PGTS '08

Proceedings of the 5th Polish-German Teletraffic Symposium

October 6-7, 2008
Berlin, Germany

Editors
Anja Feldmann
Paul J. Kühn
Michał Pióro
Adam Wolisz

Editors:

Anja Feldmann
Deutsche Telekom Laboratories, Germany
Technische Universität Berlin, Germany
Email: anja@net.t-labs.tu-berlin.de

Paul J.Kühn
Universität Stuttgart, Germany
Email: paul.j.kuehn@ikr.uni-stuttgart.de

Michał Pióro
Warsaw University of Technology, Poland
Lund University, Sweden
Email: mpp@tele.pw.edu.pl

Adam Wolisz
Technische Universität Berlin, Germany
Email: awo@ieee.org

Bibliografische Information der Deutschen Nationalbibliothek

Die Deutsche Nationalbibliothek verzeichnet diese Publikation in der Deutschen Nationalbibliografie; detaillierte bibliografische Daten sind im Internet über http://dnb.d-nb.de abrufbar.

Copyright Logos Verlag Berlin GmbH 2008
Alle Rechte vorbehalten.

ISBN 978-3-8325-2047-2

Logos Verlag Berlin GmbH
Comeniushof, Gubener Str. 47,
10243 Berlin
Tel.:+4903042851090
Fax:+4903042851092
INTERNET:http://www.logos-verlag.de

Each paper appearing in this volume of proceedings was included based on at least three favorable peer reviews. **Reviewers**:

A. Bartoszewicz, Łódź University of Technology, Poland
T. Bauschert, Chemnitz University of Technology, Germany
A. Brandt, Humboldt-Universität zu Berlin, Germany
P. Buchholz, Technische Universität Dortmund, Germany
W. Burakowski, Warsaw University of Technology, Poland
W. Bziuk, Technische Universität Braunschweig, Germany
P. Chemouil, France Telecom Research (ITC International Advisory Council), France
J. Charzinski, Nokia Siemens Networks, Munich, Germany
T. Czachórski, Polish Academy of Science, Gliwice, Poland
H. Dahms, University of Applied Science, Lübeck, Germany
J. Eberspächer, Technische Universität München, Germany
A. Feldmann, Technische Universität Berlin, Deutsche Telekom Laboratories, Germany
C. Görg, University of Bremen, Germany
A. Grzech, Wrocław University of Technology, Poland
F. Hartleb, T-Systems, Darmstadt, Germany
V.B. Iversen, Technical University of Denmark (ITC International Advisory Council), Lyngby, Denmark
A. Jajszczyk, AGH University of Science and Technology, Cracow, Poland
W. Kabaciński, Poznań University of Technology, Poland
S. Kaczmarek, Gdańsk University of Technology, Poland
A. Kasprzak, Wrocław University of Technology, Poland
M. Kiese, Technische Universität München, Germany
U. Killat, Technische Universität Hamburg-Harburg, Germany
U. Körner, Lund Institute of Technology (ITC International Advisory Council), Sweden
J. Konorski, Gdańsk University of Technology, Poland
P.J. Kühn, Universität Stuttgart (ITC International Advisory Council), Germany
R. Lehnert, Technische Universität Dresden, Germany
J. Lubacz, Warsaw University of Technology, Poland
R. Mathar, RWTH Aachen University, Germany
W. Molisz, Gdańsk University of Technology, Poland
A. Pach, AGH University of Science and Technology, Cracow, Poland
Z. Papir, AGH University of Science and Technology, Cracow, Poland
M Pióro, Warsaw University of Technology, Poland
R. Prinz, Technische Universität München, Germany
E. Rathgeb, University of Duisburg-Essen, Germany
G. Roessler, Avaya GmbH, Frankfurt/M, Germany
S. Rugel, O2, Munich, Germany
J. Sachs, Ericsson Labs, Aachen, Germany
J.B. Schmitt, Technische Universität Kaiserslautern, Germany
C. Spleiß, Technische Universität München, Germany
M. Stasiak, Poznań University of Technology, Poland
P. Tran-Gia, University of Würzburg (ITC International Advisory Council), Germany
T. Uhl, Flensburg University of Applied Sciences, Germany
C. Wietfeld, Technische Universität Dortmund, Germany
A. Wolisz, Technische Universität Berlin, Germany
J. Woźniak, Gdańsk University of Technology, Poland

ORGANIZED BY

Technische Universität Berlin
and
Deutsche Telekom Laboratories

CO-SPONSORED BY

International Teletraffic Congress
http://www.i-teletraffic.org/

The German Information Technology Society (ITG)
http://www.vde.com/

Comarch S.A.
http://www.comarch.pl

Deutsche Telekom Laboratories
http://www.laboratories.telekom.com/

Technische Universität Berlin
http://www.tu-berlin.de/

PGTS '08 COMMITTEE MEMBERS

GENERAL CONFERENCE CHAIRS

Anja Feldmann (Technische Universität Berlin, Deutsche Telekom Laboratories, Germany)
Adam Wolisz (Technische Universität Berlin, Germany)

STEERING COMMITTEE CHAIRS

Adam Grzech (Wrocław University of Technology, Poland)
Ralf Lehnert (Technische Universität Dresden, Germany)
Józef Woźniak (Gdańsk University of Technology, Poland)

TECHNICAL PROGRAM COMMITTEE CHAIRS

Paul Kühn (Universität Stuttgart, Germany)
Michał Pióro (Warsaw University of Technology, Poland)

TECHNICAL PROGRAM COMMITTEE

A. Bartoszewicz, Łódź University of Technology, Poland
T. Bauschert, Chemnitz University of Technology, Germany
A. Brandt, Humboldt-Universität zu Berlin, Germany
P. Buchholz, Technische Universität Dortmund, Germany
W. Burakowski, Warsaw University of Technology, Poland
W. Bziuk, Technische Universität Braunschweig, Germany
P. Chemouil, France Telecom Research, France
J. Charzinski, Nokia-Siemens-Networks, Munich, Germany
T. Czachórski, Polish Academy of Sciences, Gliwice, Poland
H. Dahms, University of Applied Sciences, Lübeck, Germany
Z. Dziong, École de technologie supérieure, University of Quebec, Canada
J. Eberspächer, Technische Universität München, Germany
A. Feldmann, Technische Universität Berlin, Deutsche Telekom Laboratories, Germany
J. Filipiak, Comarch S.A., Poland
C. Görg, University of Bremen, Germany
A. Grzech, Wrocław University of Technology, Poland
F. Hartleb, T-Systems, Darmstadt, Germany
C. Heinz, Mannesmann-Arcor, Germany
V.B. Iversen, Technical University of Denmark, Denmark
A. Jajszczyk, AGH University of Science and Technology, Cracow, Poland
W. Kabaciński, Poznań University of Technology, Poland
S. Kaczmarek, Gdańsk University of Technology, Poland
A. Kasprzak, Wrocław University of Technology, Poland
U. Killat, Technische Universität Hamburg-Harburg, Germany
D. Kofman, NoE Euro NGI, France
J. Konorski, Gdańsk University of Technology, Poland
U. Körner, Lund University of Technology, Sweden

R. Lehnert, Technische Universität Dresden, Germany
J. Lubacz, Warsaw University of Technology, Poland
R. Mathar, RWTH Aachen University, Germany
W. Molisz, Gdańsk University of Technology, Poland
A. Pach, AGH University of Science and Technology, Cracow, Poland
Z. Papir, AGH University of Science and Technology, Cracow, Poland
E. Rathgeb, University of Duisburg-Essen, Germany
S. Recker, Vodafone, Düsseldorf, Germany
G. Roessler, Avaya GmbH, Frankfurt/M., Germany
S. Rugel, O2, Munich, Germany
J. Sachs, Ericsson Labs, Aachen, Germany
M. Söllner, Lucent Technologies, Nürnberg, Germany
M. Stasiak, Poznań University of Technology, Poland
H. Szczerbicka, Leibniz Universität Hannover, Germany
P. Tran-Gia, University of Würzburg, Germany
T. Uhl, Flensburg University of Applied Sciences, Germany
C. Wietfeld, Technische Universität Dortmund, Germany
A. Wolisz, Technische Universität Berlin, Germany
R. Wessäly, atesio GmbH, Berlin, Germany
J. Woźniak, Gdańsk University of Technology, Poland

TABLE OF CONTENTS

PREFACE

The 5th Polish-German Teletraffic Symposium (PGTS '08) takes place in Berlin. It continues a series of events held every two years (Dresden 2000, Gdańsk 2002, Dresden 2004, Wrocław 2006). PGTS has been created to fasten the cooperation between the telecommunications and teletraffic communities of the two neighboring countries. Since 2000, PGTS has become a well-established international forum attracting contributions and participants not only from Germany and Poland but also from all over Europe.

Teletraffic Theory and Engineering has a long tradition in both countries in the academic world of universities, academies, research institutions, and also within the telecommunications industries and within network operating organizations. The Polish teletraffic community holds annual meetings (Polish Teletraffic Symposia) since many years, which are since 2000 interleaved with the PGTS meetings. The German teletraffic community is mainly organized in a Special Expert Group on System Architecture and Traffic Engineering within the German Information Technology Society (ITG) which organizes regular workshops and conferences and which considers itself as a national counterpart to the International Teletraffic Congress and its organizations.

Teletraffic problems have changed substantially during the recent years as a result of the changes in the telecommunications area towards integrated digital networks, data services, internet and mobile communications. Traffic patterns have changed giving rise to more sophisticated traffic models and more advanced analysis methods, traffic and performance management. Despite all that, the fundamental issues as resource management, quality of service, traffic control or network planning and operation are still of key importance, where stochastic processes, simulation techniques and optimization methods are applied.

PGTS '08 reflects almost all current areas of teletraffic research and engineering and, for the first time, also the upcoming field of network security. The organizers hope that PGTS '08 will contribute to the teletraffic science and be attractive for all categories of experts from theory to applications, from academic to industrial and network operators, and, especially, for promising young talents. PGTS may also act as a catalyzer for more joint research cooperations between Poland, Germany and other parts of Europe.

A symposium like PGTS would not become reality without the initiatives and help of many volunteers. We would like to thank all members of the Technical Program Committee, all reviewers of the submitted papers, our invited speakers, chairpersons, and especially the support of the staff of the Institute of Telecommunication Networks and the associated T-Labs, in particular Britta Liebscher for the website and Jürgen Malinowski for his help in editing the Proceedings. Special thanks are expressed to sponsors of PGTS '08, the International Teletraffic Congress (ITC), the German Information Technology Society (ITG), the private company Comarch S.A., the Deutsche Telekom Laboratories, and the Technische Universität Berlin.

We would like to express our thanks to the authors of all submitted papers for their contributions to make PGTS '08 a great success. We wish all participants a pleasant stay in the city of Berlin.

Berlin, September 2008

Anja Feldmann and Adam Wolisz	Conference Co-Chairs
Adam Grzech, Ralf Lehnert and Józef Woźniak	Steering Committee Co-Chairs
Paul J. Kühn and Michał Pióro	Technical Program Committee Co-Chairs

I. INVITED CONTRIBUTIONS

Anja FELDMANN*

AN OPORTUNITY FOR ISP AND P2P COLLABORATION

ABSTRACT

Peer-to-peer (P2P) systems offer astounding possibilities to their users. As such P2P users are a good source of revenue for the Internet Service Providers (ISPs). But the immense volume of P2P traffic also poses a significant challenges to the ISPs. P2P systems have to either build their overlay topologies agnostic of the underlay topology or measure the path performance themself. Accordingly, routing in P2P systems is often suboptimal and largely independent of the Internet routing. In addition, the ISP looses control of its traffic. This situation is disadvantageous for both: the ISPs and the P2P users.

To overcome this, we suggest that ISPs and P2P systems collaborate. We propose and evaluate the feasibility of a solution where the ISP offers an "oracle" to the P2P users. When the P2P user supplies the oracle with a list of possible P2P nodes, the oracle ranks them according to certain criteria, like their proximity to the user or higher bandwidth links. This can be used by the P2P user to choose appropriate neighbors, and therefore improve its performance. The ISP can use this mechanism to better manage the immense P2P traffic, e. g., to keep it inside its network, or to direct it along a desired path.

[1] VINAY AGGARWAL, OBI AKONJANG, ANJA FELDMANN. Improving User and ISP Experience through ISP-aided P2P Locality. In *Proceedings of 11th IEEE Global Internet Symposium 2008 (GI '08)*, (Location: Phoenix, AZ, USA), IEEE Computer Society, Washington, DC, USA, April 2008.

[2] VINAY AGGARWAL, ANJA FELDMANN, CHRISTIAN SCHEIDELER. *Can ISPs and P2P systems co-operate for improved performance?* ACM SIGCOMM Computer Communications Review (CCR), 37(3):29–40, July 2007.

*Deutsche Telekom Laboratories, TU Berlin, anja.feldmann@telekom.de

Key words – Science of information, spatio-temporal, semantic and structural information.

Wojciech SZPANKOWSKI*

FACETS OF INFORMATION IN COMMUNICATIONS

Information permeates every corner of our lives and shapes our universe. Understanding and harnessing information holds the potential for significant advances. Information is communicated in various forms: from business information measured in dollars, to chemical information contained in shapes of molecules, and to biological information stored and processed in living cells. Our current understanding of the formal underpinnings of information date back to Claude Shannon's seminal work in 1948 resulting in a general mathematical theory for reliable communication in the presence of noise. This theory enabled much of the current day storage and communication infrastructure. However, a wide spread application of information theory to economics, biology, life science and complex networks seems to be still awaiting us. We shall argue that a new *science of information* is to rekindle for extraction, comprehension, and manipulation of *structural, spatio-temporal* and *semantic* information. We conclude this essay with a list of challenges for future research.

1. INTRODUCTION

Information is still the distinctive mark and arguably the basic commodity of our era, so that the need for deeper reflection and study is intensifying. The notion and theory of *information* introduced by Claude Shannon in 1948 have served as the backbone to a now classical paradigm of digital communication. Television, internet, voyage to the Moon, unbreakable ciphers, computers, NASA's planetary probes, home compact disk audio systems, they all benefited from Shannon theory. Shannon's notion of information quantifies the extent to which a recipient of data can reduce its statistical uncertainty when "semantic aspects of communication are irrelevant" [23]. Unfortunately, that formalization of information hardly captures all of the needed nuances, and the accompanying theory has not lent itself to non-trivial applications outside the native context. We yet have to develop theory that provides a satisfactory formalism and overarching answers for extraction, comprehension, and manipulation of *structural, spatio-temporal* and *semantic* information in scientific and social domains. Consequently, one can argue the necessity for the next information revolution and perhaps launching a new *Science of Information* that integrates research and teaching activities from all angles: from the fundamental theoretical underpinnings of information to the science and engineering of novel substrates, biological networks, chemistry, communication networks, economics, physics, and complex social systems.

Surprisingly enough, advances in information technology and widespread availability of information systems and services have largely obscured the fact that *information* remains undefined in its generality, though considerable collective effort has been invested into its understanding [2, 3, 18, 22, 24, 27]. Shannon wrote in his 1953 paper [24]: "The word "information"

*Department of Computer Science, Purdue University, W. Lafayette, IN 47907, U.S.A., spa@cs.purdue.edu.

has been given many different meanings . . . it is likely that at least a number of these will prove sufficiently useful in certain applications and deserve further study and permanent recognition." C. F. von Weiszsäcker argued [27] against an absolute definition of information, claiming that: "Information is only that which produces information" (relativity) and "Information is only that which is understood" (rationality). One also observes that information extraction depends on available resources (e.g., think of guessing an integer with one bit storage available [21, 26]). It follows therefore, that in its generality, *information is that which can impact a recipient's ability to achieve the objective of some activity in a given context within limited available resources* [17]. We shall adopt this definition for the purpose of this essay.

In passing we should add that we view here "communication" very broadly. Living cells do communicate and process information as well as users in communication/wireless networks, information is communicated in arbitrage of financial markets, spike trains between neurons carry information to the brain, and so forth. We are aiming at understanding "communicated information" in its generality.

2. SIENCE OF INFORMATION

Information, understood broadly, is capable of unifying seemingly unrelated areas such as information theory, physics of information, value of information, Kolmogorov complexity and information flow in life sciences [19]. This is illustrated in Figure 1. One can argue, however, as the participants of the first workshop of *Information Beyond Shannon*, Orlando, FL, 2005 did, that the following aspects of information were never adequately addressed in the past and therefore threaten to raise severe impediments to diverse applications of science of information:

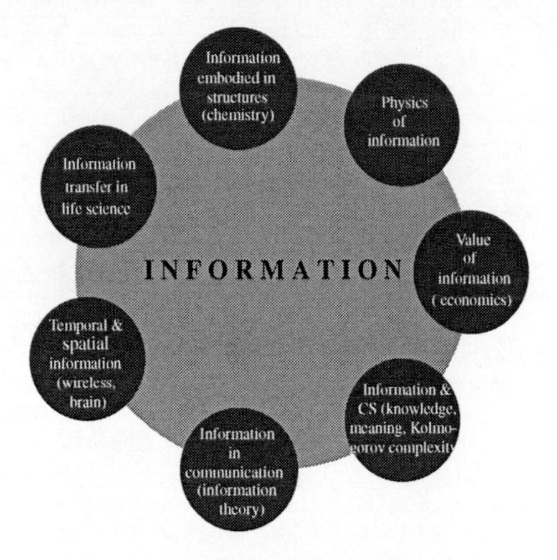

Figure 1: Information Synergy.

- **Delay**: In many communication problems, especially when information is transmitted over a network, the amount of delay incurred is an important and nontrivial factor [10]. For example, complete information arriving late maybe useless to the receiver, whereas incomplete information arriving early may be valuable (e.g., in a signaling cascade associated with a specific cell function, delay or loss of signals can be lethal, timeliness is the key for arbitrage of financial markets, timely intelligence is essential to virtually all security and defense related applications). This is not simply a question of understanding the classical delay-rate trade-off (via the reliability function [3]), but a complex issue involving our choice of how and what to transmit, as well as the actually utility of the information being transmitted.
- **Space**: In interacting systems, spatial localization often limits information exchange, with obvious disadvantages as well as benefits. These benefits typically result from reduction in interference as well as ability of system to modulate and react to stimulus (common examples range from wireless systems to immune response).
- **Information and Control**: In the above delay-bandwidth example we have a conflict between two objectives: Attaining a high transmission rate, and sending the information with small delay. But information is exchanged in space and time for decision making, thus timeliness of information delivery along with reliability and complexity constitute basic objectives. More generally, there are cases where we have some control not only over the coding part, but also about other design aspects of the communication setting (e.g., the network topology, power distribution, routing, etc). How can the two tasks be optimally combined?
- **Utility**: We often find that the utility of what is transmitted depends on different factors, such as the time at which it arrives, and perhaps even the actual contents of the message. How can such utility considerations be incorporated into the classical coding problem?
- **Semantics**: In many scientific contexts experimenters are interested in signals, without knowing precisely what these signals represent. Examples of this situation are very common in biology: DNA sequences and spike trains between neurons are certainly used to convey information, but little more than that can be assumed a priori. Often, one of the first steps in the analysis is to try and estimate the amount of information contained in these signals - how should that be done? Estimating the entropy is typically not appropriate: It offers a measure of the structural complexity of the signal, but it does not measure its actual information content. For example, it ignores the fact that there may be noise present and it does not take into account that certain parts of the signal may be irrelevant to the receiver. Is there a general way to account for the actual "meaning" of signals in a given context?
- **Optimal versus Real Communication Systems**: In the last example of observing, say, a spike train between two neurons in an animal's brain, the difficulty in extracting information from the signal (or even measuring the amount of information present) in part arises from the fact that the scientist is not in Shannon's position of designing an optimal system given known communication constraints. Instead, she is analyzing an existing and typically sub-optimal communication system. We can ask whether there is a general methodology for the study of such systems, so that the actual context within which they operate is taken into account

- **Dynamic information**. In a complex network, information is not just communicated but also processed and even generated along the way (e.g., response to stimuli is processed at various stages – with immediate response processed at site of stimulus, higher-level response processed in the brain, response to emergency events is coordinated at various levels, ranging from first responders to command and control centers). How can such considerations of dynamic sources be incorporated into an information-theoretic model?
- **Learnable Information**: One can argue (and some have) that in all scientific endeavors, the only task is to extract information from data. How much information can actually be learned from a given data set? In Shannon theory, one starts from a (possibly unknown) model for the data-generating mechanism and calculates its entropy, but in practice the starting point is only the data. Is there a general theory that provides natural model classes for the data at hand? What is the cost of learning the model, and how does it compare to the cost of actually describing the data? Risannen's MDL theory offers guidelines in this direction [20].
- **Structure and Organization**: We still lack measures and meters to define and quantify information embodied in structure and organization (e.g., information in nanostructures, biomolecules, gene regulatory and protein interaction networks, social networks, networks of financial transactions, etc.). Typically, these measures must account for associated context, and incorporate diverse (physical, social, economic) dynamic observables and system state).
- **Limited Resources**: In many scenarios, information is limited by available resources (e.g., computing devices, bandwidth of signaling channels). How much information can be extracted and processed with limited resources? This relates to **complexity and information** where different representations of the same distribution may vary dramatically when complexity is taken into account (e.g., computing a number from its prime factors is easy but factoring it is known to be much harder).

It is our belief that further advances in science of information and its diverse applications depend on answering the above challenges.

3. TWO EXAMPLES

In this section we discuss two examples: one illustrating spatio-temporal aspect of information in wireless massive networks, and the other related to information transfer in biological systems (the so called Darwin channel).

3.1. SPACE-TIME PARADOX IN STABLE WIRELESS NETWORKS

Wireless local area networks (WLANs), multihop mobile ad hoc networks (MANETs), and social networks are a logical next step towards an ubiquitous computing environment. The related technological challenges, e.g., volatile connectivity, power awareness, and increased node autonomy, have also become scientific ones. Recent research [8, 9] in MANETs has led to the discovery of the "space capacity paradox" and "time capacity paradox." The theoretical capacity of a multihop wireless network increases with node density and node mobility in spite of the apparently devastating effect of transmission interference. A deeper information-theoretic understanding of these properties is likely to bring about a breakthrough in MANET technology and deployment, on condition that more realistic network operation models are adopted.

Classical information theory studies capacity of channels connecting two endpoints. This approach is hard to adopt to mobile nodes which relay information in a multihop manner and time-varying topology. Therefore, some authors (cf. [14]) introduce the concept of the *spatio-temporal relaying.* A relay in a space-time situation carries information from a mobile transmitter (space) in its past (time) to a mobile receiver (space) in its future (space). Here, the past and future are defined with respect to the causal physical trajectory of nodes that forms a path in a spatio-temporal space of information transfer. The quality of the transmission depends on the respective spatio-temporal positions of the transmitter and receiver. Thus the concept of a space-time relay transcends classical information theory.

The challenge is to extend the celebrated Shannon capacity formula $\log_2(1+\frac{S}{N})$ per Hertz (S signal, N noise) to multi-source wireless networks. Recently, Jacquet [13] proved that the maximum information rate I per second per Hertz in a network of dimension d (e.g., $d=1$ for a line, $d=2$ for a plane, $d=3$ for a three-dimensional space) with the Rayleigh factor α is

$$I = \frac{\alpha}{d}(\log 2)^{-1}.$$

This formula is remarkable since it connects three main ingredients: space (d), physics of wave propagation (α), and information theory ($\log 2$).

Furthermore, in [9] it is shown that the theoretical capacity of a multihop wireless network is proportional to the square root of the network size (number of nodes). This remarkable result promises enormous wireless capacity for ultra-dense networks (e.g., one million nodes with available bandwidth of 1 Mb/s can reach a total capacity on order of gigabits per second, unprecedented for mobile networks). However, attempts to verify these predictions in a network with WiFi nodes bring unsatisfactory results: the space capacity has a tendency to decrease with the number of nodes, rather than increase as theoretically predicted. This reflects the well-known fact that the WiFi medium access protocol, primarily designed for wireless LANs, does not scale to multihop networks. Several analytical models using random placement of nodes were subsequently developed. By using wave attenuation and packet capture rules, minimal (slotted Aloha-like) medium access protocols were designed to fit into the theoretically predicted scaling property. Closed formulae for the probability of packet capture versus distance were found that can be used to identify the performance bottleneck for any routing protocol via certain equilibrium equations [12]; this promises further advances in the Science of Information. For example, it turns out that the Optimized Link State Routing (OLSR) protocol in its basic version cannot support more than a few thousand nodes. For networks this large the bottleneck is the *overhead*, since every node must inform all other nodes about its local connectivity, creating control traffic flows that percolate through the whole of the network (even if broadcast optimally, as prescribed by OLSR). Of interest become routing strategies whereby only significant connectivity changes are broadcast.

Recently, a time counterpart of the space paradox was demonstrated. Under the hypothesis of ergodic node mobility it was proved that the capacity of a mobile wireless network can be linear in the number of nodes. Thus, one million nodes can reach a capacity on the order of a thousand Gb/s. However, this comes with a price tag of growing delay. One can therefore ask how much useful information is really passed.

More generally there are fascinating relations between spatio-temporal properties and information propagation (speed) in wireless networks. In [14] an upper bounds is derived for the propagation speed of one bit in a wireless mobile network embedded in a map of dimension d. The question still remains how to estimate the propagation speed as function of the information rate. Clearly, the speed decreases with the requested capacity so that information theoretical upper-bounds need to be extracted and realistic models must be developed.

3.2. INFORMATION TRANSFER IN BIOLOGICAL SYSTEMS

We now switch to communications in biology and discuss information flow in biological systems by introducing the *mutation channel* and the *Darwin channel*. The mutation channel is a classical insertion/deletion channel [5, 4], while the Darwin channel is a novel information-theoretic channel, described in details below, that models preferential Darwinian selection.

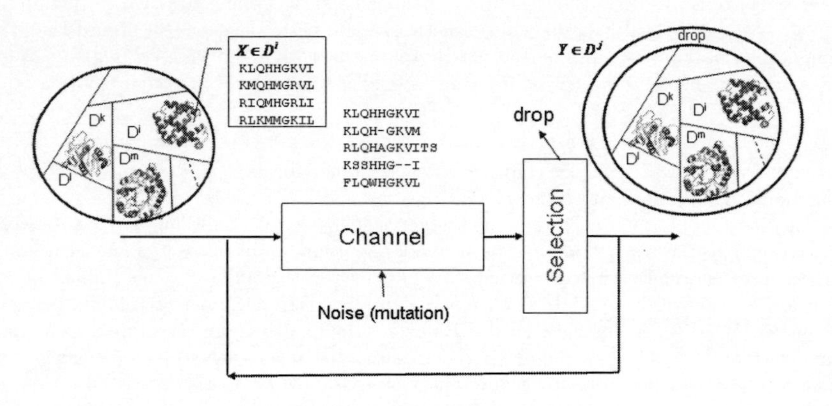

Figure 2: Darwin Channel

Darwin channel described in Figure 2 is designed to model the flow of genetic information through *temporal constrained* channel with feedback (i.e., surviving sequences/genes re-enter the evolutionary process in time and space preserving functionality). More precisely, the original input (biological) sequence $X_1^n = X_1 \dots X_n$ is restricted to a *constrained* (Darwinian preselected) set $\mathcal{D}_n$ that is a proper subset of all possible input sequences. This set is partitioned into subsets $\mathcal{D}^i$ of sequences of the same *functionality* (e.g., by using a scoring function often used in biology). This partition is represented by a function

$$F : \mathcal{D}_n \to \{0, 1, \dots, M-1\}$$

for some M (that may depend on n) such that all $X_1^n \in \mathcal{D}^I$ are assigned to index $I := F(X_1^n)$. In the Darwin channel *mutation* constitutes the *noise*, and it can take the form of insertion, deletion or substitution, where is the latter is the most dominant factor. The output sequence $Z_1^m = Z_1 \dots Z_m$ is in general of random length m. An output sequence Z_1^m is either assigned to one of the subsets $\mathcal{D}^i$ through F or erased (declare "dead"). In such an information transfer scenario an input sequence is declared to be "functionally surviving" *if both X_1^n and its corresponding output sequence Z_1^m belong to the same functional subset $\mathcal{D}^I$*; otherwise an error occurs leading to a non-surviving sequence. Furthermore, to model temporal behavior, we follow

Eigen's observations (i.e., "there are correlations between error rate and genome length") and allow error rate to be a function of n. The main open problem is how to determine the amount of "information" being transfer through the channel. As a matter of fact, we first need to address more fundamental questions, namely what is information in such systems and how to measure it?

Let us now consider two special cases to illustrate some difficulties we may encounter analyzing the Darwin channel. Assume for now there is no feedback, noise is just a substitution with probability of error ε being fixed and very small, say $\varepsilon = 10^{-8}$. Furthermore, to focus we postulate that the constrained set $\mathcal{D}_n$ consists of (d, k) $(d < k)$ binary sequences in which any run of zeros must be of length at least d and at most k (e.g., such sequences model spike trains of neurons). We make one more important assumption, namely we measure the information transfer in such a system by the statistical dependency between the output sequence Z^n and the input sequence X^n through the *mutual information* $I(X; Z)$. Then one may ask what values $I(X; Z)$ takes and perhaps what is the maximum value of $I(X; Z)$ over all possible input distribution. The latter question leads to the *noisy constrained capacity* problem that has been unsolved since Shannon [7].

Let us first look at $I(X; Z)$. Observe that

$$I(X; Z) = H(Z) - H(Z|X)$$

where $H(Z)$ and $H(Z|X)$ are entropy and conditional entropy, respectively. But, as easy to see, $H(Z|X)$ is the entropy of the noise, that is, $H(Z|X) = H(\varepsilon) = -\varepsilon \log \varepsilon - (1-\varepsilon)\log(1-\varepsilon)$. Just we are left with the problem of estimating the entropy of Z.

In our case, X is a (d, k) sequence and Z a noisy version of X. But a (d, k) sequence can be generated as an output of an automaton, thus X is a Markov sequence and then Z is a *hidden Markov process* (HMP). Unfortunately, entropy of a HMP process is not easy to estimate [6]. Fortunately, recently in [11, 15] it was proved that for $\varepsilon \to 0$ the following holds

$$H(Z) = H(P) - f_0(P)\varepsilon \log \varepsilon + f_1(P)\varepsilon + o(\varepsilon)$$

for some explicitly computable coefficients $f_0(P)$ and $f_1(P)$ where P is the distribution of the underlying Markov process of X and $H(P)$ its known Markov entropy [11, 16]. Even more interestingly, if we are interested in the maximum mutual information, that is, the noisy constrained capacity

$$C(\mathcal{D}, \varepsilon) = \sup_{X \in \mathcal{D}} I(X; Z) \lim_{n\to\infty} \frac{1}{n} \sup_{X_1^n \in \mathcal{D}_n} I(X_1^n, Z_1^n)$$

then the situation becomes more complicated. Recently, in [11, 16] it was proved that

$$C(\mathcal{D}, \varepsilon) = C(\mathcal{D}) - (1 - f_0(P^{\max}))\varepsilon \log \varepsilon + (f_1(P^{\max}) - 1)\varepsilon + o(\varepsilon)$$

where $C(\mathcal{D})$ is the noiseless capacity, that is, $C(\mathcal{D}) = -\log \rho_0$, where ρ_0 is the smallest real root of (cf. [25])

$$\sum_{\ell=d}^{k} \rho_0^{\ell+1} = 1.$$

Thus, even under these simplifying (and rather biologically naive) assumptions, the Darwin channel is a "hard nut to crack".

Let us finally consider the temporal aspect of the Darwin channel. To illustrate our point, we make the most simplifying assumptions, namely the Darwin channel is plainly a simple binary symmetric channel in which each bit incurs a random delay T before it reaches the receiver [17]. A bit that reaches the destination after a given deadline τ is dropped. Furthermore, let ε be the probability of error. We assume that the longer a bit takes to reach the receiver, the lower the probability of a successful transmission (which is an accurate model in certain biological situations as observed by M. Eigen). For $t \leq \tau$ the probability of a successful transmission is $\Phi(\varepsilon, t)$ (e.g., $\Phi(\varepsilon, t) = (1 - \varepsilon)^t$), and hence the probability of error is $1 - \Phi(\varepsilon, t)$ for some $\varepsilon > 0$. What is the capacity of such a *temporal channel*? If the delay is exponentially distributed, then one easily finds for $\Phi(\varepsilon, t) = (1 - \varepsilon)^t$ that

$$C(\tau) = [(1 - P(T > \tau))][1 - H(\rho)],$$

where $\rho = P(x|x)/(1-e^{-\tau})$ with $P(x|x) = (1-(1-\varepsilon)^{\tau}e^{-\tau})/(1-\ln(1-\varepsilon))$ being the probability of a successful transmission. Observe that with a stringent delay bound, the capacity of the channel is adversely affected by frequent erasure and the capacity drops due to temporal errors.

4. CONCLUDING REMARKS

In the upcoming workshop *Information Beyond Shannon*, Venice, Italy, December 29-30, 2008 (`http://mobilfuture.com/venice/`) the organizers list the following challenges facing us:

- Frederick P. Brooks, Jr., wrote in "The Great Challenges for Half Century Old Computer Science" [1]: "Shannon performed an inestimable service by giving us a definition of Information and a metric for Information as communicated from place to place. We have no theory however that gives us a metric for the Information embodied in structure. ... This is the most fundamental gap in the theoretical underpinning of Information and computer science. A young information theory scholar willing to spend years on a deeply fundamental problem need look no further."
- Information accumulates at a rate faster than it can be sifted through, accessed and digested by humans, so that the bottleneck, traditionally represented by the medium, is drifting towards the receiving end of the channel.
- In a growing number of situations, the overhead in accessing Information prevails over that of fruition, which makes information itself practically unattainable or obsolete.
- Capabilities akin to contents addressing and semantic access and transmission are not even in sight, while computing and communication infrastructures of the new Millennium induce drastic mutations on the conventional notions of Knowing and Learning, Guessing and Discovering.
- Microscopic systems seem not to obey Shannon postulates of information [2]. In the quantum world and on the level of living cells traditional Information often fails to accurately describe reality.

The cross-disciplinary research in science of information advocated here hopefully will lead to the development of an active and thriving community of students and scholars to pursue its goals. As the first step, we have launched recently at Purdue the *Institute for Science of Information* http://www.isi.purdue.edu that should serve as home for such activities. We certainly hope that similar centers will soon emerge in Europe and around the world.

5. ACKNOWLEDGMENT

The author thanks Drs. J. Konorski (TUG), C. Jedrzejek (PUT), and participants of the first *Workshop Information Beyond Shannon*, Orlando, 2005 for sharing their views on information.

This work was supported in part by the NSF Grants CCF-0513636, DMS-0503742, DMS-0800568, and CCF -0830140, NIH Grant R01 GM068959-01, NSA Grant H98230-08-1-0092, EU Project No. 224218 through Poznan University of Technology, and the AFOSR Grant FA8655-08-1-3018.

REFERENCES

[1] F. Brooks. Three great challenges for half-century-old computer science. *J. the ACM*, 50:25–26, 2003.

[2] C. Brukner, A. Zeilinger, Conceptual Inadequacy of the Shannon Information in Quantum Measurements. *Phys. Rev.* A 63, 2001.

[3] T.M. Cover and J.A. Thomas, *Elements of Information Theory*, Second Edition, John Wiley & Sons, New York, 2006.

[4] R. L. Dobrushin. Shannon's theorem for channels with synchronization errors. *Problems of Information Transmissions*, 3:18–36, 1967.

[5] E. Drinea and M. Mitzenmacher. On lower bounds for the capacity of deletion channels. *IEEE Transaction of Information Theory*, 52:4648–4657, 2006.

[6] Y. Ephraim and N. Merhav, "Hidden Markov processes," *IEEE Trans. Inform. Theory*, 48, 1518–1569, 2002.

[7] J. Fan, T. L. Poo, and B. Marcus. Constraint gain. *IEEE Transaction of Information Theory*, 50:1989–1999, 2004.

[8] M. Grossglauser and D. Tse, Mobility Increases the Capacity of ad-hoc Wireless Networks, *IEEE/ACM Trans. Networking*, 48, 477-486, 2002.

[9] P. Gupta and P.R. Kumar, Capacity of Wireless Networks, *IEEE Trans. Information Theory*, 46, 388-404, 2000.

[10] B. Hajek and A. Ephremides, Information Theory and Communication Networks: An Unconsummated Union, *IEEE Trans. Information Theory*, 44, 2416-2434, 1998.

[11] G. Han and B. Marcus, Capacity of Noisy Constrained Channels, *Information Theory Symposium*, Nice, 2007.

[12] P. Jacquet, Space-time Information Propagation in Mobile ad hoc Wireless Networks, http://ee-wcl.tamu.edu/itw2004/program/jacquet_inv.pdf/, *ITW 2004*, San Antonio, 2004

[13] P. Jacquet. Realistic wireless network model with explicit capacity evaluation. Technical Report INRIA RR-6407, INRIA, 2008.

[14] P. Jacquet, Bernard Mans and G. Rodolakis, Information Propagation Speed in Delay Tolerant Networks: Analytic Upper Bound. *Proc. ISIT*, 6-10, Toronto, 2008.

[15] P. Jacquet, G. Seroussi, and W. Szpankowski. On the Entropy of a Hidden Markov Process. *Theoretical Computer Science*, 395, 203-219, 2008.

[16] P. Jacquet, G. Seroussi, and W. Szpankowski. Noisy constrained capacity. In *Information Theory Symposium*, volume IEEE, pages 986–990, 2007.

[17] J. Konorski and W. Szpankowski, What is Information?, Festschrift in Honor of Jorma Rissanen, 154-172, 2008.

[18] D.G. Luenberger, *Information Science*, Princeton Univ. Press, 2006.

[19] P. Nurse, Life, Logic, and Information, *Nature*, 454, 424-426, 2008.

[20] J. Rissanen, *Stochastic Complexity in Statistical Inquiry*, World Scientific, Singapore, 1998.

[21] J. Roche. Gambling the Mnemonically Impaired, *IEEE Trans. Information Theory.*, 48:1379–1392, 2002.

[22] J. Seidler, *The Science of Information*, WNT, Warszawa, 1982 (in Polish).

[23] C. Shannon, A Mathematical Theory of Communication, *Bell System Technical Journal*, 27, 379-423 and 623-656, 1948.

[24] C. Shannon. The Lattice Theory of Information. *IEEE Transaction on Information Theory*, 1:105–107, 1953.

[25] W. Szpankowski, *Average Case Analysis of Algorithms on Sequences.* New York: John Wiley & Sons, Inc., 2001.

[26] S. Venkatesh and J. Franklin. How Much Information Can One Bit of Memory Retain About Bernoulli Sequence, *IEEE Trans. Information Theory.*, 37:1595–1604, 1991.

[27] C.F. von Weiszsäcker and E. von Weiszsäcker, Wideraufname der begrifflichen Frage: Was ist Information?, Nova Acta Leopoldina, 206, 1972.

Key words – Wireless communication, delay tolerant networking, peer-to-peer

Gunnar KARLSSON*

OPPORTUNISTIC WIRELESS CONTENT DISTRIBUTION

ABSTRACT

In recent years, there have been systems proposed for opportunistic content distribution in the wireless ad-hoc domain. Examples are those targeted towards people in metropolitan areas [1][2][3][4], vehicular environments [5][6], and wildlife sensor networks [7][8]. In common to these systems is that mobile nodes cooperate in distributing content by sharing contents in a peer-to-peer fashion over radio contacts that are established when two nodes enter within communication range of one another. Still there is much work to be done in understanding the behavior and performance of these, so called, delay-tolerant systems and how they are affected by mobility, node resources, transmission rate, cooperation degree and service discovery.

Stochastic modeling is commonly used to evaluate the performance of mobile wireless networking systems of the nature we consider in this work. Theoretical models can give important insight under simplifying assumptions about fundamental issues such as system capacity and scaling properties. However, the analytic models become intractable in many cases when the simplifying assumptions are lifted, and simulations are then commonly used. But even with the advanced simulators used today, it is not trivial to capture all necessary parameters affecting the performance of mobile wireless systems. In addition, it is difficult to know which parameter sets to use when designing and dimensioning a real system. Evaluating a system's efficiency in real-life is therefore still very important, even though wireless experiments are exposed to many parameters that cannot be controlled. Finally, relating and comparing results from theory, simulations and experiments is a significant challenge.

This talk will describe the architecture of an opportunistic content distribution system and a performance evaluation using two approaches, a stochastic Markovian model and simulations based on experimental traces. The analytical model and the trace-driven simulations are complementary methods and the goal is to obtain qualitative results on the system behavior and performance rather than giving an exact quantitative comparison. Specifically, we study how the content distribution is affected by node cooperation. We consider different levels of cooperation where the willingness of the nodes to share contents with peers differs. Since devices, such as mobile phones, audio and video players and PDAs, are small and resource limited, we presume that the capability of the nodes to cooperate will be restricted. In particular, limited battery

* KTH School of Electrical Engineering and ACCESS Linnaeus Center, 10044 Stockholm, Sweden. Email: gk@ee.kth.se

lifetime will discourage cooperative transmissions of data. The impact of limited sharing on the system's performance is therefore of particular interest.

The talk will conclude with some suggestions for further research in the area of opportunistic wireless content distribution.

REFERENCES

[1] M. Papadopouli and H. Schulzrinne, "Seven degrees of separation in mobile ad hoc networks," in Proceedings of the IEEE GLOBECOM, 2000.

[2] W. H. Yuen, R. D. Yates, and S. C. Mau, "Exploiting data diversity and multiuser diversity in noncooperative mobile infostation networks," in Proceedings of IEEE INFOCOM, San Francisco, USA, March 2003.

[3] G. Karlsson, V. Lenders, and M. May, "Delay-tolerant broadcasting," IEEE Transactions on Broadcasting, vol. 53, no. 2, pp. 369 – 381, Mar. 2007.

[4] M. Motani, V. Srinivasan, and P. Nuggehalli, "Peoplenet: Engineering a wireless virtual social network," in Proceedings of ACM MobiCom, 2005.

[5] J. Burgess, B. Gallagher, D. Jensen, and B. Levine, "MaxProp: Routing for Vehicle-Based Disruption-Tolerant Networks," in Proceeding of IEEE INFOCOM, Barcelona, Spain, April 2006.

[6] J. Ott and D. Kutscher, "A Disconnection-Tolerant Transport for Drivethru Internet Environments," in Proceedings of IEEE INFOCOM, Miami, USA, March 2005.

[7] P. Juang, H. Oki, Y. Wang, M. Martonosi, L.-S. Peh, and D. Rubenstein, "Energy-Efficient Computing for Wildlife Tracking: Design Tradeoffs and Early Experiences with ZebraNet," in Proceedings of the Tenth International Conference on Architectural Support for Programming Languages and Operating Systems (ASPLOS-X), San Jose, CA, USA, October 2002.

[8] T. Small and Z. J. Haas, "The shared wireless infostation model: a new ad hoc networking paradigm (or where there is a whale, there is a way)," in Proc. ACM MobiHoc, Annapolis, Maryland, USA, 2003, pp. 233–244.

II. TELETRAFFIC MODELS AND TRAFFIC THEORY

keywords – multirate models, multirate threshold systems, PCT1, PCT2

Mariusz GŁĄBOWSKI [1]
Maciej SOBIERAJ [2]
Maciej STASIAK [3]

EVALUATION OF QoS DETERIORATION IN THRESHOLD SYSTEMS

The paper proposes an analytical method of evaluation of calls quality deterioration in multiservice threshold systems servicing elastic and adaptive traffic. The idea of the proposed method consists in determination of the number of serviced calls of all traffic classes in particular threshold areas – influencing the calls holding time and admitted bandwidth – on the basis of the occupancy distribution in multithreshold state-dependent systems with PCT1 and PCT2 traffic. The method is based on an approximation of multidimensional Markov process occurring in the considered system by the one-dimensional Markov chain. The proposed analytical model is used to determine the traffic characteristics in UMTS networks with streaming, adaptive and elastic services. In order to determine the accuracy of the proposed model, the analytical results of percentage share of particular traffic class calls serviced in pre- and post-threshold areas with the assigned qualitative parameters are compared with simulation results.

1. INTRODUCTION

In Universal Mobile Telecommunication System (UMTS), which is one of the standards proposed for third generation cellular technologies (3G), four service classes are defined: conversation, streaming, interactive and background. The conversation class is characterised by the highest sensitivity to delays. Within this class voice transmission and video telephony services are offered. The streaming class concerns multimedia services while the interactive class is applied for web sites browsing. The advantage of background class is the least sensitivity to delays. This class offers services of transfers of electronic mail.

Each of four UMTS service classes can be assigned to one of three – considered in teletraffic theory – service classes, i.e. to the streaming, elastic or adaptive class. For a streaming service the required resources are constant and cannot be decreased with the increase in load of the system. The adaptive class allows to decrease the bandwidth available for its calls, remaining the same service duration. In the case of elastic class it is important to transfer all data, regardless of available throughput. Consequently, along with the increase in the load of the systems, the admitted bandwidth decreases and the holding time increases proportionally.

[1] Poznań University of Technology, ul. Piotrowo 3A, 60-965 Poznań, e-mail: mariusz.glabowski@et.put.poznan.pl
[2] Poznań University of Technology, ul. Piotrowo 3A, 60-965 Poznań, e-mail: msobiera@et.put.poznan.pl
[3] Poznań University of Technology, ul. Piotrowo 3A, 60-965 Poznań, e-mail: stasiak@et.put.poznan.pl

In teletraffic theory, for modelling the systems with elastic and adaptive services, the so-called threshold models are used [9, 13]. The threshold models belong to the large group of well-known *multirate models* [1, 5, 6, 8, 13, 15, 16]. In these models the system services call demands having an integer number of the so-called Basic Bandwidth Units (BBUs). When constructing multirate models for broadband systems, it is assumed that the BBU is the greatest common divisor of equivalent bandwidths of all call streams offered to the system [11, 15].

The threshold models can be applied to determine traffic characteristics of UMTS. One of the first attempt of application of the threshold models for analysis of UMTS networks was made in [8]. In [2] the analytical method of the occupancy distribution calculation was proposed for the state-dependant threshold systems servicing traffic generated by finite and infinite population of sources. The main focus of conducted researches was the determination of occupancy distribution and blocking probability in threshold systems. According to the authors knowledge, there has not been made any attempt so far to analytically determine the level of degradation of service quality in multiservice threshold systems. This deterioration is a consequence of limitation of assigned bandwidth for calls of particular traffic classes depending on system load.

In this paper it is proposed a new analytical method which allows us to determine the percentage share of serviced calls of particular traffic classes which the QoS parameters are deteriorated proportionally to the load of a system. The remaining part of the paper is organized as follows. Section 2 describes an analytical method of occupancy distribution calculation in threshold systems with finite and infinite source population. Section 3 discusses the allocation of resources in WCDMA radio interface whereas in Section 4 the proposed analytical method of evaluation of percentage share of calls serviced in particular threshold areas is presented. In Section 5 the results of calculations for an exemplary system with streaming, adaptive and elastic services were compared with simulation data. Section 6 is a conclusion.

2. THRESHOLD SYSTEMS

2.1. BASIC RECURRENCE DEPENDENCIES

Threshold systems are examples of state-dependent multiservice systems considered recently in teletraffic theory. In these systems the parameters of offered traffic can change depending on the load of the system. In most cases, an increase in the load above a certain occupancy state Q in such a system may result in a change of the number of admitted BBUs for calls of a given class (i.e. in the parameter t_i) and/or a change in the mean holding time of calls of a given class, i.e. in the parameter ${\mu_i}^{-1}$). The literature of the subject includes considerations on single-threshold systems (STS) and single-retry systems (SRS) [9, 10], in which demands of particular classes can change once, and multithreshold systems (MTS), multiretry systems (MRS) and connection-dependent threshold systems (CDTS) [8, 13, 14], where, depending on the degree of the load of the system, calls can repeatedly change their demands.

Let us consider a multithreshold system (MTS) with capacity of V BBUs (Basic Bandwidth Units) [13] and with Poisson call streams. In this model we assume that for class i calls a set of thresholds $\{Q_{i,1}, Q_{i,2}, \ldots, Q_{i,q}\}$ is individually introduced, where the first index indicates the class of a call, while the second index the number of a threshold. Additionally, it is assumed that $\{Q_{i,1} \leq Q_{i,2} \leq \ldots \leq Q_{i,q}\}$.

The operation of the MTS can be presented in the following way: in each post-threshold area k of class i a traffic stream of class i, defined by the own set of parameters $\{\lambda_i, t_{i,k}, \mu_{i,k}\}$, is offered, where λ_i is the intensity of call stream of class i, $t_{i,k}$ is the number of the demanded BBUs for the calls of class i in post-threshold area k, $\mu_{i,k}$ is the intensity of service stream (the inverse of mean holding time) of class i in post-threshold area k. Additionally, it is assumed that: $t_{i,0} > t_{i,1} > \ldots > t_{i,k} > \ldots > t_{i,q}$ and $\mu_{i,0}^{-1} \leq \mu_{i,1}^{-1} \leq \ldots \leq \mu_{i,k}^{-1} \leq \ldots \leq \mu_{i,q}^{-1}$. This means that with the increase in the load of the group, the number of BBUs assigned to service calls of particular classes decreases and, at the same time, the average holding time of the calls can be extended. In the literature [2, 9, 13] it is proved, that occupancy distribution in MTS can be determined by the Generalized Kaufman-Roberts Recursion (GKRR):

$$n\,[P_n]_V = \sum\nolimits_{i=1}^{M} A_i t_i \sigma_i(n - t_i)\,[P_{n-t_i}]_V, \tag{1}$$

where $[P_n]_V$ is the probability of n BBUs being busy in the system with capacity of V BBUs, $A_i = \lambda_i/\mu_i$ is the mean traffic offered by class i calls, M is the number of traffic classes, and $\sigma_i(n)$ is the so-called conditional probability of passing between adjacent states of the service process in the considered system.

Let us consider the local balance equation for the pre-threshold area ($0 \leq n \leq Q_{i,1}$). Consequently we get equation, which takes the following form:

$$n\,[P_n]_V = \sum_{i=1}^{M} A_{i,0} t_{i,0} \sigma_{i,0}(n - t_{i,0}) \left[P_{n-t_{i,0}}\right]_V, \tag{2}$$

where $\sigma_{i,0}(n)$ is conditional probability of passing which describes the pre-threshold area:

$$\sigma_{i,0}(n) = \begin{cases} 1 & \text{for } n \leq Q_{i,1}, \\ 0 & \text{for } n > Q_{i,1}, \end{cases} \tag{3}$$

whereas $A_{i,0}$ is the traffic intensity of class i, offered to the group within the pre-threshold area $0 \leq n \leq Q_{i,1}$:

$$A_{i,0} = \frac{\lambda_i}{\mu_{i,0}}. \tag{4}$$

By applying analogous considerations for each of the post-threshold areas k, we get:

$$n\,[P_n]_V = \sum_{i=1}^{M} A_{i,k} t_{i,k} \sigma_{i,k}(n - t_{i,k}) \left[P_{n-t_{i,k}}\right]_V, \tag{5}$$

where:

$$\sigma_{i,k}(n) = \begin{cases} 1 & \text{for } Q_{i,k} < n \leq Q_{i,k+1}, \\ 0 & \text{for the remaining } n, \end{cases} \tag{6}$$

whereas the traffic intensity $A_{i,k}$, offered to the group within the area $Q_{i,k} < n \leq Q_{i,k+1}$, is equal to:

$$A_{i,k} = \frac{\lambda_i}{\mu_{i,k}}. \tag{7}$$

In Formula (6) we adopt $Q_{i,0} = 0$.

Since individual inter-threshold areas are separable, we can sum up Eqs. (2) and (5) to obtain the occupancy distribution in the MTS system:

$$n\,[P_n]_V = \sum_{i=1}^{M} \sum_{k=0}^{q} A_{i,k} t_{i,k} \sigma_{i,k}(n - t_{i,k}) \left[P_{n-t_{i,k}}\right]_V. \tag{8}$$

In the paper [2] it was proved, that the blocking probability for class i calls can be determined with the following formula:

$$E_i = \sum_{k=0}^{q} E_{i,k}, \tag{9}$$

where $E_{i,k}$, i.e. the blocking probability for class i calls in the threshold k, takes the following form:

$$E_{i,k} = \begin{cases} 0 & \text{for } \begin{cases} V - t_{i,k} \geq Q_{i,k+1}, \\ V - t_{i,k} > Q_{i,k}, \end{cases} \\ \sum\limits_{n=V-t_{i,k}+1}^{Q_{i,k+1}} [P_n]_V & \text{for } \begin{cases} V - t_{i,k} < Q_{i,k+1}, \\ V - t_{i,k} > Q_{i,k}, \end{cases} \\ \sum\limits_{n=Q_{i,k}+1}^{Q_{i,k+1}} [P_n]_V & \text{for } \begin{cases} V - t_{i,k} < Q_{i,k+1}, \\ V - t_{i,k} \leq Q_{i,k}, \end{cases} \end{cases} \tag{10}$$

and $Q_{i,q+1} = V$.

For a given occupancy state n of the system, we can determine the values $y_{i,k}(n)$ of the service streams of class i in the post-threshold area k. The values correspond to the number of class i calls being serviced in the occupancy state n. The parameters $y_{i,k}(n)$ for MTS system can be determined by the following formula [2, 17]:

$$y_{i,k}(n) = A_{i,k}\, \sigma_{i,k}(n - t_{i,k}) \left[P_{n-t_{i,k}}\right]_V / \left[P_n\right]_V$$
$$\text{for } 0 \leq k \leq q \text{ and } Q_{i,k} + t_{i,k} < n \leq Q_{i,k+1} + t_{i,k}. \tag{11}$$

2.2. THRESHOLD SYSTEMS WITH THE PCT1 AND PCT2 TRAFFIC

Consider now a threshold system with the capacity of V BBUs, which is offered two types of traffic streams: M_1 Erlang streams (PCT1[4]streams) and M_2 Engset streams (PCT2[5] streams). The call intensity of the class i Erlang stream in the post-threshold area k is $\lambda_{i,k}$, whereas the call intensity of the class j Engset stream is $\lambda_j(y_{j,k}(n))$ and is dependant on the number of $y_{j,k}(n)$ of currently serviced calls of the stream in the state n in the post-threshold area k:

$$\lambda(y_{j,k}(n)) = (N_j - y_{j,k}(n))\gamma_j, \tag{12}$$

[4]PCT1 – Pure Chance Traffic Type One – type of traffic in which we assume that the service times are exponentially distributed and the arrival process is a Poisson process [7]. This type of traffic is also known as Erlang traffic.

[5]PCT2 – Pure Chance Traffic type Two – type of traffic in which we assume that the service times are exponentially distributed and the arrival process is formed by the limited number of sources. This type of traffic is known as Engset traffic.

where γ_j is the call intensity of a single idle PCT2 traffic source, while the number of PCT2 traffic sources of class j is equal to N_j. The PCT2 traffic intensity, offered by one idle single source of class j in the post-threshold area k, is then equal to:

$$\alpha_{j,k} = \gamma_j / \mu_{j,k}. \tag{13}$$

The basis of the analytical modeling of multithreshold systems that are offered different multi-rate traffic streams of the type PCT1 and PCT2 is the GKRR, expressed – for threshold systems – by Equation (8). It is assumed in the method that the number $y_{j,k}(n)$ of Engset class j calls being serviced in state n is the same as the number of calls being serviced in the equivalent Erlang stream that generates the offered traffic with the intensity $A_{j,k} = N_j \alpha_{j,k}$, equal to the traffic offered by all sources of class j [3,4]. If the number of calls of Engset stream being serviced in subsequent macrostates is known then Equation (8) can be rewritten in the form which include nature of PCT2 stream:

$$n\left[P_n\right]_V = \sum_{i=1}^{M_1}\sum_{k=0}^{q_i} A_{i,k} t_{i,k} \sigma_{i,k}(n - t_{i,k}) \left[P_{n-t_{i,k}}\right]_V + \\ + \sum_{j=1}^{M_2}\sum_{k=0}^{q_j} A_{j,k}(n - t_{j,k}) t_{j,k} \sigma_{j,k}(n - t_{j,k}) \left[P_{n-t_{j,k}}\right]_V, \tag{14}$$

where $A_{j,k}(n)$ is the PCT2 traffic offered by the class j Engset stream in the state n, in the post-threshold area k:

$$A_{j,k}(n) = [N_{j,k} - y_{j,k}(n)]\alpha_{j,k}, \tag{15}$$

whereas the coefficients $\sigma_{i,k}(n)$ and $\sigma_{j,k}$(n) are determined according to (6). It should be emphasized that the parameter $y_{j,k}$ is determined on the basis of (11) with the initial assumption that the value of the offered traffic $A_{j,k}$ does not depend on the state and is $N_j \alpha_{j,k}$ [3].

Having the occupancy distribution (14) we can determine the blocking probability for particular traffic classes on the basis of Equation (9) in multiservice systems, including UMTS.

3. RESOURCE ALLOCATION IN WCDMA

System UMTS is a soft-capacity system using the WCDMA radio interface (*Wideband Code Division Multiple Access*). This interface has a great theoretical capacity in case of an isolated cell. Simultaneously, the available capacity is limited due to different types of interference [12]: co-channel interference within a cell, outer co-channel interference, adjacent channels interference, all possible noise and interference coming from other systems and sources, both broadband and narrowband. The occurrence of those types of interference means that in the WCDMA radio interface, together with increasing traffic load, there is also an increase in noise generated by other users who are serviced by the same cell or by different cells. In order to ensure an appropriate service level, it is necessary to limit the number of allocated resources. Therefore, the soft capacity of the WCDMA radio interface is called the noise limited capacity. Thus, in the radio interface of the UMTS system allocation does not consist in adding bit rates, however, it consists in adding noise loads.

The noise load factor for a single traffic source of class i can be determined on the basis of

Table 1. Exemplary loads of radio interface WCDMA for the service of calls of different classes

Service (i)	I	II	III	IV
W [Mchip/s]	3.84			
R_i [kb/s]	12.2	64	144	384
ν_i	0.67	1	1	1
E_b/N_0 [db]	4	2	1.5	1
L_i	0.005	0.026	0.050	0.112
I - Speech				
II, III, IV - Data transmission				

a formula proposed in [12], which for multirate traffic takes the following form:

$$L_c = \frac{1}{1 + \frac{W}{\left(\frac{E_b}{N_0}\right)_c R_c \nu_c}}. \tag{16}$$

In formula (16) we adopted the following notation:

W – chip rate of spreading signal,

R_c – bit rate of data signal coming from one traffic source of class c[6],

ν_c – activity factor of traffic source of class c, which defines percentage of holding time of transmitting channel, in which source is active, i.e. transmits signal with bit rate R_c,

E_b/N_0 – ratio of energy per bit to noise spectral density,

L_c – load factor of radio interface for the service of a call of class c.

Table 1 shows exemplary WCDMA radio interface loads for the service of calls of different classes.

Let us notice that the load factor L_c is dimensionless and defines a fraction of possible interface load. This factor shows also the non-linear relation between the percentage interface load and the bit rate of a given traffic source. Based on known load factors of single traffic sources, it is possible to define the total load η_{UL} for the uplink:

$$\eta_{UL} = (1+\delta)\left(\sum_{i=1}^{M_1} N_i L_i + \sum_{j=1}^{M_2} N_j L_j\right), \tag{17}$$

where N_i and N_j is the number of serviced traffic sources of class i and j, respectively, in the considered uplink, δ is a ratio of the interference from other cells to the interference within a given cell.

The total load for the downlink can be written based on a formula proposed in [12] which in the case of dividing traffic sources into appropriate classes takes the following form:

$$\eta_{DL} = (1-\xi+\delta)\left(\sum_{i=1}^{M_1} N_i L_i + \sum_{j=1}^{M_2} N_j L_j\right), \tag{18}$$

[6]In the article the letter i denotes a PTC1 traffic class, the letter j – a PTC2 traffic class, and the letter c – an arbitrary traffic class (PCT1 or PCT2).

where ξ indicates the influence of the usage of channel codes based on the technique of Orthogonal Variable Spreading Factors.

Due to multiservice character of the UMTS network the radio interface services a few traffic types and each of them requires different bit rates for servicing a call. Universal Mobile Telecommunication System – in view of the throughput of performed services – can be considered as a discrete network with integrated services, in which the so called Basic Bandwidth Unit can be expressed with a fraction of the link load factor. In multirate systems the value of BBU is assumed to be less or equal to the greatest common divisor (GCD) of the resources required by particular call streams [15]. For the WCDMA radio interface we can write:

$$L_{\text{BBU}} = \text{GCD}(L_1, L_2, \ldots, L_{M_1+M_2}). \tag{19}$$

Now, the capacity of a system (interface) can be expressed in the number of BBUs determined above:

$$V = \lfloor \eta / L_{\text{BBU}} \rfloor, \tag{20}$$

where η is the radio interface capacity for the uplink or the downlink. Similarly, we can express the number of BBUs required by a call of a given class:

$$t_c = \lceil L_c / L_{\text{BBU}} \rceil, \tag{21}$$

4. PERCENTAGE SHARE OF CALLS IN THRESHOLD AREAS

Let us reconsider the multithreshold system which is offered two types of traffic streams: M_1 Erlang streams and M_2 Engset streams. The capacity of the considered system is equal to V BBUs. For each class c calls, a set of thresholds $\{Q_{c,1}, Q_{c,2}, \ldots, Q_{c,q}\}$, is individually introduced. Each set related to class c calls includes q_c elements. For each threshold k and for each traffic class c, a couple of parameters $t_{c,k}$ and $\mu_{c,k}^{-1}$ is determined.

Let us determine now the percentage share $d_{c,k}$ of class c calls, accepted in the post-threshold area k, in the number of all accepted calls in the system. The parameter $d_{c,k}$ allows us – at the stage of system dimensioning – to estimate the number of calls accepted to the service with decreased qualitative parameters, i.e. with reduced bandwidth, from the initially required value $t_{c,0}$ to – depending on the systems state – the value $t_{c,k}$.

The basis of the estimation of the parameter $d_{c,k}$ in MTS system with elastic and adaptive services is the occupancy distribution (14). This distribution allows us to determine the average number $y_{c,k}(n)$ of class c calls being serviced in the pre-threshold area and in the post-threshold areas on the basis of Equation (11). Having the occupancy distribution in MTS system and the values of parameters $y_{c,k}(n)$, we can determine the percentage share of class c calls in the threshold area k on the basis of the following equation:

$$d_{c,k} = \frac{\sum_{n=0}^{V} y_{c,k}(n) \mu_{c,k} [P_n]_V}{\sum_{n=0}^{V} \sum_{w=1}^{M_1+M_2} \sum_{k=0}^{q_w} y_{w,k}(n) \mu_{w,k} [P_n]_V} \cdot 100\%, \tag{22}$$

where

$$\sum_{c=1}^{M_1+M_2} \sum_{k=0}^{q_c} d_{c,k} = 100\%. \tag{23}$$

The numerator of Equation (22) defines the average number of class c calls, being serviced in time unit in the post-threshold area k, whereas the denominator defines the average number of calls of all classes, being serviced in time unit in the system (in all threshold areas).

5. NUMERICAL RESULTS

The presented method for determining the percentage share of class c calls serviced in the post-threshold area k in the system with PCT1 and PCT2 multirate traffic is an approximate method. In order to determine the accuracy of the proposed model, the results of the calculations were compared with the simulation data.

The study was carried out for a single UTMS cell which was offered 4 classes of service. It was assumed that the number of subscribers using the voice service was much higher than the capacity of the system, which allowed us to model the call stream of the service by the stream of the type PCT1. For the remaining services, serviced in the UMTS cell, it was assumed that their call streams were PCT2 streams and that the intensity of the streams decreases with the number of serviced users.

The study was carried out for a cell with the radio interface capacity amounting to 120 BBUs. The resources demanded by particular traffic classes were expressed in the number of basic bandwidth units (noise units). It was assumed that the voice service with the required bandwidth $L_1 = 1$ BBU was a streaming service and the required resources would not be decreased with the increase in load of the system. For the second and the third traffic classes it was assumed that they were elastic services for which along with the decrease in a demanded bandwidth the holding time increases proportionally. For the fourth traffic class, it was assumed that it was an adaptive service, for which, with the decrease in the available bandwidth, the holding time did not increase. For each traffic class the number of thresholds, the corresponding number of required BBUs and the holding times in each threshold area, is defined:

- first class: $q_1 = 0$; $t_{1,0} = 1$ BBU, $\mu_{1,0}^{-1} = 1$
- second class: $q_2 = 1$; $Q_{2,1} = 105$ BBUs; $t_{2,0} = 4$ BBUs, $\mu_{2,0}^{-1} = 1$; $t_{2,1} = 2$ BBUs, $\mu_{2,1}^{-1} = 2$; $N_2 = 40$
- third class: $q_3 = 2$; $Q_{3,1} = 92$ BBUs, $Q_{3,2} = 107$ BBUs; $t_{3,0} = 7$ BBUs, $\mu_{3,0}^{-1} = 1$; $t_{3,1} = 5$ BBUs, $\mu_{3,1}^{-1} = 1.4$; $t_{3,2} = 3$ BBUs, $\mu_{3,2}^{-1} = 2.333$; $N_3 = 50$
- fourth class: $q_4 = 2$; $Q_{4,1} = 90$ BBUs, $Q_{4,2} = 101$ BBUs; $t_{4,0} = 10$ BBUs, $\mu_{4,0}^{-1} = 1$; $t_{4,1} = 8$ BBUs, $\mu_{4,1}^{-1} = 1.25$; $t_{4,2} = 6$ BBUs, $\mu_{4,2}^{-1} = 1.667$; $N_4 = 50$

The percentage share of calls of each traffic class in pre-threshold area and post-threshold areas is presented in Figures 1–4. The simulation results are shown in the figures in the form of appropriately denoted points with 95-percent confidence interval, calculated according to the t-Student distribution for 5 series, with 1000000 calls (of the class generating the least number of calls) in each series. In many cases the value of the confidence interval is lower than the height of the sign used to indicate the value of the simulation experiment. The results of blocking probability are presented depending on the average value of traffic offered to a single bandwidth unit of the system: $a = \frac{\sum_{i=1}^{M_1} A_i t_i + \sum_{j=1}^{M_2} N_j \alpha_j t_j}{V}$. The system was offered traffic in the following proportions of $\frac{\lambda_1}{\mu_1} t_1 : \frac{\lambda_2}{\mu_2} t_2 : \ldots : \frac{\lambda_m}{\mu_m} t_m = 1 : 1 : \ldots : 1$.

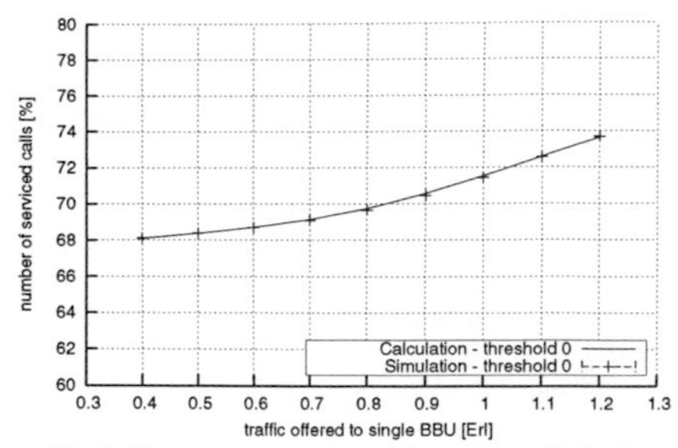

Fig. 1. The average number of first class calls in the pre-threshold area

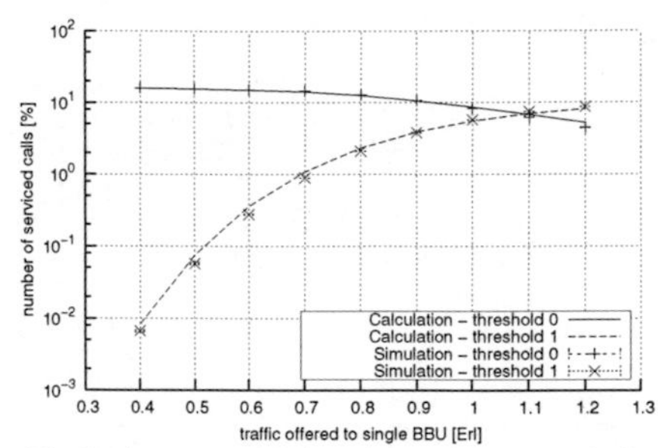

Fig. 2. The average number of second class calls in the pre-threshold area and post-threshold area

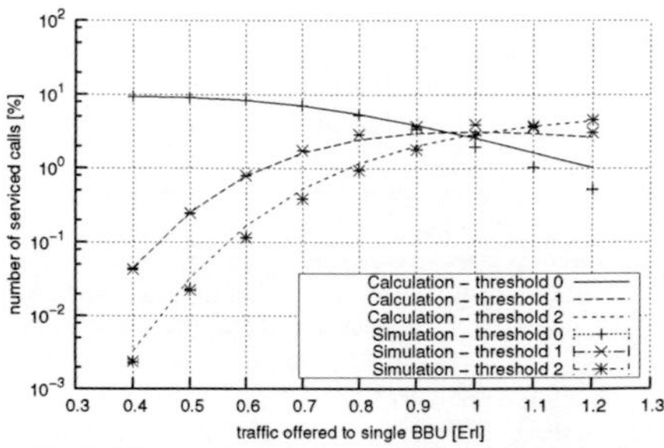

Fig. 3. The average number of third class calls in the pre-threshold area and post-threshold areas

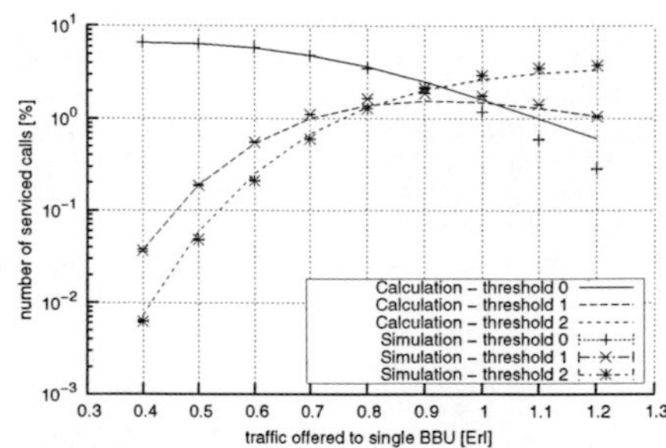

Fig. 4. The average number of fourth class calls in the pre-threshold area and post-threshold areas

Figure 1 allows us to observe the share of streaming class calls for which the required resources cannot be reduced along with an increase in load of the system. Consequently, no post-threshold areas are defined for this class. The share of the first class calls remains approximately constant for the whole range of traffic intensity. In the case of the other classes, we observe (Figures 2–4) that the share of serviced calls in particular post-threshold areas changes along with an increase in the value of offered traffic. For light load of the system the calls serviced in the pre-threshold area have the greater share. An increase in the value of offered traffic causes an increase in the number of calls serviced in the post-threshold areas.

6. CONCLUSIONS

In the paper it is proposed a new analytical method of estimation of percentage share of calls serviced in a pre-threshold area and post-threshold areas with specific quality of service parameters. The proposed method is based on the model of multithreshold systems which is offered a mixture of multirate traffic streams generated by infinite and finite source population. The proposed method allows us to estimate – at the stage of network dimensioning – the degree of degradation of quality of service parameters of services offered to the network.

REFERENCES

[1] BZIUK W., *Approximate state probabilities in large shared multi-rate loss systems with an application to trunk reservation*, in Proceedings of 2nd Polish-German Teletraffic Symposium (9th Polish Teletraffic Symposium), Gdańsk, Poland, 2002, pp. 145–152.

[2] GŁĄBOWSKI M., *Continuous threshold model for multi-service wireless systems with PCT1 and PCT2 traffic*, in Proceedings of the 7th International Symposium on Communications and Information Technologies, Sydney, Australia, Oct. 2007.

[3] GŁĄBOWSKI M., *Modelling of state-dependent multi-rate systems carrying BPP traffic*, Annals of Telecommunications, 63 (2008), pp. 393–407. DOI: 10.1007/s12243-008-0034-5.

[4] GŁĄBOWSKI M. AND STASIAK M., *An approximate model of the full-availability group with multi-rate traffic and a finite source population*, in Proceedings of 3rd Polish-German Teletraffic Symposium, P. Buchholtz, R. Lehnert, and M. Pióro, eds., Dresden, Germany, Sept. 2004, VDE Verlag, pp. 195–204.

[5] HARTMANN H. L. AND KNOKE M., *The one-level functional equation of multi-rate loss systems*, European Transactions on Telecommunications, 14 (2003), pp. 107–118.

[6] IVERSEN V., *The exact evaluation of multi-service loss systems with access control*, in Seventh Nordic Teletraffic Seminar (NTS-7), Lund, Sweden, Aug. 1987, pp. 56–61.

[7] IVERSEN V., ED., *Teletraffic Engineering Handbook*, ITU-D, Study Group 2, Question 16/2, Geneva, Dec. 2003.

[8] KALLOS G. A., VASSILAKIS V. G., MOSCHOLIOS I. D., AND LOGOTHETIS M. D., *Performance modelling of W-CDMA networks supporting elastic and adaptive trafic*, in Proc. 4th International Working Conference on Performance Modelling and Evaluation of Heterogeneous Networks (HET-NETs '06), Ilkley, 2006.

[9] KAUFMAN J. S., *Blocking with retrials in a completly shared recource environment*, Journal of Performance Evaluation, 15 (1992), pp. 99–113.

[10] KAUFMAN J. S., *Blocking in a completely shared resource environment with state dependent resource and residency requirements*, in IEEE INFOCOM '92: Proceedings of the eleventh annual joint conference of the IEEE computer and communications societies on One world through communications (Vol. 3), Los Alamitos, CA, USA, 1992, IEEE Computer Society Press, pp. 2224–2232.

[11] KELLY F., *Loss networks*, The Annals of Applied Probability, 1 (1991), pp. 319–378.

[12] LAIHO J., WACKER A., AND NOVOSAD T., *Radio Network Planning and Optimization for UMTS*, John Wiley & Sons, Ltd., 2006.

[13] MOSCHOLIOS I. D., LOGOTHETIS M. D., AND KOKKINAKIS G. K., *Connection-dependent threshold model: a generalization of the Erlang multiple rate loss model*, Journal of Performance Evaluation, 48 (2002), pp. 177–200.

[14] MOSCHOLIOS I. D., LOGOTHETIS M. D., AND KOKKINAKIS G. K., *Call-burst blocking of on-off traffic sources with retrials under the complete sharing policy.*, Perform. Eval., 59 (2005), pp. 279–312.

[15] ROBERTS J., MOCCI V., AND VIRTAMO I., eds., *Broadband Network Teletraffic, Final Report of Action COST 242*, Commission of the European Communities, Springer, Berlin, 1996.

[16] STAEHLE D. AND MÄDER A., *An analytic approximation of the uplink capacity in a UMTS network with heterogeneous traffic*, in 18th International Teletraffic Congress (ITC18), Berlin, 2003, pp. 81–91.

[17] STASIAK M. AND GŁĄBOWSKI M., *A simple approximation of the link model with reservation by a one-dimensional Markov chain*, Journal of Performance Evaluation, 41 (2000), pp. 195–208.

Key words - Network Calculus, Effective Bandwidth, Effective Capacity, Statistical Multiplexing

Kishore ANGRISHI, Ulrich KILLAT [1]

ANALYSIS OF STOCHASTIC PERFORMANCE BOUNDS IN FEED-FORWARD NETWORKS

The statistical network calculus is an evolving methodology for backlog and delay analysis of networks that can account for statistical multiplexing gain. However, the multiplexing gain accomplished is limited to the network ingress due to the lack of information about the statistical characteristics of flows inside the network. This paper suggests the usage of the well-known notion of effective bandwidth to describe the arrival process of a flow along its path in a feed-forward network. As a result, two disparate methods for the analysis of end-to-end stochastic performance bounds in such a network can be defined. The first method is based on a per-hop analysis benefiting from the (approximate) invariance property of effective bandwidth. The second method uses the concept of effective capacity together with effective bandwidth in the framework of statistical network calculus for an efficient end-to-end analysis of the feed-forward network. The significance of these two novel approaches is that, apart from achieving efficient utilization of statistical multiplexing of independent flows inside the network, the computed stochastic performance bounds of a feed-forward network scale linearly with the number of nodes.

1. Introduction

The increasing number of real time applications over the Internet has motivated the study of Quality of Service (QoS) guarantees. One of the popular developments in recent times is the theory of network calculus [2, 3]. The network calculus is a framework for analyzing worst-case delays and backlog in a network where the traffic, and sometimes also the service, is characterized in terms of envelope functions. However, the worst case deterministic performance bounds are often conservative and far away from practical use since they lead to underutilization of the network resources. This has motivated the search for a probabilistic extension of the network calculus, commonly referred to as "statistical network calculus", which describes arrivals and services probabilistically while preserving the elegance and expressiveness of the original framework. A significant step towards statistical network calculus is presented in [4,5,9], where the concept of effective envelope and effective service envelope were introduced to describe arrivals and service in a network node respectively. The main advantage of statistical network calculus over its deterministic counterpart is the inclusion of statistical multiplexing gain. However this utilization of statistical multiplexing is limited to network ingress. This is because, the stochastic information about the arrival flows are lost once the effective envelope

[1]Hamburg University of Technology, Institute for Communication Networks, E-4, D-21071, Hamburg, Germany, {kishore.angrishi,killat}@tu-harburg.de

is derived at the network ingress [12, 16]. Accordingly, no statistical multiplexing can be observed inside the network. To this end, effective bandwidth which is a better statistical descriptor of an arrival process can be used inside the network instead of effective envelope. It captures the effect of the rate fluctuation, burstiness and represents the rate at which the considered flow needs to be served to satisfy certain quality of service demands. One of the important properties of effective bandwidth is its *(approximate) invariance* [7, 11] while passing a network node even for a small number of flows. The relationship between network calculus and effective bandwidth has been first investigated by Chang [2]. Later a formal relationship between effective bandwidth and statistical network calculus was proposed in [1], which was used to derive effective envelope at the network ingress from the effective bandwidth of the aggregate flow.

Another important aspect of end-to-end performance bound analysis is its scalability in the number of traversed nodes n. The standards set by deterministic network calculus are a scaling in $O(n^2)$ for bounds that are derived iteratively compared to $O(n)$ obtained from end-to-end convoluted service curves [3]. In a probabilistic environment $O(n^2)$ has been achieved for concatenated effective service curves [1] where the quadratic scaling is due to the increasing length of end-to-end busy periods. Recently, the requirement for busy period bounds was eliminated using sophisticated rate correction terms which enable a scaling in $O(n{\cdot}log\, n)$ as shown for traffic with exponentially bounded burstiness in [9]. In [16], authors have developed an intuitive, system-theoretic formulation of a probabilistic network calculus with moment generating functions which achieves the target scaling in $O(n)$.

In this paper, we introduce two approaches for an efficient end-to-end performance bound analysis in a feed-forward network, which uses effective bandwidth to describe the arrival process inside the network within the framework of statistical network calculus. The significance of these two novel approaches is that, apart from achieving efficient utilization of statistical multiplexing of independent flows inside the network, the computed stochastic performance bounds of a feed-forward network scale linearly with the number of nodes. The first approach uses the (approximate) invariance property of effective bandwidth [7, 11] within the framework of statistical network calculus as discussed in [12], to perform the iterative per-hop performance bound (e.g., delay bound) analysis in a feed-forward network. This approach requires a minimum number of independent flows, referred as effective threshold [14], to be multiplexed at each node inside the network to observe (approximate) invariance property of effective bandwidth. The second approach uses the framework of statistical network calculus with effective bandwidth and effective capacity [13] to perform the end-to-end performance bound analysis in a feed-forward network. In our earlier work in [13], we have used the effective bandwidth and effective capacity within statistical network calculus to perform per-hop delay bound analysis. However, the per-hop delay bound analysis described in [13] is found not to yield a linear scaling of delay bounds as can be obtained by the end-to-end delay bound analysis presented in this paper. The second approach differs from the probabilistic network calculus using moment generating functions from [16] in the following two aspects: Firstly, the moment generating functions of arrivals have the unpleasant property that they grow to infinity very quickly, especially the product rule for multiplexing is challenging for a large number of flows. It can however be solved efficiently for effective bandwidths, where the logarithm converts the product into a sum. Secondly, the bound for the output effective bandwidth is conservative due to the conventional de-convolution operation of univariate moment generating functions evaluated over the time interval of $(0, \infty)$. This bound can be improved by restricting the evaluation time interval to busy period bound [4] or to dominant time scale (DTS) [15]. These improvements are included into our new end-to-end statistical network calculus framework.

In the remainder of this paper, we analyze only flows that have stationary independent increments. The discrete time $t \in N = \{0, 1, 2, \dots\}$ is used. We assume the flows are independent at the network ingress. The analysis described in this paper assumes the network to be feed-forward with respect to source-destination pairs.

The sequence of this paper is organized as follows: In the next section, we will recapitulate the required background on statistical network calculus. The approach using (approximate) invariance property of effective bandwidth within statistical network calculus is presented in section 3. In section 4, we introduce our end-to-end statistical network calculus with effective bandwidth and effective capacity. We present the numerical evaluation and the conclusions in the last section.

2. Statistical Network Calculus

This section provides a brief overview on statistical network calculus [4, 5]. Arrival and departure processes of a system are described by real-valued cumulative functions $A(t)$ and $B(t)$ respectively, which represent the amount of data observed in the interval $(0, t]$. Clearly, $A(t)$ and $B(t)$ are nonnegative and increasing in t. An effective envelope for an arrival process A with a violation probability of $\varepsilon_g \in [0, 1]$ is defined as a non-negative function $\mathcal{G}^{\varepsilon_g}$ such that for all $t \geq 0$

$$P\{A(t+\tau) - A(t) \leq \mathcal{G}^{\varepsilon_g}(\tau)\} \geq 1 - \varepsilon_g \tag{1}$$

In other words, an effective envelope provides a stationary bound for an arrival process. Effective envelopes can be obtained for individual flows, as well as for multiplexed arrivals [1].

Assume $S(t)$ is a random process that denotes the service offered by the system in the interval $(0, t]$ where $S(t)$ is nonnegative, increasing in t for all $t \geq 0$. A (minimum) effective service envelope with a violation probability of $\varepsilon_s \in [0, 1]$ for an arrival process A is defined as a non-negative function $\mathcal{S}^{\varepsilon_s}$ that satisfies for all $t \geq 0$,

$$P\{B(t) \geq A \otimes \mathcal{S}^{\varepsilon_s}(t)\} \geq 1 - \varepsilon_s \tag{2}$$

where $\otimes$ is the min-plus convolution operator [3]. By letting $\varepsilon_g \to 0$ in equation (1) and $\varepsilon_s \to 0$ in equation (2), we recover the arrival and service curves of the deterministic calculus [2, 3] with probability one.

For the calculus we make the assumption that there exists a number $T^{\varepsilon_b} < \infty$ such that for all $t \geq 0$,

$$P\{\exists \tau \leq T^{\varepsilon_b} : B(t) \geq A(t-\tau) + \mathcal{S}^{\varepsilon_s}(\tau)\} \geq 1 - (\varepsilon_s + \varepsilon_g) \tag{3}$$

T^{ε_b} is bound on the range of the convolution in equation (2) which holds with violation probability ε_b. Thus, equation (3) is a probabilistic bound on the largest relevant time scale that related arrivals and departures. In a work conserving scheduler, such a bound can be established in terms of a probabilistic bound of the busy period of the scheduler, where a busy period for a given time t is the maximal time interval containing t during which the backlog from the flows remains positive [1].

The performance bounds can be established in terms of min-plus algebra operations on effective envelopes and effective service curves. Note that we are dealing with three violation probabilities: ε_g is the probability that arrivals violate the effective envelope, ε_s is the probability that the service violates the effective service curve, and ε_b is the probability that the bound on the time scale T^{ε_b} is violated.

Consider a node that offers service $S(t)$ with effective service envelope $\mathcal{S}^{\varepsilon_s}$ to the arrival process $A(t)$ with effective envelope $\mathcal{G}^{\varepsilon_g}$, and assume T^{ε_b} satisfies equation (3). Assume $S(t)$ and $A(t)$ are statistically independent, stationary, then the backlog bound (b) and delay bound (d) that are violated at most with probability $\varepsilon(= \varepsilon_b + \varepsilon_s + \varepsilon_g \cdot T^{\varepsilon_b}) \in [0, 1]$ for any time $t \geq 0$ are,

$$b = \mathcal{G}^{\varepsilon_g} \oslash \mathcal{S}^{\varepsilon_s}(0)\} \quad (4)$$
$$d = \inf\{\tau : \{\mathcal{G}^{\varepsilon_g} \oslash \mathcal{S}^{\varepsilon_s}(-\tau)\} \leq 0\} \quad (5)$$

where $\oslash$ is the min-plus de-convolution operator [3].

The main reason for a statistical network calculus is the utilization of statistical multiplexing within a framework for performance analysis that enables an effective concatenation of network elements. Note, however, that once effective envelopes are fixed, the statistical information about the flow is lost and no further multiplexing gain is feasible inside the network. Further, a stochastic network service envelope leads to end-to-end bounds that scale with $O(n^2)$ for n network elements in sequence, compared to $O(n)$ in deterministic case.

In the following sections we present two approaches to perform end-to-end performance bound analysis, where we propose traffic's effective bandwidth is used to describe its arrival process in the network along its path. This has two advantages: (i) Visibility of the statistical characteristics of the flows inside the network. (ii) Easy handling of traffic characterization due to inherent properties of effective bandwidth e.g, additivity property [6], invariance property and decoupling [7].

3. Per-Hop Analysis with (Approximate) Invariance Property of Effective Bandwidth

In this section, we use the (approximate) invariance property of effective bandwidth to perform iterative per-hop analysis in a feed-forward network to identify end-to-end probabilistic performance bound e.g, delay bound.

The usage of the traffic's effective bandwidth as its descriptor inside the network will help to retain information about its statistical characteristics inside the network. This will help in additional statistical multiplexing gain along the path of the traffic, when multiplexed with other independent cross flows which share the resources with the flow of interest along its path.

Consider a system with finite buffer size B and the buffer is drained at constant rate of C amount of data per unit time (i.e., $S(t) = C \cdot t$). The aggregate arrival traffic is made of L similar, independent, stationary flows, each with effective bandwidth of $\alpha^{(L)}(\theta, t)$. In a stable system, the mean rate of the aggregated arrival process is always less than the service rate (C) at a node. It is shown in [7] that, under many sources limiting regime (infinite sources), if the condition $L \cdot \alpha^{(L)}(0, t) < C$ is strictly valid, the effective bandwidth associated with flows passing a network node does not change, i.e. all flow effective bandwidth stay the same on their path through the network. The proof relies on the fact that when there are many independent sources the queue empties regularly with a high probability. So, how many input processes are needed for this limiting result to be accurate? Numerical simulations suggest that in some cases only a small number of independent inputs are needed to make the input and output look nearly identical. The real question though is how many input processes are needed for reasonable convergence over the scale of interest. We define "Effective Threshold" as the minimum number input processes required at the node to achieve *(approximate) invariance* of the effective bandwidth of output flows. Effective threshold can be identified using the information about the departure process from a node. The following theorem [14] provides a good bound on the effective bandwidth of the departure process. It is a extended version of the output bound studied in [16].

***Theorem 1* [14]** *Let $A(t)$ and $S(t) = C \cdot t$ be two independent processes with stationary increments, representing aggregate arrival and service processes at a node, respectively. Let $\alpha(\theta, t)$ $(= L \cdot \alpha^{(L)}(\theta, t))$ be the effective bandwidth of the aggregate arrival process and $\tilde{\alpha}(\theta, t)$ be the effective bandwidth of the departure process for any $t \geq 0$ and $\theta \geq 0$. Then an upper bound ($\hat{\alpha}_L(\theta, t)$) on the effective bandwidth of the departure process with violation probability $\varepsilon \in (0, 1]$ for any $t \geq 0$ and $\theta \geq 0$ is given by,*

$$P\left\{\tilde{\alpha}(\theta,t) \leq \hat{\alpha}_L(\theta,t) = \frac{1}{\theta t}\left[log\left(\sum_{s=0}^{T^{\varepsilon_b}} e^{\theta((t+s)\cdot\alpha(\theta,t+s)-C\cdot s)}\right)\right]\right\} \geq 1-\varepsilon \tag{6}$$

where T^{ε_b} is the probabilistic busy period bound at the node which, with a small violation probability ε_b, is the smallest value of t complying with

$$\inf_{\theta \geq 0}\{\theta t(\alpha(\theta,t) - C)\} = log\varepsilon_b \tag{7}$$

The existence of probabilistic busy period bound for a constant rate server ($S(t) = C \cdot t$) has been proved in [1].

The above theorem can be used to identify the effective threshold. The effective threshold ($\overline{L}$) is the smallest value being a solution for L of the expression,

$$|\gamma(L,\theta,t) - \gamma(\theta,t)| \leq \zeta \qquad \forall\, t, \theta \geq 0 \tag{8}$$

where,

$$\gamma(\theta,t) = \lim_{L\to\infty} \gamma(L,\theta,t) = \lim_{L\to\infty}\left|\frac{\hat{\alpha}_L(\theta,t) - L\cdot\alpha^{(L)}(\theta,t)}{L\cdot\alpha^{(L)}(\theta,t)}\right| \tag{9}$$

The term $\gamma(L, \theta, t)$ is the relative error of departure process effective bandwidth with respect to input process effective bandwidth and ζ is the tolerance limit for the observance of approximate invariance of effective bandwidth. The value of ζ should be normally less than 1%, preferably, less than $0.1\%\ \forall\, t$. Since the upper bound ($\hat{\alpha}_L$) of effective bandwidth of departure process is used instead of the actual effective bandwidth ($\tilde{\alpha}$) to determine the effective threshold, the value of relative error percentage $\gamma(L, \theta, t)$ may never converge to 0 as the number of arrival flows L tends to ∞. The term $\gamma(\theta, t)$ in equation (8) is used for correction.

Though, the effective threshold obtained is valid for any θ and $t \geq 0$, if we are interested in the probability of overflow at a downstream node we want reasonable convergence of the effective bandwidth at the operating point (θ, t) of that node. The effective threshold at a network node depends directly on the operating point of the node (relevant space parameter θ and time parameter t at a given node) [10].

Efficient models for statistical multiplexing are known from the theory of effective bandwidth. Using effective bandwidth to represent the arrival process inside the network within statistical network calculus, allows inheriting some useful large deviation results for an efficient network analysis. The easy handling of effective bandwidth is mostly due to its additivity. For example, in [6], it has been proved that the effective bandwidth of an aggregate flow consisting of different independent flows of different service classes multiplexed together is simply the sum of their independent single flow effective bandwidth. If $\alpha_x(\theta, t)$ and $\alpha_y(\theta, t)$ are the effective bandwidth of two independent arrival process $X(t)$ and $Y(t)$ for all $t \geq 0$ and $\theta \geq 0$, then the effective bandwidth $\alpha_{x+y}(\theta, t)$ of these aggregated flows is given by

$$\alpha_{x+y}(\theta, t) = \alpha_x(\theta, t) + \alpha_y(\theta, t) \tag{10}$$

From *(approximate) invariance* property, the effective bandwidth of a traffic is the same at all points in the network (though the different nodes will typically have different operating points so the values of the function will be different). Hence, if we can construct an effective envelope for the incoming traffic at each network node using the formal relationship established between these two concepts in [1], the traffic's effective envelope does not deteriorate along the path of the flow inside the network. The constructed effective envelope and effective service curve of the network node are then used to analyze the delay and backlog observed at each node using the equations (5) and (4). The end-to-end performance bound can be obtained by the summation of the observed per-node bound found iteratively along the path of the flow of interest. Since the effective envelope does not deteriorate due to approximate invariance property of effective bandwidth, the end-to-end backlog and delay bounds scale linearly in the number of servers in series, denoted by n.

4. End-to-End Analysis with Effective Bandwidth and Effective Capacity

This section introduces the concept of effective capacity which summarizes the time varying resource availability to the arrival process at a node and derives end-to-end probabilistic performance bounds e.g, delay bound, using statistical network calculus with effective bandwidth and effective capacity.

In order to develop a framework for end-to-end analysis using statistical network calculus with effective bandwidth, we require an analogous notion called effective capacity to represent the stochastic behavior of the service process at a network node.

Definition : *Effective capacity* describes the maximum constant arrival rate of the traffic to a node with a given stochastic service rate to provide an expected QoS. The *effective capacity function* of service process $S(t)$ is defined as

$$\beta(\theta, t) = -\frac{1}{\theta t} logE\left[e^{-\theta \cdot S(t)}\right] \tag{11}$$

where θ, t are the parameters dependent on the multiplexing link, i.e., the characteristics of the stochastic service. The time parameter t corresponds to the most probable duration of the buffer busy period prior to overflow. The space parameter θ indicates the bandwidth fluctuation. The effective capacity of a server determines the constant arrival rate of the traffic somewhere between the average and minimum service rate of the server.

The effective capacities of the nodes in series can be effectively transformed into an equivalent, single effective capacity of the network using the following theorem. This theorem can be seen as the generalization of the Lemma 5 in [13].

Theorem 2 *Let $S^1(t)$,$S^2(t)$ ... $S^N(t)$ be N independent and stationary service processes of N nodes in series and $\beta^1(\theta, t)$, $\beta^2(\theta, t)$ ... $\beta^N(\theta, t)$ be their corresponding effective capacity functions for any $t \geq 0$ and $\theta \geq 0$. Then the effective capacity function of the network of N nodes in series satisfies*

$$t\beta^{(S^1 \otimes S^2 \otimes \cdots \otimes S^N)(t)}(\theta, t) \geq -\frac{(N-1)}{\theta} log(t+1) + (t\beta^1(\theta) \otimes t\beta^2(\theta) \otimes \cdots \otimes t\beta^N(\theta))(t) \tag{12}$$

For a proof of Theorem 2, see the Appendix.

The two performance measures, namely backlog and delay are of particular interest in the context of networking and networked applications. The following theorem is necessary to compute backlog and delay experienced in the network of tandem nodes. It gives a a good bound on the output flow effective bandwidth from a network. This theorem can be seen as the generalization of the Lemma 4 in [13].

Theorem 3 *Let $A(t)$ and $S^1(t)$,$S^2(t)$ ... $S^N(t)$ be independent and stationary processes representing aggregate arrival and service processes of the N tandem nodes, respectively. Let $\alpha(\theta,t)$, $\beta^1(\theta,t)$, $\beta^2(\theta,t)$ ... $\beta^N(\theta,t)$ be their corresponding effective bandwidth and effective capacity functions for any $t \geq 0$ and $\theta \geq 0$. Then the effective bandwidth function of the output process from the network satisfies*

$$t\alpha^{(A\oslash(S^1\otimes S^2\otimes\cdots\otimes S^N))(t)}(\theta,t) \leq \frac{N}{\theta}log(T^{\varepsilon_b}+1)+(t\alpha(\theta)\oslash t\beta^1(\theta)\oslash t\beta^2(\theta)\oslash\cdots\oslash t\beta^N(\theta))(t) \quad (13)$$

where T^{ε_b} is the probabilistic upper bound for the length of any busy period with a small violation probability ϵ_b at the node, given by,

$$T^{\varepsilon_b} = inf\,\{t>0:\inf_{\theta\geq 0}\{\theta(t\alpha(\theta,t)-(t\beta^1(\theta)\otimes t\beta^2(\theta)\otimes\cdots\otimes t\beta^N(\theta))(t)) \\ +(N-1)log(t+1)\} = log\epsilon_b\} \quad (14)$$

The backlog is upper bounded with a violation probability of ε according to

$$b \leq t\alpha^{(A\oslash(S^1\otimes S^2\otimes\cdots\otimes S^N))(t)}(\theta,0) - \frac{log\varepsilon}{\theta} \quad (15)$$

The delay is upper bounded with a violation probability of ε according to

$$t\alpha^{(A\oslash(S^1\otimes S^2\otimes\cdots\otimes S^N))(t)}(\theta,-d) - \frac{log\varepsilon}{\theta} \leq 0 \quad (16)$$

For a proof of Theorem 3, see the Appendix.

The framework presented utilizes the potential of the concatenation property one usually gets from composing systems, which is considered as one of the raisons d'etre of network calculus. It can be observed from Theorem 2 and Theorem 3, that the end-to-end performance bounds derived using equations (15) and (16) scale with $O(n)$, where n is the number of servers in series. This result is of considerable importance, because it shows that probabilistic service curves can actually comply with the renowned end-to-end scaling property of deterministic network calculus.

5. Numerical Result and Conclusion

As an example of our finding we provide numerical delay bounds for n servers with cross traffic in series as shown in Fig.1. The servers each have capacity C and leftover service is determined under the general scheduling model. We use a variant of on-off traffic with independent increments as traffic sources. On-Off traffic models are frequently used to model the behavior of (unregulated) compressed voice sources. As illustrated in Fig. 2, we describe an on-off traffic source as a two-state memoryless process. In the 'On' state, traffic is produced at the peak rate P, and in the 'Off' state, no traffic is

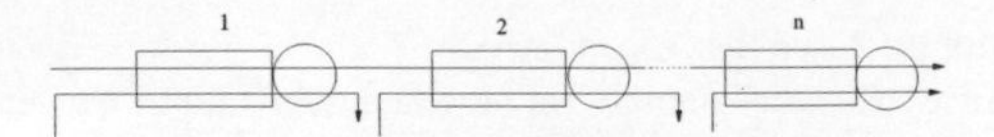

Fig. 1. Tandem servers with cross traffic

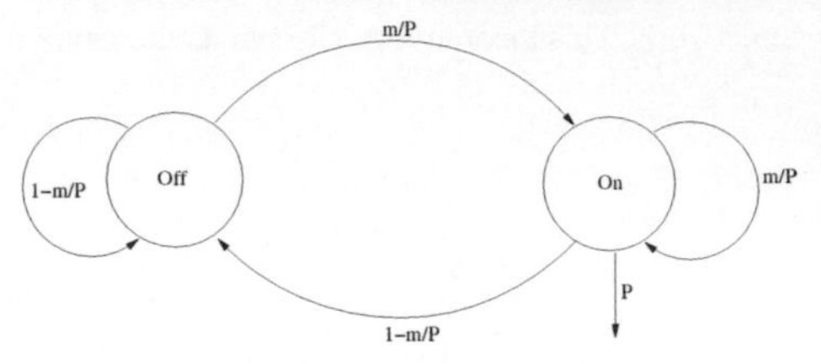

Fig. 2. On-Off Transition Model

produced, with an overall average traffic rate $m < P$. The effective bandwidth of such traffic is given by [2],

$$\alpha(\theta) = \frac{1}{\theta}\left\{log\left(1 + \frac{m}{P}\left(e^{\theta P} - 1\right)\right)\right\} \quad (17)$$

We compute stationary end-to-end delay bound with violation probability $\varepsilon = 10^{-6}$ using equation (16) and compare the results to respective bounds obtained using iterative per-hop delay bound analysis discussed in Section 3. The effective threshold is found to be 33 flows using equation (8) for $\zeta = 1\%$. At each server the current cross traffic is de-multiplexed and fresh, independent cross traffic is multiplexed. Assume the servers are constant rate servers, each with capacity $C = 2Mb/s$. We consider only the situation where $N \cdot m < C =< N \cdot P$, since otherwise the expected backlog is either infinite or zero. Delay bounds are derived for an aggregate of 33 independent cross flows and 12 independent through flows at each server, with on-off traffic parameters $P = 64Kb/s$ and $m = 44.3Kb/s$. For the iterative per-hop method discussed in Section 3, we assume $\varepsilon_g = \varepsilon_s = \varepsilon_b = \frac{10^{-6}}{2-T^{\varepsilon_b}}$ and T^{ε_b} can be found using the equation (3). Fig. 3 shows scaling of delay bounds for n servers in series with 33 cross flows and 12 through flows. The results confirm the scalability of the approach. The bounds from both the methods scale in $O(n)$, where n is the number of servers in series. It can be observed from the graph that the single server delay bound is same for both the methods, where as end-to-end delay bound obtain using equation (16) provides a more conservative bound than the iterative per-hop method.Thus, the important end-to-end scalability of deterministic network calculus has been achieved using probabilistic calculus which effectuates a noticeable statistical multiplexing gain.

REFERENCES

[1] Li, C., Burchard, A., and Liebeherr, J.,: A Network Calculus with Effective Bandwidth, Technical Report CS-2003-20, University of Virginia, 2003.

[2] Chang, C.-S.,: Performance Guarantees in Communication Networks, Springer-Verlag, 2000.

[3] Le Boudec J.-Y., and Thiran P.,: Network Calculus A Theory of Deterministic Queuing Systems for the Internet, ser. LNCS. Springer-Verlag, 2001, no. 2050.

[4] Boorstyn, R.-R., Burchard, A., Liebeherr, J., and Oottamakorn C.,: Statistical Service Assurances for Traffic Scheduling Algorithms, IEEE Journal on Selected Areas in Communications, 18(12):2651-2664, 2000.

[5] Burchard, A., Liebeherr, J., and Patek, S. D.,: A calculus for end-to-end statistical service guarantees (revised), Technical Report CS-2001-19, University of Virginia, Computer Science Department, May 2002.

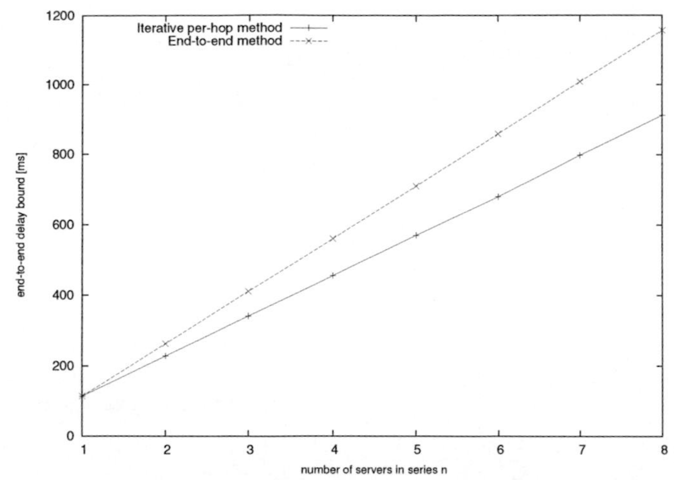

Fig. 3. End-to-end concatenation of servers in series

[6] Chang, C.-S., and Thomas, J.A.,: Effective bandwidth in high speed digital networks, IEEE Journal on Selected Areas in Communications, vol. 13, no. 6, 1995.

[7] Wischik, D.,: The output of a switch, or, effective bandwidths for networks, Queueing Systems, Volume 32, 1999.

[8] Kelly,F. P.,: Notes on effective bandwidths, ser. Royal Statistical Society Lecture Notes. Oxford University, 1996, no. 4.

[9] Ciucu, F., Burchard, A., and Liebeherr, J.,: A network service curve approach for the stochastic analysis of networks, in Proc. ACM SIGMETRICS, 279-290, June 2005.

[10] Aspirot, L., Belzarena, P., Bermolen, P., Ferragut, A., Perera, G., SimÃşn, M.: Quality of service parameters and link operating point estimation based on effective bandwidths, Performance Evaluation 59 (2005), 103-120.

[11] Abendroth, D., and Killat, U.,: An advanced traffic engineering approach based on the approximate invariance of effective bandwidths, Telecommunication Systems Journal, Kluwer Publishers, 2004, vol 27 (2-4).

[12] Angrishi, K., Zhang, S., Killat, U.,: Analysis of a Real-Time Network using Statistical Network Calculus with Approximate Invariance of Effective Bandwidth, 15.ITG/GI - Fachtagung Kommunikation in Verteilten Systemen (KiVS 2007), Bern, Switzerland, March 2007

[13] Angrishi, K., Killat, U.,: Analysis of a Real-Time Network using Statistical Network Calculus with Effective Bandwidth and Effective Capacity, 14. GI/ITG Konferenz Messung, Modellierung und Bewertung von Rechen- und Kommunikationssystemen (MMB 2008), Dortmund, Germany, March 2008

[14] Angrishi, K., Killat, U.,: On the Threshold for Observing Approximate Invariance of Effective Bandwidth, Proceedings of the 11th International Symposium on Performance Evaluation of Computer and Telecommunication Systems (SPECTS'08), Edinburgh, UK, June 2008,

[15] Eun, D.Y., and Shroff, N.,: A Measurement-Analytic Framework for QoS Estimation Based on the Dominant Time Scale, Proc. of IEEE INFOCOM 2001, Anchorage, Alaska, 2001.

[16] Fidler, M.,: An End-to-End Probabilistic Network Calculus with Moment Generating Functions, Proceedings of IWQoS, June 2006.

A. Appendix

Proof of Theorem 2

$$t\beta^{(S^1 \otimes S^2 \otimes \cdots \otimes S^N)(t)}(\theta, t) = -\frac{1}{\theta} logE\left[e^{-\theta (S^1 \otimes S^2 \otimes \cdots \otimes S^N)(t)}\right]$$

$$
\begin{aligned}
&= -\frac{1}{\theta}log(\sum_{z=-\infty}^{\infty} e^{\theta z} P(\sup_{0\le s_n \le \dots s_3 \le s_2 \le t} [-S^1(t-s_2) - S^2(s_2-s_3)\cdots - S^N(s_N)] = z)) \\
&\ge -\frac{1}{\theta}log(\sum_{z=-\infty}^{\infty} \sum_{s_2=0}^{t} \sum_{s_3=0}^{s_2} \cdots \sum_{s_N=0}^{s_{N-1}} e^{\theta z} P([-S^1(t-s_2) - S^2(s_2-s_3)\cdots - S^N(s_N)] = z)) \\
&\ge -\frac{1}{\theta}log\left((t+1)^{N-1} \sup_{0\le s_n \le \dots s_3 \le s_2 \le t} E\left[e^{-\theta[S^1(t-s_2)+S^2(s_2-s_3)\cdots + S^N(s_N)]}\right]\right) \\
&= -\frac{N-1}{\theta}log(t+1) + \inf_{0\le s_n \le \dots s_3 \le s_2 \le t} (t-s_2)\beta^1(\theta, t-s_2) + (s_2-s_3)\beta^2(\theta, s_2-s_3) + \dots \\
&\quad + s_N \beta^N(\theta, s_N) \\
&= -\frac{N-1}{\theta}log(t+1) + (t\beta^1(\theta) \otimes t\beta^2(\theta) \otimes \cdots \otimes t\beta^N(\theta))(t)
\end{aligned}
$$

Hence, proves the claim.

Proof of Theorem 3 if $S^{net} = (S^1 \otimes S^2 \otimes \cdots \otimes S^N)(t)$ and the corresponding effective capacity is β^{net}

$$
\begin{aligned}
t\alpha^{(A\oslash S^{net}(t)}(\theta, t) &= \frac{1}{\theta}logE\left[e^{\theta(A\oslash S^{net})(t)}\right] \\
&= \frac{1}{\theta}log(\sum_{z=-\infty}^{\infty} e^{\theta z} P(\sup_{0\le s}[A(t+s) - S^{net}(s)] = z)) \\
&\le \frac{1}{\theta}log(\sum_{z=-\infty}^{\infty} \sum_{s=0}^{\tau^*} e^{\theta z} P([A(t+s) - S^{net}(s)] = z)) \\
&\le \frac{1}{\theta}log\left((\tau^*+1) \sup_{0\le s\le \tau^*} E\left[e^{\theta[A(t+s)-S^{net}(s)]}\right]\right) \\
&= \frac{log(\tau^*+1)}{\theta} + \sup_{0\le s\le \tau^*} [(t+s)\alpha(\theta, t+s) - s\beta^{net}(\theta, s)]
\end{aligned}
$$

from Theorem 2, we get,

$$
\begin{aligned}
&\le \frac{log(\tau^*+1)}{\theta} + \sup_{0\le s\le \tau^*} \Big[(t+s)\alpha(\theta, t+s) - (-\frac{N-1}{\theta}log(s+1) + \\
&\quad (s\beta^1(\theta) \otimes s\beta^2(\theta) \otimes \cdots \otimes s\beta^N(\theta))(s))\Big] \\
&= \frac{log(\tau^*+1)}{\theta} + \frac{N-1}{\theta}log(\tau^*+1) + \\
&\quad \sup_{0\le s\le \tau^*} [(t+s)\alpha(\theta, t+s) - (s\beta^1(\theta) \otimes s\beta^2(\theta) \otimes \cdots \otimes s\beta^N(\theta))(s)] \\
&= \frac{N}{\theta}log(\tau^*+1) + [t\alpha(\theta) \oslash (t\beta^1(\theta) \otimes t\beta^2(\theta) \otimes \cdots \otimes t\beta^N(\theta))(t)] \\
&= \frac{N}{\theta}log(\tau^*+1) + [t\alpha(\theta) \oslash t\beta^1(\theta) \oslash t\beta^2(\theta) \oslash \cdots \oslash t\beta^N(\theta))(t)]
\end{aligned}
$$

Probabilistic upper bound for the length of any busy period with a small violation probability ϵ at the node, backlog and delay bound which are violated at most with probability ε can be obtained by direct application of Theorem 2 on to their corresponding definitions found in [1].

Hence, proves the claim.

Key words – reservation, Erlang's ideal grading

Sławomir HANCZEWSKI*
Maciej STASIAK*

MODELING OF SYSTEMS WITH RESERVATION BY ERLANG'S IDEAL GRADING

The paper presents a new method for modeling systems with reservation. The method is based on the generalized ideal grading model servicing multirate traffic. The paper proposes an algorithm for determining such an availability value in the ideal grading that causes blocking equalization of different classes of calls. The proposed algorithm was worked out for integer values of the availability parameters. A comparison of the analytical results with the simulation results proves good accuracy of the proposed method.

1. INTRODUCTION

The ideal grading model with single-service traffic is one of the oldest traffic engineering models. The appropriate formula to determine the blocking probability in the group was worked out by A. K. Erlang as early as 1917 [2]. The formula is called *Erlang's interconnection formula.* Even though the ideal grading did not find any practical applications in the past due to a large number of load groups, the system was used for many years for approximate modeling of other telecommunications systems [6]. As it turned out that the characteristics and properties of the majority of homogeneous grading were similar to those of ideal limited-availabilty groups. For example, the Erlang's ideal grading was used for modeling single-service and multi-service outgoing groups in switching networks [3], [11]. The ideal grading was also used to model switching networks with multicast connections [4]. In [12], an approximate ideal grading model servicing multirate traffic, which assumes identical availability value for all classes of calls, is proposed.
The present paper proposes a generalized model of the ideal grading servicing multirate traffic in which each call class is characterized by a different availability. Such an approach has made it possible to model the values of the blocking probabilities of particular classes depending on the changes in the value of the availability parameter. The paper shows that with appropriately matched availability parameters in the ideal grading it is possible to equalize all blocking probabilities, which is, in fact, equivalent to the operation of the resevation mechanism in the full-availability group with multirate traffic. The paper also proposes an appropriate algorithm for determining such an availability value that effects in the blocking equalization in different classes of calls.

* Chair of Telecommunication and Computer Networks, Poznan University of Technology, Poland

The obtained results are promising and indicate further possibilities in implementing the generalized formula of the ideal grading model in determining characteristics and properties of other systems. This approach can also be very effective in modeling systems for 3G mobile networks.

2. STATE-DEPENDENT MULTI-RATE SYSTEMS

The occupancy distribution in state dependent multi-rate systems can be determined on the basis of the so-called generalized Kaufman-Roberts recursion [1],[13]:

$$nP(n) = \sum_{i=1}^{M} a_i t_i \sigma_i (n - t_i) P(n - t_i), \tag{1}$$

where:
$P(n)$ - the state probability, i.e. the probability of an event that there are n busy BBUs in the system,
t_i - the number of demanded BBUs (Basic Bandwidth Unit) by a i class calls. BBU is defined as the greatest common divisor of equivalent bandwidths of all call streams offered to the system [10],
a_i - the traffic offered by the i class calls,
M - the number of classes of calls serviced by the system,
$\sigma_i(n)$ - the conditional probability of passing between the adjacent states of the process associated with the class i call stream. The way of determining the value of this parameter depends on the kind of a considered state-dependent system. If the probabilities of passing $\sigma_i(n)$ are equal to one for all states, the equation (1) is reduced to the Kaufman-Roberts recursion [5], [7]:

$$nP(n) = \sum_{i=1}^{M} a_i t_i P(n - t_i). \tag{2}$$

Formula (2) determines the occupancy distribution in the state independent system i.e. full-availability group with multi-rate traffic streams.

The conditional probability of passing $\sigma_i(n)$ in the state depended system (equation (1)) can be written as follows:

$$\sigma_i(n) = 1 - \gamma_i(n), \tag{3}$$

where $\gamma_i(n)$ is the conditional blocking probability for class i calls in a considered system, calculated under the assumption that in this system n BBUs are busy. Therefore, the total blocking probability in the state-depended multi-rate system for class i calls can be expressed by the following formula:

$$E_i = \sum_{n=0}^{V} \gamma_i(n) P(n). \tag{4}$$

3. FULL-AVAILABILITY GROUP WITH RESERVATION (FAGR)

The aim of the introduction of the reservation mechanism in telecommunications systems is to ensure similar values of the parameters of the quality of service for calls of different classes. For this purpose, the reservation threshold Q_i for each traffic class is designated. The parameter Q_i determines the borderline state of a system, in which servicing class i calls is still possible. All states higher than Q_i belong to the so called reservation space Q_i, in which class i calls will be blocked:

$$R = V - Q_i . \tag{5}$$

According to the equalisation rule [9],[8],[14] the blocking probability in the full-availability group will be the same for all call stream classes if the reservation threshold for all traffic classes is identical and equal to the difference between the total capacity of a group and the value of resources required by the call of maximum demands ($t_M = t_{\max}$):

$$Q = V - t_M . \tag{6}$$

The occupancy distribution in the full-availability group with reservation (FAGR) can be calculated on the basis of equation (1) in which conditional probabilities of passing $\sigma_i(n)$ are determined in the following way:

$$\sigma_i(n) = \begin{cases} 0 & \text{for } n > Q \\ 1 & \text{for } n \le Q \end{cases} . \tag{7}$$

Such a definition of the parameter $\sigma_i(n)$ means that in states higher than Q (reservation space), calls of all traffic classes will be blocked.

The equalized blocking probability for all traffic classes in the FAGR can be determined as follows:

$$E_i = \sum_{n=Q+1}^{V} P(n) . \tag{8}$$

4. ERLANG'S IDEAL GRADING (EIG)

The structure of an Erlang's ideal grading (EIG) is characterized by three parameters: availability d, capacity V and the number of grading groups g. Erlang's ideal grading is a system in which the conditional probability of passing between the adjacent states does not depend on the call intensity, and can be determined combinatorially [2]. The number of grading groups of an Erlang's ideal grading is equal to the number of possible ways of choosing the d channels (BBUs) from their general number V, whereby two grading groups differ from each other by at least one channel. This means that, a separate grading group is intended for each possible combination of d channels. With the same traffic offered to all grading groups and the random hunting strategy, the load of each channel of an Erlang's ideal grading is identical. Moreover, for each combination of busy channels, the occupancy distribution in each grading group is the same. This means, that for an arbitrary "busy" state in n channels ($0 \le n \le V$) in the considered grading, no matter how many n channels from among the possible V channels are busy, the probability of busy j channels of a given grading group ($0 \le j \le d$) is equal to the probability of busy j channels in each other grading group.

4.1 GENERALIZED MODEL OF ERLANG'S IDEAL GRADING

In [12], a model of the ideal grading with multi-rate traffic is proposed. The model assumes that the availability for all classes of calls is the same.

Let us consider now the generalized model in which each call class is characterized by a different availability. This means that a different number of grading groups is related to each of the call class. Figure 1 shows a simple model of the ideal grading with the capacity $V = 3$ BBU's. The group services two classes of calls with the availability $d_1 = 2$, $d_2 = 3$. Hence, the number of load groups for relevant call classes is equal to:

$$g_1 = \binom{V}{d_1} = \binom{3}{2} = 3; \quad g_2 = \binom{V}{d_2} = \binom{3}{3} = 1$$

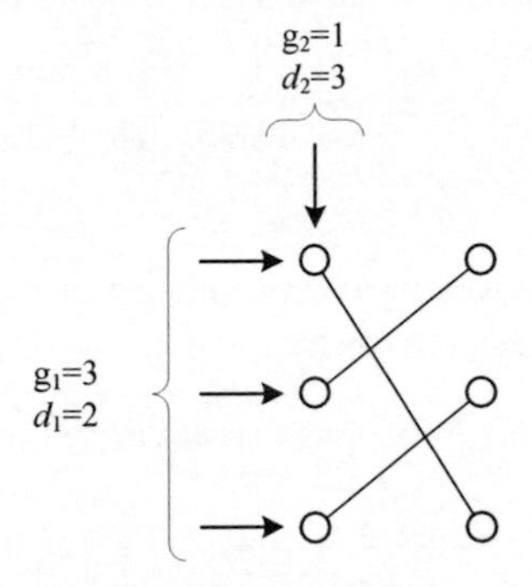

Fig.1. Model of the ideal grading with different availabilities

The occupancy distribution in an Erlang's ideal grading with multi-rate traffic can be determined on the basis of the recursive formula (1). To determine conditional blocking probabilities in the ideal grading with different availabilities for different call classes, model [12] is used, in which the parameters related to a given call of class i are made dependent on the availability d_i attributed to this class (in model [12] these parameters were dependent on the availability d). Due to ideal grading definition, the distribution of busy outgoing BBU's is identical in each group and the conditional probability of blocking of class i calls is equal to the blocking probability in one (given) grading group:

$$\gamma_i(n) = P_{i,n} = \sum_{x=d_i-t_i+1}^{k_i} P_{V,d_i}(n,x). \quad (9)$$

where:
$k_i = n$, if $(d_i - t_i + 1) \le n < d_i$,
$k_i = d_i$, if $n \ge d_i$,
The parameter $P_{V,d_i}(n,x)$ is the conditional probability of x BBU's being busy in one grading group, when the total number of busy BBU's in the system is equal to n. The probability $P_{V,d_i}(n,x)$ can be described by the hypergeometrical distribution [12]:

$$P_{V,d_i}(n,x) = \binom{d_i}{x}\binom{V-d_i}{n-x} \bigg/ \binom{V}{n}. \quad (10)$$

After determining all probabilities $\gamma_i(n)$, the blocking probability for class i calls in the ideal grading carrying a mixture of different multichannel traffic streams can be calculated according to the following formula:

$$E_i = \sum_{n=d_i-t_i+1}^{V} \gamma_i(n) P(n). \quad (11)$$

5. BLOCKING PROBABILITY EQUALIZATION IN ERLANG'S IDEAL GRADING WITH MULTIRATE TRAFFIC

Let us consider an ideal grading servicing two classes of calls. The calls of particular classes require respectively t_1 and t_2 BBUs ($t_1 < t_2$). The capacity of the group is equal to V, while the availability for each of the classes varies and equals respectively d_1 and d_2.

In the considered group, changes in the value of the parameters d_1 and d_2 will result in changes in respective values of blocking probabilities of individual classes. In a particular case, it is possible to equalize the blocking probability of all serviced classes of calls. The study carried out by the authors has proved that the lowest value of the equalized blocking probability is obtained under the assumption that the availability of the second class of calls is constant and is equal to the capacity of the system, i.e. $d_2 = V$. Additionally, it turned out that the equalized blocking probability in the ideal grading following the above assumption is equal to the equalized blocking probability in the full-availability group with reservation (assuming the same capacity of the systems and the identical structure of the offered traffic). In order to determine the value of the parameter d_1 that effects in the equalization of blocking probabilities of serviced calls, the following algorithm can be employed.

In the first step of the algorithm, the initial value of the parameter d_1 is assumed (for example, $d_1^{(0)} = t_1$) and the blocking probability $E_1^{(1)}$ and $E_2^{(1)}$ is determined:

$$(E_1^{(1)}, E_2^{(1)}) = F(A_1, t_1, d_1^{(0)}, A_2, t_2, d_2 = V), \quad (12)$$

where the function F determines blocking probabilities of calls in the ideal grading on the basis of Eq. (9) – (11). The adoption of the initial value $d_1^{(0)} = t_1$ effects in a situation where the blocking probability of the first class takes on higher values than those of the blocking probability for class 2, i.e. $E_1^{(1)} > E_2^{(1)}$.

Each step of the algorithm determines the successive values of the blocking probability on the basis of the value of the parameter d_1, determined in the previous step, while the remaining parameters of the function F do not change:

$$(E_1^{(n)}, E_2^{(n)}) = F(A_1, t_1, d_1^{(n-1)}, A_2, t_2, d_2 = V). \tag{13}$$

The availability parameter in each of the steps of the process is determined in the following way:

$$d_1^{(n)} = d_1^{(n-1)} + 1. \tag{14}$$

The algorithm terminates when the following condition is met:

$$E_1^{(n)} < E_2^{(n)}. \tag{15}$$

Such an approach means that each step of the algorithm decreases the difference between the values of blocking probabilities of the considered classes and the algorithm terminates its operation if the value $E_2^{(n)}$ exceeds the value $E_1^{(n)}$. The result of the operation of the algorithm is the value of the parameter d in $(n-1)$ or in the n-th step of the algorithm according to the following conditions:

$$d_1 = d_1^{(n)} \text{ if } \left|E_1^{(n)} - E_2^{(n)}\right| < \left|E_1^{(n-1)} < E_2^{(n-1)}\right| \quad , \tag{16}$$

$$d_1 = d_1^{(n-1)} \text{ if } \left|E_1^{(n)} - E_2^{(n)}\right| > \left|E_1^{(n-1)} < E_2^{(n-1)}\right| \quad . \tag{17}$$

6. NUMERICAL EXAMPLES

To prove the accuracy and correctness of the proposed analytical model, the results of the analytical calculations were compared with the results of the simulation experiments with sample groups. The operation of the proposed model is presented with the example of the Erlang's ideal grading with the capacity of V=30 BBU's servicing two classes of calls. Further diagrams show the results for the following mixture of classes: $t_1 = 1$, $t_2 = 2$, a_1:a_2=1:1 (Figure 2), $t_1 = 1$, $t_2 = 9$, a_1:a_2=1:1 (Figure 3), $t_1 = 2$, $t_2 = 5$, a_1:a_2=1:1 (Figure 4) in relation to the traffic offered to one BBU: $a = \sum_{i=1}^{M} \frac{a_i t_i}{V}$. The diagrams also show the results of the calculations obtained for the full-availability group with multirate traffic and reservation that equalize losses, i.e. the reservation space R=t_2. When the differences in the number of demanded BBU's are appropriately large in relation to the capacity of the group (for example, $t_2 - t_1 = 8$ with V=30), the results of the probability E_1 take the form of a "steplike curve" (Fig. 4). The above example clearly shows the possibilities of increasing the accuracy of the method by the introduction of non-integer values of the availability to the model, which will be subject to further studies. A comparison of the results for the Erlang's ideal grading and the full-availability group with reservation shows that the blocking probability equalization in both systems takes on identical values. Thus, systems with reservation can be modeled by the Erlang's ideal grading. Figure 5 presents the changes in the parameter d_1 (in the function of traffic offered to one BBU of the group) that effect in the equalization of the blocking probability in the ideal grading. It is noticeable that the value of the parameter d_1, with which the equalization of losses ensues, is not constant and decreases with the increase of the offered traffic. All the results presented in the diagrams confirm the accuracy of the adopted assumptions for the analytical model.

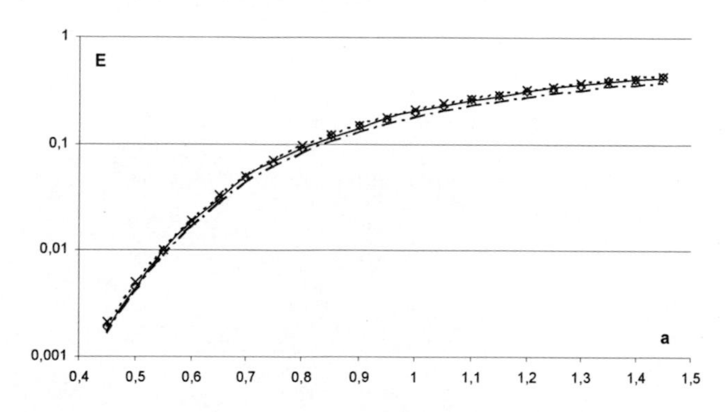

Fig. 2. Blocking probability in Erlang's ideal grading
Simulations: EIG ◇ t_1=1, × t_2=2
Calculations: EIG —— t_1=1, - - - t_2=2
Calculations: FAGR -·- t_1=1, —··— t_2=2

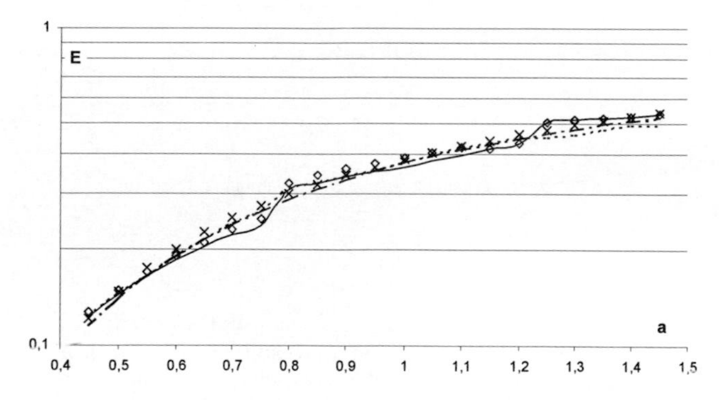

Fig. 3. Blocking probability in Erlang's ideal grading
Simulations: EIG ◇ t_1=1, × t_2=9
Calculations: EIG —— t_1=1, - - - t_2=9
Calculations: FAGR -·- t_1=1, —··— t_2=9

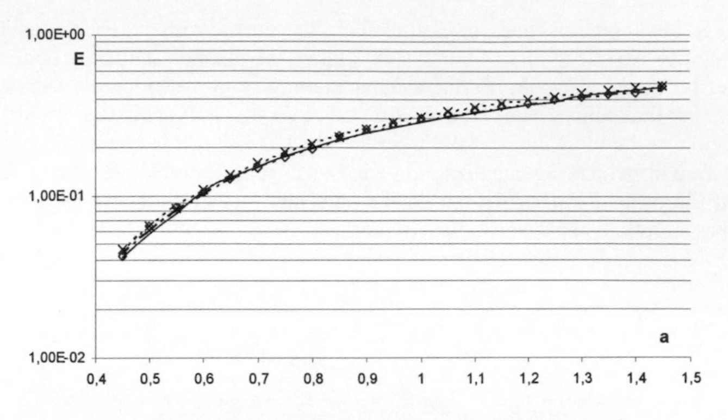

Fig. 4. Blocking probability in Erlang's ideal grading
Simulations: EIG ◇ t_1=2, × t_2=5
Calculations: EIG ——— t_1=2, - - - - t_2=5
Calculations: FAGR - · - t_1=2, — · · — t_2=5

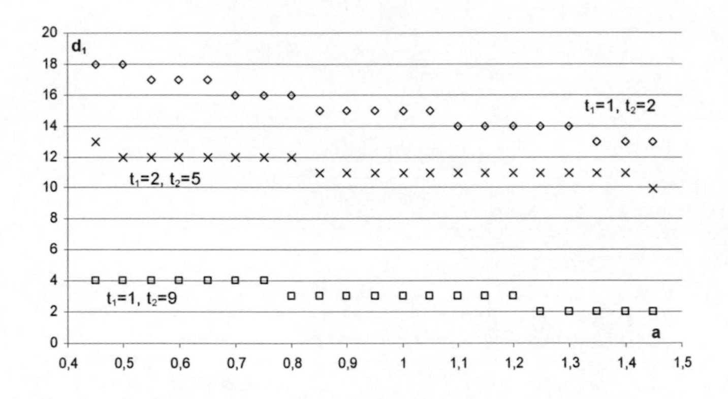

Fig. 5. Changes in parameter d_1 in relation to offered traffic

7. CONCLUSIONS

The paper presents a method for modeling systems with reservation by limited-availability systems. The operation of the method is based on a generalized Erlang's ideal grading servicing multirate traffic. A comparison of the analytical results with the simulation results proves good accuracy of the proposed method. The accuracy of the proposed method can be even greater through further works on developing a method for blocking probability calculation in Erlang's ideal grading for non-integer values of the parameter d_1. The obtained results indicate a possibility of implementing the generalized model of Erlang's ideal grading in determining characteristics of other systems. Such an approach can prove very beneficial in the case of modeling systems in 3G mobile networks.

REFERENCES

[1] BESHAI, M., MANFIELD, D., Multichannel services performance of switching networks. In: Proc. 12th ITC, Torino, Italy (1988) p.5.1A.7

[2] BROCKMEYER E., HALSTROM H., JENSEN A., The life and works of A. K. Erlang, *Acta Polytechnica Scandinavica*, No. 287, 1960.

[3] ERSHOVA E.B., ERSHOV V.A., Cifrowyje sistiemy raspriedielenia informacji, Radio i swiaz, Moskwa, 1983

[4] HANCZEWSKI S., STASIAK M., Point-to-group blocking in 3-stage switching networks with multicast traffic streams. In: Proceedings of SAPIR, Lecture Notes in Computer Science, P. Dini, P. Lorenz, J.N. de Souza (red) 3126:219– 230, Fortaleza, August 2004. Springer.

[5] KAUFMAN J.S., *Blocking in shared resource environment*, IEEE Transactions on Communications, vol. COM-29, No. 10, 1981, pp. 1474-1481.

[6] LOTZE A., History and development of grading theory, Proc. 5th Int. Teletraffic Congress, New York, 1967, s. 148-161.

[7] ROBERTS J. W., *A service system with heterogeneous user requirements*, Performance of data communications systems and their applications, Nort Holland Pub Co, Amsterdam, 1981.

[8] ROBERTS J.W., Teletraffic models for the Telcom 1 Integrated Services Network, Proc. 10th Int. Teletraffic Congress, Montreal, 1983, paper 1.1. 2.

[9] ROBERTS J.W., Ed., *Performance Evaluation and Design of Multiservice Networks, Final Report COST 224*, Commission of the European Communities, Brussels, Holland, 1992.

[10] ROBERTS J.W., MOCCI V. and VIRTAMO I., Eds., *Broadband Network Teletraffic, Final Report of Action COST 242*, Commission of the European Communities, Springer Verlag, Berlin, Germany, 1996.

[11] STASIAK M., Blocage interne point a point dans les reseaux de connexion. *Annals of Telecommunications.*, vol. 43, No. 9-10, 1988, pp. 561-575

[12] STASIAK M., An approximate model of a switching network carrying mixture of different multichannel traffic streams, *IEEE Transactions on Communications,* 1993, 41, 6, pp. 836-840

[13] STASIAK M., Blocking probability in a limited-availability group carrying mixture of different multichannel traffic streams, *Annals of Telecommunications*, 1993, vol. 48, No. 1-2, pp. 71-76

[14] TRAN-GIA P., HUBNER F., An analysis of trunk reservation and grade of service balancing mechanisms in multiservice broadband networks, Proc. Int. Teletraffic Congress Seminar: Modeling and Performance evaluation of ATM technology, La Martynique, 1993.

III. SWITCHING NODE ARCHITECTURES AND PERFORMANCE

Key words – Clos-network, Dispatching Algorithm, Packet Switching, Packet Scheduling

Janusz KLEBAN, Maciej SOBIERAJ*

DELAYED RESPONSE OF CENTRAL ARBITER IN THREE-STAGE BUFFERLESS CLOS-NETWORK SWITCHES

Multiple-stage architectures of switching fabrics are very attractive because they may be used for the implementation of large-capacity switches and routers due to their modularity and scalability. In particular, the three-stage Clos switching fabric is an attractive solution to the next generation packet switching nodes. In a Clos-network switch packet scheduling is needed as there is a large number of points where contention may occur. In order to solve contention problems it is possible to use a central arbiter, or distributed arbiters. In the distributed manner, each output has its own arbiter operating independently from others where it is necessary to send many request-grant-accept signals. It is very difficult to implement such arbitration in the real environment because of time constraints. A central arbiter may also create a bottleneck because of time complexity. In this paper we detail the influence of delayed response of a central arbiter on the performance of the three-stage bufferless Clos switching fabric. The new input port - output port matching scheme called the longest VOQ matching (LVM) is proposed and evaluated. The algorithm sends information to the central arbiter about VOQs size. The arbiter can analyze the information during several time slots, and after that time the input-output matching and route assignment patterns are sent to the Clos switching fabric. We show via simulation that the arbiter may respond after several time slots, and is still possible to obtain acceptable results for uniform and nonuniform traffic distribution patterns.

1. INTRODUCTION

The continuing growth of the Internet traffic implies a demand for scalable switches/routers with high speed interfaces and large switching capacity. The main part of each network node is switching fabric which transfers a packet from its input link to its output link. The high-performance switches internally operate on fixed-size data units, called cells from the ATM jargon. This means that in the case of variable-size packets on transmission lines, as it is normally the case in the Internet, packets must be segmented into cells at switch inputs, and cells must be reassembled into packets at switch outputs [3]. There are mainly two approaches to the implementation of high-speed packet switching systems. One approach is the single-stage switch architecture, the other one is the multiple-stage switch architecture, such as the Clos-network switch. Most high-speed packet switching systems in the backbone of the Internet are currently built on the basis of a single-stage switching fabric with a centralized scheduler. An example of the single-stage architecture is a crossbar switching fabric which is widely used with the appropriate scheduling in single-stage

*Poznan University of Technology, janusz.kleban@et.put.poznan.pl, maciej.sobieraj@et.put.poznan.pl

switches. Single-stage crossbar switches are quite expensive as the switches become larger because the number of switching elements is proportional to the square of the number of switch ports. For the number of ports greater than 32 or 64 the multiple-stage switching fabrics become preferable due to less costs (less number of crosspoints), much higher switch capacity and scalability. The multiple-stage switching fabrics are made of smaller-size switching elements, where each such element is usually a crossbar.

In 1953, Clos proposed a class of space-division three-stage switching networks and proved strictly non-blocking conditions of such networks [4]. These kind of switching fabrics are widely used and extensively studied as a scalable and modular architecture for the next generation switches/routers. The Clos switching fabric can achieve a nonblocking property with the smaller number of total crosspoints in the switching elements than crossbar switches. Nonblocking switching fabrics are divided into four classes: strictly nonblocking (SSNB), wide-sense nonblocking (WSNB), rearrageable nonblocking (RRNB) and repackably nonblocking (RPNB) [6], [8]. SSNB and WSNB ensures, that any pair of idle input and output can be connected without changing any existing connections, but a special path set-up strategy must be used in WSNB networks. In RRNB and RPNB any such pair can be also connected, but it may be necessary to re-switch existing connections to other connecting paths. The difference is in time these reswitchings take place. In RRNB, when a new request arrives, and is blocked, an appropriate control algorithm is used to reswitch some of existing connections to unblock the new call. In RPNB, a new call can always be set up without reswitching of existing connections, but reswitching takes place when any of existing call is terminated. These reswitchings are done to prevent a switching fabric from blocking states before a new connection arrives.

The three-stage Clos-network switch architecture may be categorized into two types: bufferless and buffered. The former one has no memory in any stage, and it is also referred to as the space-space-space (S^3) Clos-network switch, while the latter one employs shared memory modules in the first and third stages, and is referred to as the memory-space-memory (MSM) Clos-network switch. The buffers in the second stage modules cause an out-of-sequence problem, so a re-sequencing function unit in the third stage modules is necessary but difficult to implement when the port speed increases. One disadvantage of the MSM architecture is that the first and third stages are both composed of shared-memory modules. It is necessary to use different kind of switching modules in each stage of the switching fabric. Lack of uniqueness in the structure of module may increase cost of the total system. In this paper we provide a packet dispatching scheme for the bufferless (S3) Clos network switches. In this architecture all shared memory modules are replaced by bufferless crossbar switches. All cells are stored outside the switching fabric in the input port cards. This solution can overcome the memory speedup problem and is similar to the Virtual Output Queuing (VOQ) structure in the single-stage crossbar switches.

The virtual output queuing is widely considered in the literature as a good solution for IQ (Input Queued) switches, to avoid the Head-Of-Line (HOL) blocking problem encountered in the input-buffered switches [1], [19]. HOL blocking causes the idle output to remain idle even if at an idle input there is a cell waiting to be sent to an (idle) output. In VOQ switches every input provides a single and separate FIFO for each output. Such a FIFO is called a Virtual Output Queue (VOQ). When a new cell arrives at the input port, it is stored in the designated queue and waits for transmission through a switching fabric. Cells in a VOQ do not block cells in any other VOQs except for contention. In VOQ switches internal blocking and output port contention problems are resolved by fast arbitration schemes. The arbitration scheme decides which items of information should be passed from inputs to arbiters, and – based on that decision – how each arbiter picks one cell from among all input cells destined for the output.

The switching process in the S^3 bufferless Clos network switch consists of two activities, namely port-to-port matching (scheduling), and route assignment between the first and third stages. The scheduling phase is necessary for determining a set of input-output matches for cell switching. During the route assignment phase a set of internally conflict-free path for the above matches is assigned. It is possible to go through these two phases separately or simultaneously. Due to these two phases the scheduling in the bufferless Clos network switch is more challenging than in the MSM architecture. The main switching problem in the MSM architecture lies in the route assignment between input and output modules. Algorithms which can solve this problem are usually called packet dispatching schemes. Considerable work has been done on scheduling algorithms for the MSM Clos-network switches. Most of them achieve 100% throughput under the uniform traffic, but the throughput is usually reduced under the nonuniform traffic [10], [11], [12], [15], [17], [18]. A switch can achieve 100% throughput under the uniform or nonuniform traffic if the switch is stable, as it was defined in [16]. In general, a switch is stable for a particular arrival process if the expected length of the input queues does not grow without bound.

Most of schemes proposed for VOQ switches accomplish request, grant and accept phases. This approach is difficult to implement when too many contention points exists. Matching schemes to configure S3 Clos network switches have been proposed in [2], [13], [14], [18]. In these papers performance evaluation results obtained only for the uniform traffic distribution patterns were shown. Many of the proposed schemes solve the configuration process in two phases: port matching first and routing thereafter, as routing uses the results of the port matching phase.

In this paper, instead of request, grant and accept phases, we propose to use the central arbiter for port-to-port matching and route assignment simultaneously. To solve port matching issue the new packet dispatching algorithm called the longest VOQ matching (LVM) is proposed. The route assignment problem, for a given set of input-output matches, in the S3 Clos network can be solved using the well known algorithms for simultaneous connections [5], [7]. The algorithms can be classified into two categories [2]:

- optimized algorithms provided guaranteed routes for all matches, but with a higher complexity in time or in implementation;
- heuristic algorithms provided all or part of the matches with routes in much lower complexity.

As we desired to investigate the influence of delayed response of the central arbiter on cell delay and maximum VOQ size an optimized algorithm was used in our simulation. To relax time constraints the central arbiter responded after D (D = 0, 2, 4, 6, 8, 10) time slots. We found that using proposed matching algorithm it is possible to postpone the matching decision by several time slots, and that the performance of the switching fabric in terms of the average cell delay and maximum queue size is acceptable for both the uniform and the nonuniform traffic distribution patterns. The port-to-port matching pattern and connecting paths in the switching fabric are kept up to the new decision of the central arbiter.

The remainder of this paper is organized as follows. Section 2 introduces some background knowledge concerning the S^3 Clos switching fabric that we refer to throughout this paper. Section 3 presents the LVM scheme and connecting path assignment algorithm. Section 4 is devoted to performance evaluation of the proposed IP–OP matching scheme under delayed response of the central arbiter. Two possible variants of the central arbiter responding where investigated, namely the LVMa and LVMb. We conclude this paper in section 5.

2. S³ CLOS SWITCHING FABRIC

The three-stage Clos-network architecture is denoted by $C(m, n, k)$, where parameters m, n, and k entirely determine the structure of the network. There are k input switches of capacity $n \times m$ in the first stage, m switches of capacity $k \times k$ in the second stage, and k output switches of capacity $m \times n$ in the third stage. The capacity of this switching system is $N \times N$, where $N = nk$. The three-stage Clos switching fabric is strictly nonblocking if $m \geq 2n-1$ and rearrangeable nonblocking if $m \geq n$.

We define the S³ Clos switching fabric based on the terminology used in [17] (see Fig. 1 and Table I).

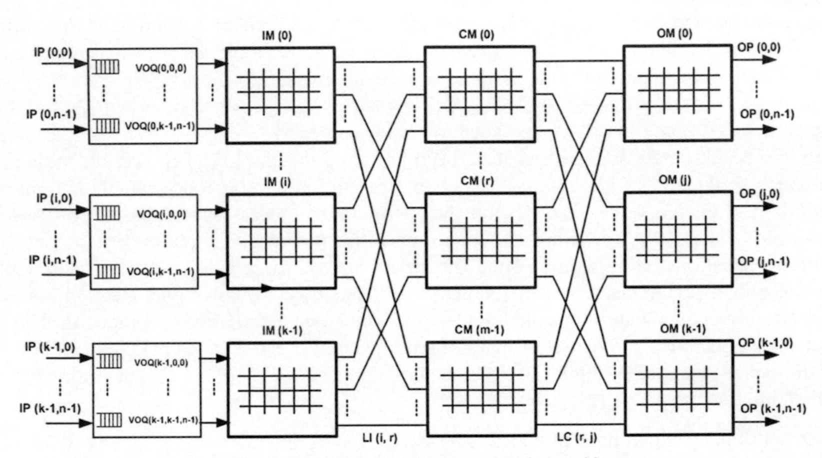

Fig. 1. The bufferless Clos-network switch architecture

Table I. The terminology for S³ Clos switching fabric

IM	Input module at the first stage
CM	Central module at the second stage
OM	Output module at the third stage
i	IM number, where $0 \leq i \leq k-1$
j	OM number, where $0 \leq j \leq k-1$
h_i	Input port number in each IM, where $0 \leq h_i \leq n-1$
h_j	Output port number in each OM, where $0 \leq h_j \leq n-1$
r	CM number, where $0 \leq r \leq m-1$
$IM(i)$	The $(i+1)$th input module
$CM(r)$	The $(r+1)$th central module
$OM(j)$	The $(j+1)$th output module
$IP(i, h_i)$	The (h_i+1)th input port at IM(i)
$OP(j, h_j)$	The (h_j+1)th output port at OM(j)
$LI(i, r)$	Output link at $IM(i)$ that is connected to $CM(r)$
$LC(r, j)$	Output link at $CM(r)$ that is connected to $OM(j)$
$VOQ(i, j, h_j)$	Virtual output queue that stores cells from $IM(i)$ to $OP(j, h_j)$

In the S^3 Clos switching fabric architecture the first stage consists of k IMs, and each of them has an $n \times m$ dimension. The second stage consists of m CMs, and each of them has a $k \times k$ dimension. The third stage consists of k OMs of capacity $m \times n$. In this kind of network all cells are stored in the input port cards and the virtual output queuing VOQ structure is used to alleviate the HOL blocking phenomenon that occurs in input-buffered switches. Input buffers located in IMs may be arranged as follows:

- An input buffer in each input port is divided into N parallel queues, each of them storing cells directed to different output ports. Each IM has nN VOQs, no memory speedup is required.
- There is one input buffer for each IM and it is divided into N parallel queues, each of them storing cells destined to different output ports. Memory speedup of n is necessary here. This kind of buffer arrangement is shown in Fig. 1.

In our simulation model of the three-stage bufferless Clos switching fabric we have implemented the latter case of input buffers arrangement.

3. PORT-TO-PORT MATCHING AND CONNECTING PATH ASSIGNMENT

The proposed LVM matching scheme makes a matching between $IP(i, h_i)$ and $OP(j, h_j)$ taking into account the number of cells waiting in VOQs. Each VOQ has its own counter $PV(j, h_j)$, which shows the number of cells destined to $OP(j, h_j)$. The value of $PV(j, h_j)$ is increased by 1, when a new cell is written into a memory and decreased by 1, when the cell is sent out to $OP(j, h_j)$. The algorithm uses the central arbiter to indicate the matched pairs of $IP(i, h_i)$ and $OP(j, h_j)$.

In detail, the LVM scheme works as follows:

Step 1 (each IM): Sort the values of $PV(j, h_j)$ in descending order. Send to the central arbiter the request, which contains a list of $OP(j, h_j)$. The number of output port $OP(j_a, h_{ja})$ to which $VOQ(i_a, j_a, h_{ja})$ stores the most number of cells should be placed on the list as the first one, and the number of output port $OP(j_z, h_{jz})$ to which $VOQ(i_z, j_z, h_{jz})$ stores the least number of cells should be placed on the list as the last one.

Step 2 (central arbiter): The central arbiter divides a time period necessary for making matching (e.g. four time slots) into n mini-slots. In every mini-slot for each $IM(i)$, the central arbiter makes only one the most requested and possible matching between input port of $IM(i)$ and $OP(j, h_j)$. In a mini-slot the central arbiter analyzes the request received from $IM(i)$ and checks if it is possible to match the input port $IP(i, h_i)$ of this IM with output port $OP(j_a, h_{ja})$ the number of which was sent as the first one on the list in the request. If the matching is not possible, because $OP(j_a, h_{ja})$ is matched with the input port of other IM, the arbiter selects the next output port number $OP(j_b, h_{jb})$ on the list. The round-robin routine is employed for selection of $IM(i)$ the request of which is analyzed in the first mini-slot.

Step 3 (central arbiter): After matching phase (step 2) the central arbiter runs the path assignment scheme.

Step 4 (central arbiter): Send to each IM the IP – OP matching and path assignment patterns. The message include up to n matched output port numbers $OP(j, h_j)$ and associated central module numbers $CM(r)$.

Step 5 (each IM): Match all input links of $IM(i)$ with cells from selected $VOQ(i, j, h_j)$. If there is less than n cells to be sent to output ports, some input links remain unmatched. Set connecting paths between input links and output links $LI(i, r)$ according to path assignment pattern sent by the central arbiter.

Step 6 (each IM): For each cell send to $OP(j, h_j)$ decrease the value of $PV(j, h_j)$ by one.

Step 7 (each IM): In the next time slot send the cells from the matched $VOQ(i, j, h_j)$ to the $OP(j, h_j)$ selected by the central arbiter.

Several control algorithms for set up simultaneous connections in Clos rearrangeable networks have been proposed [5], [7]. For solving the path assignment problem Neiman has proposed an algorithm based on the Hungarian method. This algorithm is interpreted as a matrix decomposition procedure consists of two phases: simple preparatory phase and complex iteration phase. The simple structure of the algorithm was proposed by A. Jajszczyk in [7], please refer to that paper for details.

To alleviate time constraints we have investigated in simulation experiments two possible variants of the central arbiter responding. In the first case (denoted by LVMa), the central arbiter responds in the same time slot it receives the requests from IMs (the central arbiter has to go through the IP-OP matching and paths assignment phases during one time slot), but the proposed connection pattern is hold in the Clos switching fabric during D time slots, where D = 2, 4, 6, 8, 10; D=0 means that the central arbiter sends the response in each time slot. In the real equipment this situation may appear when we have very fast central arbiter, but more than one time slot is needed to transmit requests - sorted lists of $OP\ (j, h_j)$ - from IMs to the central arbiter.

In the second case (denoted by LVMb), the central arbiter responds after D time slots, D = 0, 2, 4, 6, 8, 10 (D=0 means that the central arbiter responds without delay, in the same time slot). Here the arbiter can go through the two phases during D time slots. For instance, the IMs send the requests to the central arbiter in time slot i and they get the response in time slot $i+4$. During the time slots: $i+1$, $i+2$, $i+3$, $i+4$, the connection pattern from the previous response of the central arbiter is held in the switching fabric. In the real equipment, it may take place when the central arbiter is not fast enough to go through the two phases during one time slot.

4. SIMULATION EXPERIMENTS

A computer simulation was used for performance evaluation of the proposed LVM matching scheme and the influence of delayed response of the central arbiter on the cell delay and the maximum VOQs size under the uniform and the nonuniform traffic distribution patterns. In this section, conditions of the simulation experiments and simulation results are studied.

4.1 PACKET ARRIVAL MODELS AND TRAFFIC DISTRIBUTION MODELS

Two packet arrival models are considered in the paper: the Bernoulli model and the bursty traffic model. In the Bernoulli arrival model cells arrive at each input in slot-by-slot manner. The probability that a cell may arrive in a time slot is denoted by p and is referred to as the load of the input. This type of traffic defines a memoryless random arrival pattern.

In the bursty traffic model, each input alternates between active and idle periods. During active periods, cells destined for the same output arrive continuously in consecutive time slots. The average burst (active period) length is set to 16 cells.

We consider several traffic distribution models which determine the probability that a cell which arrives at an input will be directed to a certain output. The same traffic distribution models are used in many papers for evaluation of the performance of packet distribution schemes. The considered traffic models are:

Uniform traffic – this type of traffic is the most commonly used traffic profile. In the uniformly distributed traffic probability p_{ij} that a packet from input i will be directed to output j is uniformly distributed through all outputs, i.e.:

$$p_{ij} = p / N \quad \forall i, j \tag{1}$$

Trans-diagonal traffic – in this traffic model some outputs have a higher probability of being selected, and respective probability p_{ij} was calculated according to the following equation:

$$p_{ij} = \begin{cases} \frac{p}{2} & for \quad i = j \\ \frac{p}{2(N-1)} & for \quad i \neq j \end{cases} \tag{2}$$

Bi - diagonal traffic – is very similar to the nonuniform traffic but packets are directed to one of two outputs, and respective probability p_{ij} was calculated according to the following equation:

$$p_{ij} = \begin{cases} \frac{2}{3} p & for \quad i = j \\ \frac{p}{3} & for \quad j = (i+1) \bmod N \\ 0 & otherwise \end{cases} \tag{3}$$

Chang's traffic – this model is defined as:

$$p_{ij} = \begin{cases} 0 & for \quad i = j \\ \frac{p}{N-1} & otherwise \end{cases} \tag{4}$$

4.2 RESULTS OF SIMULATION EXPERIMENTS

The experiments have been carried out for the S^3 Clos switching fabric of size 64×64 and for a wide range of traffic load per input port: from p = 0.05 to p = 1, with the step 0.05. The 95% confidence intervals that have been calculated after t-student distribution for five series with 200,000 cycles (after the starting phase comprising 10,000 cycles, which enables to reach the stable state of the switching fabric) are at least one order lower than the mean value of the simulation results, therefore they are not shown in the figures. We have evaluated two performance measures: average cell delay in time slots and maximum VOQs size (we have investigated the worst case). The size of the buffers at the input port cards is not limited, so cells are not discarded. However, cells encounter the delay instead. The results of the simulation are shown in the charts (Fig. 2-11). Fig. 2, 4, 6, 8 show the average cell delay in time slots obtained for the uniform, Chang's, trans-diagonal, and bi-diagonal traffic patterns respectively, whereas Fig. 3, 5, 7, 9 show the maximum VOQ size in number of cells. Fig. 10 and 11 show the results for the bursty traffic with the average burst size b=16 (16 is the number of cells) under the uniform traffic distribution pattern. The algorithm proposed in [7] was used for set up the connecting paths in the investigated Clos switching fabric.

We can see that the S^3 Clos switching fabric with the proposed matching scheme has 100% throughput for all kinds of investigated traffic distribution patterns and for the bursty traffic. It is very good result especially for such demanding traffic patterns as trans-diagonal and bi-diagonal.

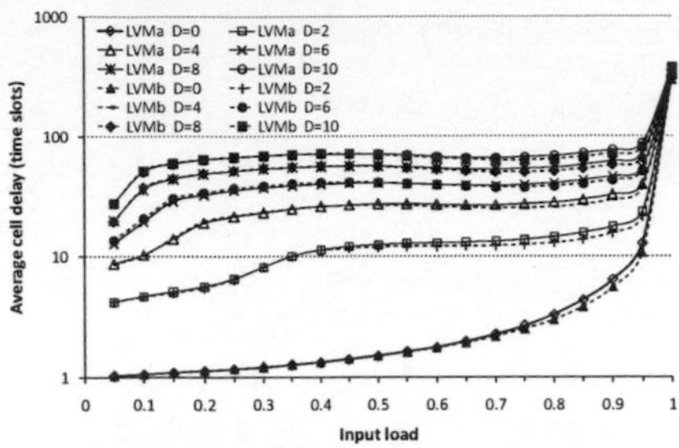

Fig. 2. Average cell delay, uniform traffic

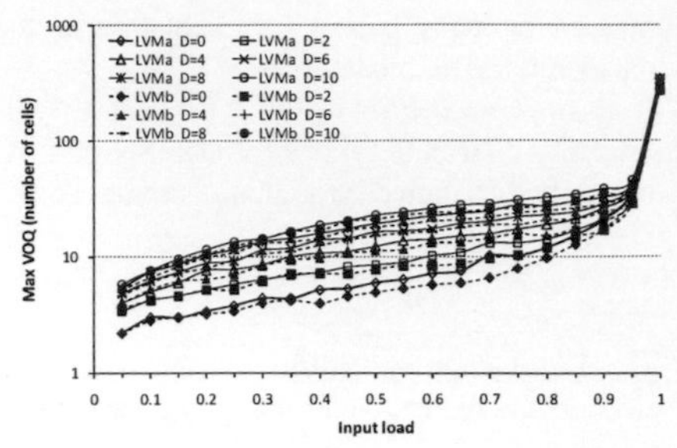

Fig. 3. The maximum VOQ size, uniform traffic

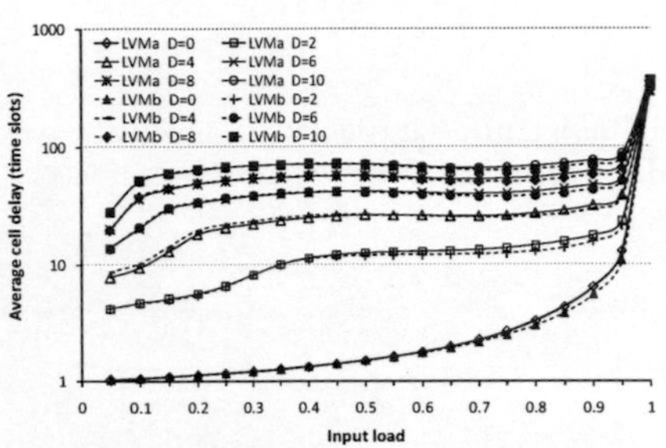

Fig. 4. Average cell delay, Chang's traffic

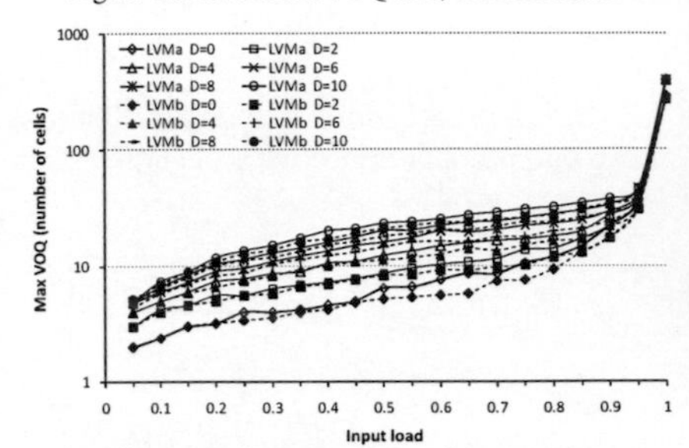

Fig. 5. The maximum VOQ size, Chang's traffic

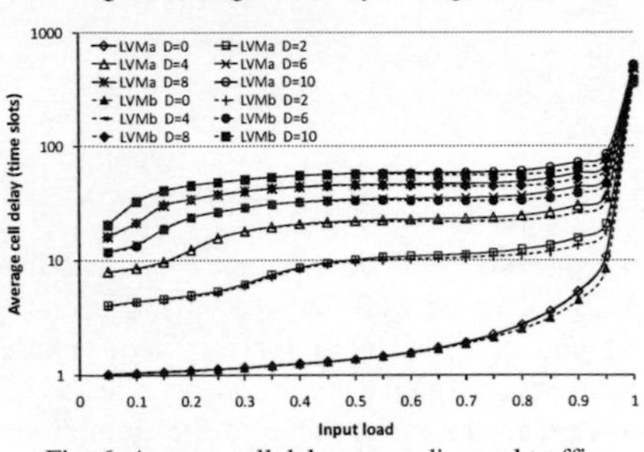

Fig. 6. Average cell delay, trans-diagonal traffic

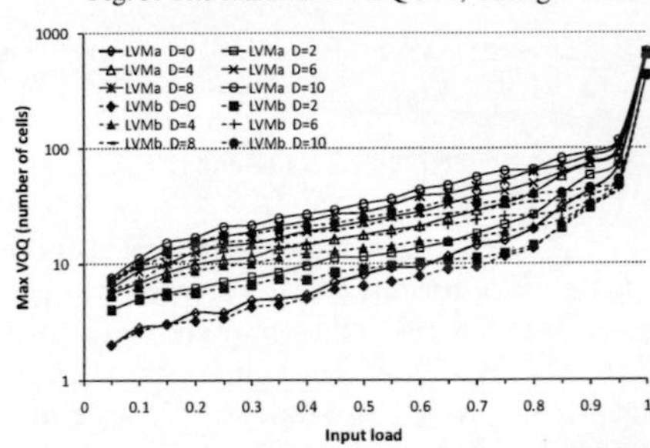

Fig. 7. The maximum VOQ size, trans-diagonal traffic

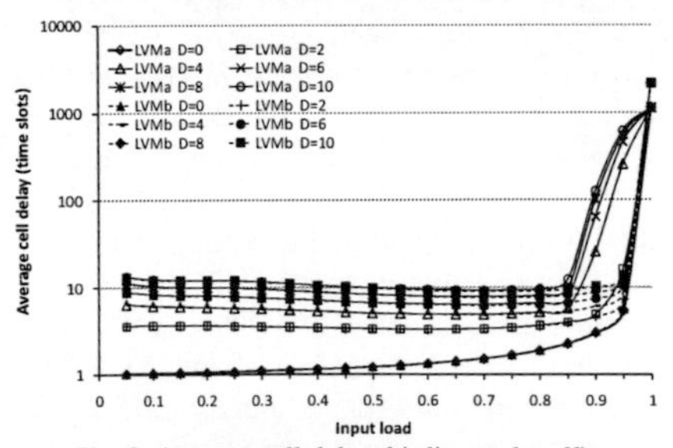

Fig. 8. Average cell delay, bi-diagonal traffic

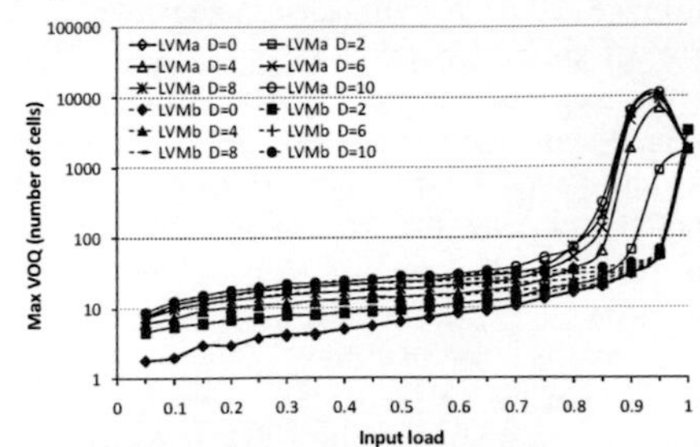

Fig. 9. The maximum VOQ size, bi-diagonal traffic

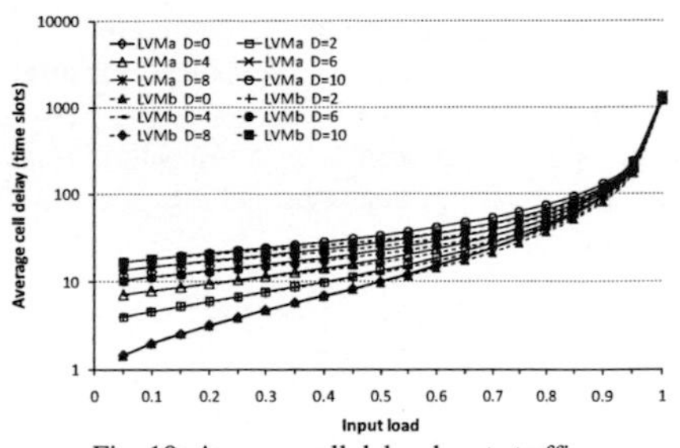

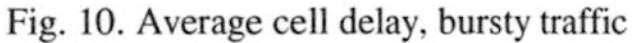
Fig. 10. Average cell delay, bursty traffic

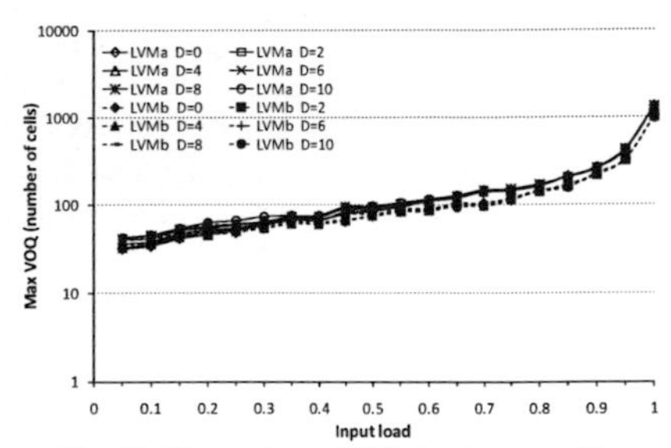

Fig. 11. The maximum VOQ size, bursty traffic

We found that there is almost no difference between results obtained for the two investigated variants of the central arbiter responding (the LVMa and LVMb), except for the bi-diagonal traffic (Fig. 8) and for the very high input load, p>0,85. In this case the LVMb algorithm gives better results than the LVMa scheme. For the highest investigated delay of the response of the central arbiter (equal to 10 time slots) the average cell delay is less than 100 time slots for wide range of input load, regardless of the traffic distribution pattern. What's more the results obtained for the uniform, Chang's and trans-diagonal traffic distribution patterns are very similar (Fig. 2, 4, 6). For the bi-diagonal traffic distribution pattern the average cell delay is lower than 15 time slots for input load below 0.95, and grow very fast for p>0.95, but do not exceed 2200 time slots (Fig. 8). For the bursty traffic, the average cell delay grows very similar to a linear function of input load for p<0.9, but for p>0.9 grows rapidly to the value of 1100 time slots.

The size of VOQ in the three-stage bufferless Clos network switch depends on the traffic distribution pattern and the delay of the central arbiter response. We comment on only the worst investigated delay, namely D=10. For the uniform (Fig. 3) and Chang's (Fig. 5) traffic distribution patterns the maximum size of VOQ is less than 50 for p<0.95; for p≥0.95 it grows to 350. For the trans-diagonal traffic distribution pattern and p<0.95 the maximum VOQ size is less than 100, and above p=0.95 it grows very fast to 500 (Fig. 7). For the bi-diagonal traffic and p<0.75 the maximum size of VOQ is less than 50 time slots (Fig. 9). For p≥0.75 we found a big difference between maximum VOQ size obtained for the LVMa and PVMb schemes. For the LVMa scheme the maximum VOQ size grows very fast to 10 000 for p>0.75 and drops to 2000 for p=1. For the LVMb scheme and 0.75<p<0.95 the maximum size of VOQ is less than 60 and grows very fast to 3000 for p≥0.95. For the bursty traffic we noticed that the maximum VOQ size is almost independed of the delay of the central arbiter response (Fig. 11). For p<0.55 the maximum VOQ size is less than 100, and grows to 1000 for 0.5<p≤1.

5. CONCLUSIONS

In this paper we have proposed the IP–OP matching scheme called LVM for the S^3 Clos switching fabric. The algorithm uses the central arbiter to perform both matching and connecting paths assignment processes. Using computer simulation we have shown that it is possible to postpone the central arbiter decision and still obtain the reasonable performance measures, namely average cell delay and maximum VOQ size for the uniform and the nonuniform traffic distribution patterns. To our best knowledge, these are first such results obtained for the nonuniform traffic distribution patterns and the S^3 Clos switching fabric. We have investigated two possible variants of the central arbiter responding called LVMa and LVMb. Due to the time complexity of connecting path assignment process, the most important is this case in which the central arbiter responds after several time slots. Here is possible to relax the requirements of the central arbiter processing power in the real equipment. Instead of an optimal connecting path assignment algorithm suboptimal algorithm can be used e.g. proposed by M. Karol in [9]. In the case of a suboptimal algorithm the average cell delay will be higher, but it will be a subject for further research.

REFERENCES

[1] Anderson T., et al., *High-speed switch scheduling for local-area networks*, ACM Trans. on Computer Systems, vol. 11, no. 4, 1993, pp. 319-352.

[2] Chao H. J., Jing Z., and Liew S. Y., *Matching algorithms for three-stage bufferless Clos network switches*, IEEE Commun. Mag., vol. 41, no. 10, pp. 46-54, Oct. 2003.

[3] Chao H. J., Lam C. H., and Oki E., *Broadband Packet Switching Technologies: A Practical Guide to ATM Switches and IP Routers*, Willey, New York, 2001.

[4] Clos C.: *A Study of Non-Blocking Switching Networks*, Bell Sys. Tech. Jour., 1953, pp. 406-424.

[5] Hwang F. K., *Control Algorithms for Rearrangeable Clos Networks*, IEEE Trans. on Communications, vol. COM-31, No. 8, August 1983.

[6] Hwang F. K., *The Mathematical Theory of Nonblocking switching Fabrics*, World Scientific, Singapore, 1998.

[7] Jajszczyk A., *A Simple Algorithm for the Control of Rearrangeable Switching Networks*, IEEE Trans. on Communications, vol. COM-33, No. 2, February 1985.

[8] Kabaciński W., *Nonblocking Electronic and Photonic Switching Fabrics*, Springer, 2005.

[9] Karol M. and I C.-L., *Performance Analysis of a Growable Architecture for Broadband Packet (ATM) Switching*, GLOBECOM '89, 1989, pp. 1173-80.

[10] Kleban J., Santos H., *Packet Dispatching Algorithms with the Static Connection Patterns Scheme for Three-Stage Buffered Clos-Network Switches*, in Proc. IEEE International Conference on Communications 2007 - ICC-2007, 24-28 June 2007 Glasgow, Scotland.

[11] Kleban J., Sobieraj M., Węclewski S., *The Modified MSM Clos Switching Fabric with Efficient Packet Dispatching Scheme*, in Proc. IEEE High Performance Switching and Routing 2007 – HPSR 2007, New York, May 30 to June 1, 2007.

[12] Kleban J., Wieczorek A., *CRRD-OG: A Packet Dispatching Algorithm with Open Grants for Three-Stage Buffered Clos-Network Switches*, in Proc. High Performance Switching and Routing 2006 – HPSR 2006, pp. 315-320.

[13] Lee T. T. and Liew S.-Y., *Parallel routing algorithm in Benes-Clos networks*, in Proc. IEEE INFOCOM 1996, pp. 279-286.

[14] Lin C. B., and Rojas-Cessa R., *Module Matching Schemes for Input-Queued Clos-Network Packet switches*, IEEE Comm. Letters, vol. 11, No. 2, February 2007, pp. 194-196.

[15] Lin C.-B. and Rojas-Cessa R., *Frame Occupancy-Based Dispatching Schemes for Buffered Three-stage Clos-Network switches*, in Proc. 13th IEEE International Conference on Networks 2005.

[16] McKeown N., Mekkittikul A., Anantharam V., Walrand J., *Achieving 100% Throughput in an Input-queued Switch*, IEEE Trans. Commun., Aug. 1999, pp. 1260-1267.

[17] Oki E., Jing Z., Rojas-Cessa R., and Chao H. J., *Concurrent Round-Robin-Based Dispatching Schemes for Clos-Network Switches*, IEEE/ACM Trans. on Networking, vol. 10, no.6, 2002, pp. 830-844.

[18] Pun K., Hamdi M.: *Dispatching schemes for Clos-network switches*, Computer Networks no. 44, 2004, pp.667-679.

[19] Tamir Y. and Frazier G.: *High performance multi-queue buffers for VLSI communications switches*, in Proc. Computer Architecture, Honolulu, Hawaii, United States, 1988, pp. 343-354.

Ethernet, QoS, Classes of Service, simulations

Piotr KRAWIEC*, Robert JANOWSKI*, Wojciech BURAKOWSKI*

SUPPORTING DIFFERENT CLASSES OF SERVICE IN ETHERNET SWITCHES WITH SHARED BUFFER ARCHITECTURE

This paper deals with the problem of assuring strict QoS guarantees for connections going through from Ethernet access network. The motivation of this paper was the problem identified during the FP6 IST EuQoS project, which showed that, despite high link capacities, Ethernet access network might be the reason of QoS deterioration in odd cases. The primary reason precluding target QoS level is the lack of appropriate QoS differentiation and traffic isolation mechanisms in some Ethernet switches. For instance, the shared buffers and priority schedulers seem to be not sufficient to guarantee strict QoS. That's why a new solution was proposed for these cases. This solution relied on additional traffic control mechanisms available in some Ethernet switches. The proposed approach was evaluated by simulations studies for TCP and UDP traffic.

1. INTRODUCTION

New demands for using multimedia applications over the Internet, such as videoconferences, tele-medicine, tele-education etc., have caused considerable research to develop the Internet into multi-service networks offering QoS (*Quality of Service*) guarantees. One of the projects focusing on these issues was the FP6 IST EuQoS [9]. The key objective of this project was to design, research, prototype and test a new system, called the EuQoS system, which assures strict end-to-end (e2e) QoS quarantees at the packet level in multi-domain, heterogeneous environment. To provide strict e2e QoS guarantees in multi-domain heterogeneous network, we require assured QoS in each part of the network: core domains as well as access domains, therefore, during the EuQoS project researches, we studied also the approach for supporting QoS in the Ethernet access network.

So far, a lot of effort was put to assure QoS guarantees in Ethernet access network. However, some solutions do not provide strict QoS, as EtheReal [18], which is throughput oriented and only supports best effort traffic, or stochastic approach presented in [4], which only yields average delay performance bound. Other approaches, as [13] and [7] require additional cost of hardware and/or software modification. Solutions proposed in [6] and [3] do not override the IEEE specifications and rely on standard Ethernet switches with priority scheduling, but require a separate queue for each priority class. Unfortunately, typical Ethernet switches currently offered by vendors own only a common buffer, which is shared by all priority classes. It may lead to the violation of QoS

* Warsaw University of Technology, Institute of Telecommunications, ul. Nowowiejska 15/19, 00-665 Warsaw, Poland, e-mail: {pkrawiec, rjanowsk, wojtek}@tele.pw.edu.pl

guarantees of high priority traffic in case when the whole Ethernet switch buffer is occupied by low priority traffic.

In this paper we present an approach to assure strict QoS guarantees in the Ethernet access network. We assume the use of currently accessible Ethernet equipment, with shared buffers, priority scheduler and traffic control mechanism similar to WRED (*Weighted Random Early Detection*). No further modification of switch software or hardware is necessary.

The rest of the paper is organized as follows: Section 2 describes the main problem treated in this paper, i.e. the assurance of target QoS level in Ethernet access network with switches, which contain shared buffers. In Section 3 the proposed solution is presented while in Section 4 it is evaluated by a series of simulations. Section 5 summarizes the paper.

2. STATEMENT OF THE PROBLEM

The approach for assuring QoS in multi-domain networks, which has been applied in the EuQoS project, bases on the implementation of end-to-end Classes of Service (e2e CoSs) [5] dedicated to handle packets generated by respective type of application, e.g. VoIP (*Voice over IP*). Roughly speaking, the e2e CoS corresponds to the network capabilities for transferring the packets belonging to selected connections with assumed QoS guarantees. These QoS guarantees are expressed by the following metrics (as defined in [10]): (1) IP packet loss ratio IPLR, (2) IP packet transfer delay IPTD and (3) IP packet delay variation IPDV.

Within the IST EuQoS project, five e2e CoSs have been defined: Telephony, RT Interactive, MM Streaming, High Throughput Data (HTD) and Standard (STD) CoSs, according to different types of traffic profiles generated by the different applications studied in EuQoS. The maximum values of QoS metrics (i.e. IPLR, IPTD and IPDV) for each e2e CoS one can find in [11].

To implement these e2e CoSs, adequate CAC (*Connection Admission Control*) algorithms were designed (to limit the QoS traffic) and appropriate QoS mechanisms like schedulers, shapers, policers etc., available in network elements were used. Only in the case of Standard CoS there is neither CAC function performed nor the QoS parameters are guaranteed since this CoS is intended to provide similar service as Best Effort network, i.e. without guarantees in the QoS parameters.

The implementation of e2e CoSs runs into different obstacles when considering each of possible access network technologies i.e. WiFi, UMTS, xDSL or Ethernet. In our paper we focus on the problem of Ethernet access network. In this technology the primary mechanism to differentiate traffic is Priority Scheduler, practically available in almost every switches. The 802.1p specification (which is a part of IEEE 802.1D [8]) defines 8 priority classes, and EuQoS project proposed the mapping between them and end-to-end EuQoS CoSs as presented in table 1.

Table 1. Mapping between end-to-end EuQoS CoSs and Ethernet priority classes.

e2e EuQoS CoS	Telephony, RT Interactive	MM Streaming, High Throughput Data	Standard
Ethernet priority class	Voice	Controlled Load	Best Effort
802.1p priority value	6 (high)	4	0 (low)

The mapping shown in table 1 implies that the traffic from STD CoS is served with the lowest priority in comparison with other e2e EuQoS CoSs. Unfortunately, typical Ethernet switches do not support per class buffer but only a common one, which is shared by all CoSs including STD CoS. Although this buffer is quite large – usually thousands of packets, the fact it is shared poses the main problem [15]. Since CAC function, by definition, does not control the amount of traffic submitted to

STD CoS, it is possible that this traffic overloads the network and fully occupies the Ethernet buffer. This situation may deteriorate IPLR metric of other CoSs, since the arriving packets from other CoSs will be dropped due to the lack of room in shared buffer space. The mechanism of shared buffer combined with the priority scheduling has been symbolically shown in fig. 1 by using model of Drop Tail queue at the entrance to the Ethernet buffer but Priority Scheduler at the exit.

It is worth mentioning that IPTD and IPDV metrics, in spite of the shared buffer, will never be influenced since once a packet enters into the shared buffer, it is scheduled to the transmission according to the priority rules and, hence the traffic from STD CoS cannot delay the packets from other CoSs [11]. Thus, the problem of assuring appropriate performance of e2e EuQoS CoSs in Ethernet network is mainly the problem of controlling traffic from STD CoS and preventing it from occupying too much buffer space avoiding, in this way, the packet losses due to shared buffer space.

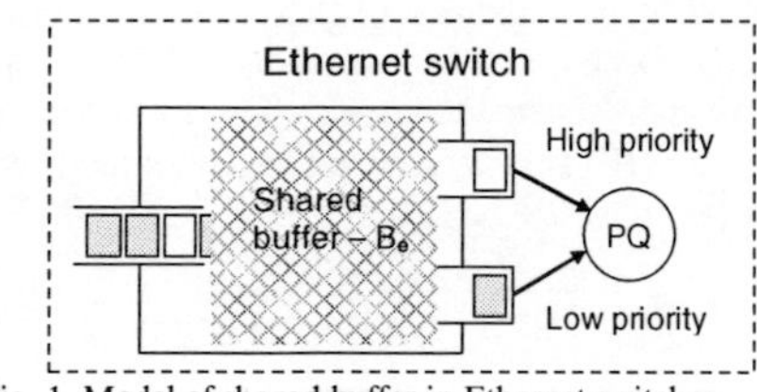

Fig. 1. Model of shared buffer in Ethernet switches.

3. THE PROPOSED SOLUTION

The main idea to alleviate the problem outlined in section 2 relies on controlling the buffer space occupied by each CoS. In this way, we assure the isolation between the different CoSs. First of all, we distinguish between the CoSs, for which we guarantee some level of QoS (we refer to them as QoS CoSs) and the STD CoS, for which we do not guarantee any level of QoS. Such a classification is caused because we may apply different methods for controlling the occupancy of the buffer space for QoS CoSs and for STD CoS. In the first case, as any of the connections belonging to the QoS CoSs must pass the CAC function, the amount of the occupied buffer space can be controlled by appropriate resource provisioning and configuration of the CAC function. However, in the case of STD CoS the approach described above is not possible because the connections belonging to this CoS are not controlled by CAC function. To control the volume of this traffic and the shared buffer occupancy due to it, we propose a solution designed to switches that support traffic control mechanism similar to WRED. The desired isolation of CoSs might be achieved when the size of the common Ethernet buffer B_e is able to accommodate the whole buffer space required by the QoS CoSs (B_{QoS}) and by the STD CoS (B_{STD}). It means that the Ethernet switch buffer size should meet the following condition: $B_e > B_{QoS} + B_{STD}$. More precisely, for each QoS CoS (i.e. for each Ethernet priority class associated with it) we can dedicate the buffer size $B_{i\text{-}j}$ (j=1,..,7) taking into account the adequate QoS requirements. For example, for Telephony CoS we design rather short buffers to guarantee low IPDV values. This buffer size $B_{i\text{-}j}$, together with dedicated capacity $C_{i\text{-}j}$ is later used by RA (*Resource Allocator*), the control module responsible for performing CAC algorithm and resource reservations, to calculate the CAC limit for the given output port i for particular CoS j. Since all originating and terminating connections which use these CoSs must pass the CAC function performed in the RA module, we can control the volume of the traffic offered to each QoS CoS within each output port (see fig. 2).

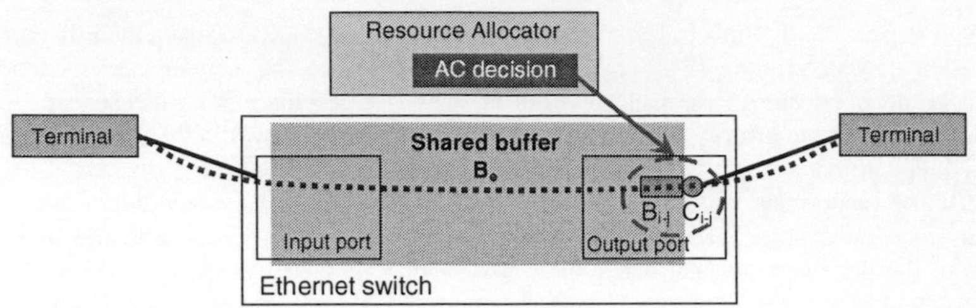

Fig. 2. Resource Allocator (RA) module controls the volume of the traffic by performing CAC algorithm for dedicated resources: buffer size $B_{i\text{-}j}$ and capacity $C_{i\text{-}j}$.

The CAC algorithm used by RA module in Ethernet access network depends on the CoS it is used for. In general, the requirements for IPLR and mean IPTD determine the maximum admissible load ρ_{max} that can be accepted in given QoS CoS. The maximum admissible load $\rho_{max,IPTD}$ that satisfies the requirement for mean IPTD is determined based on the Pollaczek-Khintchin formula:

$$\rho_{\max,\,IPTD} = \frac{2(IPTD - T_{prop} - \frac{L}{C})}{2IPTD - 2T_{prop} - \frac{L}{C}} \tag{1}$$

where T_{prop} denotes the propagation delay (which also contributes to mean IPTD), L denotes the packet length (in bits) and C denotes the fraction of link capacity dedicated to the given QoS CoS.

On the other hand, the requirement for target IPLR determines another value of the maximum admissible load, namely $\rho_{max,IPLR}$. The lower of these two values is finally considered as the maximum admissible load ρ_{max} to the given QoS CoS:

$$\rho_{\max} = \min[\ \rho_{\max,\,IPTD}\ ;\ \rho_{\max,\,IPLR}\ ] \tag{2}$$

For Telephony CoS, the value of $\rho_{max,IPLR}$ is calculated based on the dedicated buffer size $B_{i\text{-}6}$, and the target IPLR value according to the algorithm described in [17]. The value of $B_{i\text{-}6}$ is determined from the provisioned capacity $C_{i\text{-}6}$ and the requirement on the IPDV value, which is defined as an upper bound of the maximum packet queueing delay.

In the case of STD CoS the approach described above is not possible because the connections belonging to this CoS are not controlled by CAC function. To control the volume of this traffic and the shared buffer occupancy due to it we propose a solution which is designed for switches that support traffic control mechanism similar to WRED. For example, such mechanism is available in Super Stack 4 5500 Ethernet switch which was a part of EuQoS test bed. As recognized in the relevant technical documentation [2], this WRED mechanism lets us set the queue threshold $Q_{th,i\text{-}j}$ and the dropping probability $P_{drop,i\text{-}j}$ for each output port i and each Ethernet priority level j independently. Furthermore, it works in the following way: when a new packet arrives, the corresponding output port i and associated Ethernet priority level j are determined, then the size of the adequate queue $Q_{i\text{-}j}$ is compared with the earlier configured threshold $Q_{th,i\text{-}j}$. If the queue size is below than $Q_{th,i\text{-}j}$ the packet is queued in the common buffer, otherwise it is dropped with probability $P_{drop,i\text{-}j}$. In comparison with WRED mechanism known from routers, where the two queue size thresholds are specified and the dropping decisions are worked out on the basis of the avaraged queue size, it is a kind of simplified version with only one queue size threshold and the packet drop desicions based on the instantenous queue size.

The main idea of the proposed solution is that by setting the appropriate threshold $Q_{th,i-0}$ for the queue Q_{i-0} associated with STD CoS (j=0) on the output port i and the related dropping probability $P_{drop,i-0}$, we are able to drop the excessive traffic and, in this way, we tend to keep the buffer occupancy (due to STD CoS on this output port) below the value $Q_{th,i-0}$. This is suited for STD CoS since, on one hand, it has no guarantees about IPLR, IPTD nor IPDV values and, on the other hand, it carries mainly TCP controlled traffic with possibly greedy behavior tending to grab all the available capacity and the buffer space B_e. Unfortunately, Super Stack 4 5500 lets for setting the P_{drop} value only in the range <0; 92%> [1]. It means that in the case when the thresholds are exceeded, we can never drop the whole traffic incoming to the Ethernet switch but, at the most, only the 92% of it. However, as most of the STD CoS traffic uses TCP transport protocol we assume that this method is sufficient for bounding the maximum buffer occupancy.

When the above approach is applied to all N output ports of the Ethernet switch, the total resulting occupancy B_{STD} of the shared buffer due to STD CoS traffic should stay below the value:

$$B_{STD} <= \sum_{i=1}^{N} Q_{th,i-0} \qquad (3)$$

Accordingly, the remaining buffer space should be available for the traffic from QoS CoSs.

4. EVALUATION BY SIMULATIONS

The main objective of the simulation studies was to verify if the proposed solution is able to assure the target values of IPLR, IPTD, IPDV parameters for the traffic carried within particular e2e CoS when each CoS is in CAC limit and STD CoS is overloaded. An additional objective was to verify whether the QoS mechanisms available in Ethernet switch let us to control the consumption of shared buffer space, especially its occupancy due to the STD CoS traffic.

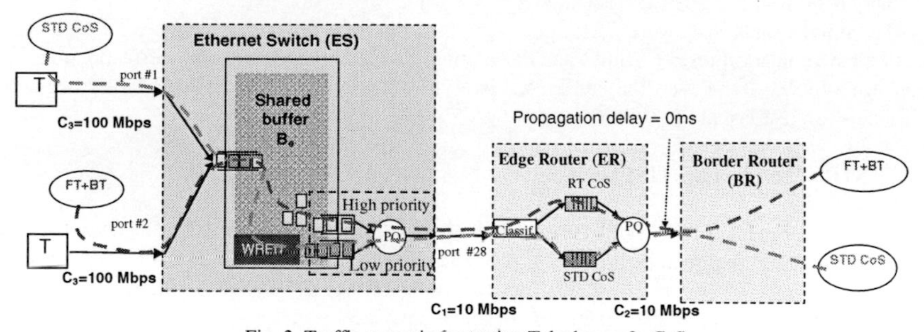

Fig. 3. Traffic scenario for testing Telephony e2e CoS

For the simulation studies we assumed the same network topology as in EuQoS test bed. Accordingly, Ethernet access network includes Ethernet Switch (ES), which connects to a number of Terminals (T) and one Edge Router (ER) which provides connectivity to the IP core (see fig. 3). ES features 28 ports, among which 27 are connected to end terminals (T) and one to ER. All the links are duplex. The Ethernet link capacities (C_1, C_3) as well as the capacity C_2 of the ER link toward

Border Router (BR) are configurable. Since we want to perform the tests in conditions when ES is a bottleneck we set the capacities C_1 and C_2 equal. To create the congestion conditions with a minimum set of terminals generating traffic (which is important when performing trials in test bed) the input links are configured with capacity 100Mbps i.e. 10 times faster than links C_1 and C_2.

We have evaluated the possibility of supporting e2e CoSs in a set of scenarios, where the traffic from only one e2e QoS CoS (Telephony, RT-Interactive, etc.) together with STD CoS traffic was present at the same time. In each test we measured relevant QoS parameters of a single traffic stream constituting so called Foreground Traffic (FT). However, for these measurements to be adequate we provoked the worst traffic conditions in the network that are allowed by CAC algorithm. It means that we loaded a tested e2e CoS to ρ_{max} value. This additional type of traffic creating CAC limit condition, we refer to as Background Traffic (BT). The FT and BT were appropriately modelled depending on the type of tested e2e QoS CoS. Due to space limitation in this paper we present only the simulation results obtained for the case of Telephony and STD CoS.

In the performed tests we differentiated between the case where the STD CoS traffic used TCP (section 4.1) or UDP (section 4.2). This differentiation is important because the applied transport protocol impacts traffic characteristics mostly due to the presence or lack of closed-loop congestion control, respectively.

The details of simulated traffic scenario are as follows. There are only two types of traffic: one representing traffic from Telephony CoS, for which we must guarantee target QoS level and the other one, which represents traffic from STD CoS. The Telephony CoS FT and BT traffic comes from the terminal connected to Ethernet port #2 and STD CoS traffic from terminal connected to port #1. The whole traffic is destined to BR across ES and ER. The propagation delay T_{prop} between ER and BR is set to 0 ms reflecting the low distances between particular elements of the Ethernet access network. The capacity dedicated to Telephony CoS on C_1 link is $C_{28\text{-}6}$=2Mbps. Packets belonging to this class are 200B long and packets belonging to STD CoS are 1500B long. FT and BT of Telephony CoS are modelled as CBR stream (parameters relevant for G.711 codec) and Poisson stream, respectively. The parameters of the WRED mechanism are the following: threshold is set to 85 pkts (and this value indicates maximum buffer space assumed for STD CoS), dropping probability is set to 0.92. The Ethernet shared buffer value is B_e=1000 pkts.

The simulation studies were performed using ns-2 platform [14]. The results were obtained by repeating the simulation tests 12 times and calculating the mean values with the corresponding 95% confidence intervals. However, the confidence intervals were not given in cases they are negligible. Each simulation test lasted for 1000 seconds.

4.1 TCP STREAMS INTO STD COS

In this test the STD CoS traffic consists of N=10 TCP greedy connections. The nominal values of QoS parameters that we assumed for the Telephony CoS in Ethernet access network are summarized in table 2 (we refer to them as to “designed values”). These values were used to determine the traffic load that we admitted into Telephony CoS.

For this test, the measured IPLR, mean IPTD and IPDV are below target designed values (see table 2). Measured IPLR equals 0 since the buffer space dedicated to Telephony CoS ($B_{28\text{-}6}$=10 packets) was much less than the whole Ethernet switch buffer size and the buffer occupancy due to STD CoS was well controlled. In fact when STD CoS traffic is TCP controlled the maximum STD CoS queue size Q_{max} (determined with probability 10^{-3}) deviates only a little from the WRED threshold which was set to 85 packets (see table 3). Thanks to TCP mechanism, after the STD CoS queue reaches WRED threshold and the new arriving packet is dropped, the TCP source slows down its sending rate letting the STD CoS queue size to decrease. The measured IPDV value was also below the designed value because Telephony CoS traffic was served with priority on the link with physical capacity 10Mbps and this value is much higher than the assumed provisioned capacity C_{28-6}=2Mbps. Also the measured mean IPTD was below the designed value but the reason was that for this simulation ρ_{max}=0.714 was determined from IPLR as a more constraining factor (see eq. 2).

Table 2. Simulation results of QoS parameters for Telephony CoS.

QoS parameter	Designed value	Measured value
IPLR	10^{-3}	0
IPTD [ms]	2.5	2.0
IPDV [ms]	8	1.3±0.1

Table 3. Simulation results of the queue size in ES.

Tested e2e CoS	Measured queue size in ES [pkts]		
	Parameter name Q	Mean	Q_{max}: Prob{$Q>Q_{max}$}<10^{-3}
Telephony	Q_{QoS}	1.1	5
STD	Q_{STD}	73±0.5	86.5±0.5

4.2 UDP STREAMS INTO STD COS

In this section we discuss two cases. Firstly, in section 4.2.1 we provide the results for the case when the WRED queue size threshold for the STD CoS is set to the desired value of queue size (we call this value as nominal threshold). Next, in section 4.2.2 we provide some guidelines how to tune the WRED threshold in order to better control the STD CoS queue size.

4.2.1 RESULTS WITH NOMINAL THRESHOLD

In this test, the STD CoS traffic consists on one CBR stream with rate equal to 100 Mbps. In these conditions, we observed that the maximum queue size Q_{max} of STD CoS traffic in Ethernet buffer is much greater than the WRED threshold at which the ES starts dropping STD CoS packets (see table 4). The reason for the long queue size, exceeded desired value of 85 pkts, is that the STD CoS packets are dropped in a probabilistic way. It means that during short periods, fewer packets than expected may be dropped and, as a consequence, it leads to uncontrolled growth of the queue. To illustrate possible IPLR deterioration of Telephony CoS traffic, we repeated the simulation keeping the size of shared buffer B_e = 95 pkts (10 pkts for Telephony CoS and 85 pkts for STD CoS), as described in table 5. In this case, the measured mean IPTD and IPDV are below target values because of the same reasons as in section 4.1, whereas IPLR is higher than designed value.

Table 4. Simulation results of queue size in ES.

Tested e2e CoS	Measured queue size in ES [pkts]		
	Parameter name Q	Mean	Q_{max}: Prob{$Q>Q_{max}$}<10^{-3}
Telephony	Q_{QoS}	0.7	5
STD	Q_{STD}	92.5±0.1	134±1.3

Table 5. Simulation results of QoS parameters for Telephony CoS (B_e=95 pkts).

QoS parameter	Designed value	Measured value
IPLR	10^{-3}	$1.2 \cdot 10^{-2}$
IPTD [ms]	2.5	2.0
IPDV [ms]	8	1.2

4.2.2 RESULTS WITH IMPROVED THRESHOLD TUNING

The simulation studies performed for Ethernet access network showed that controlling shared buffer occupancy due to STD CoS traffic by means of WRED mechanism is not a trivial task. Since the control of the occupied buffer space is crucial to provide separation between CoSs and in this way, to assure target values of QoS parameters, we must precisely control it. It means that only when STD CoS traffic is TCP controlled, we can assume that the queue size of STD CoS is well limited to the WRED threshold. In the case when this traffic is UDP we cannot assume that the STD CoS queue size is around desired WRED threshold since the simulations proved it is much above it. For this case, we need to control the queue size more precisely e.g. by setting lower value of WRED threshold to start dropping packets earlier. The question arising is the top value of WRED threshold in order not to exceed the target maximum queue size. The answer comes from the analysis of the phenomena, which is responsible for the excessively growing queue.

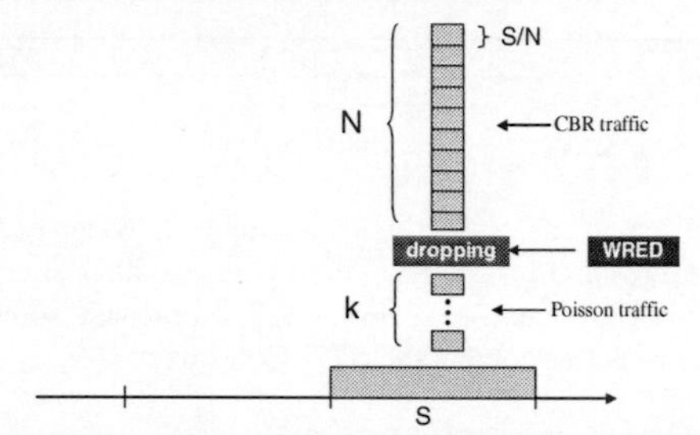

Fig. 4 Characteristics of the arriving packet stream after passing WRED mechanism

Since in the considered case, the capacity C_3 is 10 times higher than capacity C_1 and the STD CoS traffic is assumed to be CBR with rate equal to C_3 (see fig. 3), then, during the service time of each STD CoS packet on the link C_1, other 10 packets of STD CoS arrive to the output port #28. When the queue size of STD CoS packets gathered on port #28 exceeds the WRED threshold (here 85 packets), the WRED mechanism starts dropping the arriving STD CoS packets. However, it drops each of them with a probability 0.92. On average the rate of STD CoS traffic, which passes through the WRED is only 8 Mbps (i.e. 8% of incoming traffic) which guarantees that the system is stable since the total rate offered to the output port #28 equals 9.428 Mbps (8 Mbps due to STD CoS traffic and $\rho_{max}\times C_{28,6}$=0.714×2Mbps=1.428 Mbps due to Telephony CoS traffic) and stays below the service rate C_1 (10 Mbps). However, because of the probabilistic nature of WRED packet dropping, it might happen that for some period of time more packets than the foreseen average will pass through the WRED. These packets will contribute to the extensive growth of the queue beyond the WRED threshold. In order to understand how this queue grows it is necessary to characterize the stream of packets that have passed WRED. The fig. 4 illustrates the dependencies between the packet service time S on the link C_1, the original CBR packet stream (N=10 new packets during the service time S) and the stream of packets that have passed WRED (k packets out of any N arriving).

Since consecutive packets are dropped by WRED independently, the probability distribution of the number of packets that are not dropped (k out of N) is binomial:

$$\Pr ob\{k\} = \binom{N}{k} p^k (1-p)^{(N-k)}, \quad where\ k = 0,1,...,N\ and\ N = 10 \tag{4}$$

with parameter p=0.08 (probability that WRED does not drop the packet).

For our purpose, this distribution can be replaced by Poisson distribution (with mean equal to $N\times p$) as the latter one has greater variability and thus, we can consider the results obtained with the Poisson distribution as an upper bound. After characterizing the packet stream, which contributes to the STD CoS queue, we proceed with the analysis. The starting point is the following model with two CoSs: Telephony and STD and the traffic loads ρ_1 and ρ_2, respectively (see fig. 5a).

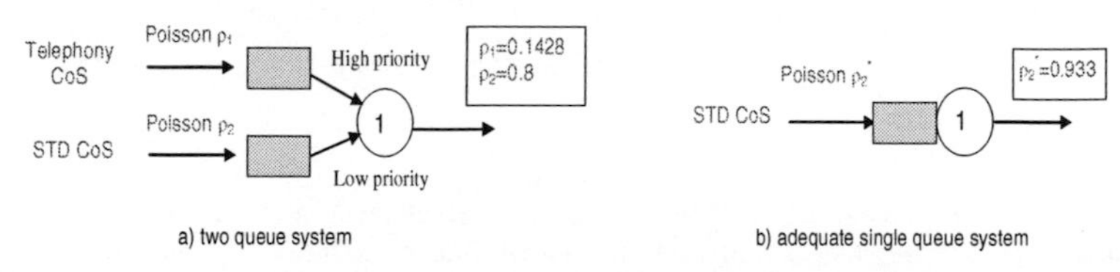

Fig. 5 The approach for analysing low priority (STD CoS) queue size

From the point of view of STD CoS traffic, we may replace the original model (fig. 5a) by a single queue system with Poisson stream as an input and appropriately recalculated ρ_2^*, which considers the impact of the high priority traffic (Telephony CoS) on the low priority one (STD CoS) [12] (see fig. 5b). The load ρ_2^* is determined basing on the following equation:

$$\rho_2^* = \frac{\rho_2}{1-\rho_1} \tag{5}$$

Next, we can approximate the STD CoS queue size probability distribution using the formula proposed in [16]:

$$\Pr ob\{Queue > x\} \approx e^{-2x\frac{(1-\rho_2^*)}{\rho_2^*}} \tag{6}$$

From (6), we can determine the target value of the queue size (X) exceeded only with some small probability P_0:

$$X = \frac{-\rho_2^* LnP_0}{2(1-\rho_2^*)} \tag{7}$$

The equation (7) gives us the information how the target value of the queue size X depends on the chosen probability P_0 and the system load ρ_2^*. If we want to assure that the STD CoS queue size in Ethernet switch will exceed the value L only with probability P_0, we have to set such a threshold T that added to additional queue growth X, offers a value not greater than L, as described in (8):

$$T + X = L \Rightarrow T = L + \frac{\rho_2 LnP_0}{2(1-\rho_1-\rho_2)} \tag{8}$$

The equation (8) provides the guideline to set the value of WRED threshold T when want the STD CoS queue size to be below L with a probability 1-P_0 in the case when the STD CoS load is ρ_2 and the load due to other CoSs (served with higher priority than STD CoS) is ρ_1.

To verify the proposed approach we repeated the simulation study described in section 4.2.1 (table 4). We assumed that the target STD CoS queue size L is 85 packets and the probability of its violation is $P_0=10^{-2}$. Since ρ_1=0.1428 and ρ_2=0.8 then, the equation (8) returns the value of the

WRED threshold T=36. Setting this value in simulation tests we obtained the results presented in table 6.

Table 6: Simulation results of queue size in ES obtained in test with improved threshold tuning.

Tested CoS	**Measured queue size in ES [pkts]**		
	Parameter name Q	**Mean**	**Q_{max}: Prob{$Q>Q_{max}$}<10^{-3}**
Telephony	Q_{QoS}	0.6	5
STD	Q_{STD}	43	85

5. SUMMARY

In this paper we showed that in some situations Ethernet access network might be a bottleneck in providing strict QoS guarantees, due to limited capabilities to assure traffic isolation in the shared buffer space. To cope with this problem we proposed a solution that engaged additional traffic control capabilities available in the considered Ethernet switch. The tested examples proved that our approach allows assuring target buffer space for given QoS CoS even if the network is congested by Standard CoS traffic.

REFERENCES

[1] 3Com Switch 5500 Family: Command Reference Guide

[2] 3Com Switch 5500 Family: Configuration Guide

[3] CHEN J., WANG Z., SUN Y., Switch Real-Time Industrial Ethernet with Mixed Scheduling Policy, IECON 02, Volume 3, 5-8 Nov. 2002 Page(s):2317 - 2321 vol.3

[4] CHOI B. Y., SONG S., BIRCH N., and HUANG J., Probabilistic approach to switched Ethernet for real-time control applications, Proc. of 7th International Conference on Real-Time Computing Systems and Applications (RTCSA'2000), Dec. 2000

[5] ENRÍQUEZ J., et al., EuQoS architecture update for Phase2, EuQoS Consortium, Deliverable D122, Dec. 2007.

[6] FAN X., JONSSON M., Guaranteed Real-Time Services over Standard Switched Ethernet, Proceedings of the IEEE Conference on Local Computer Networks 30th Anniversary (LCN'05), 15-17 Nov. 2005.

[7] HOAI H., et al., Switched real-time ethernet with earliest deadline first scheduling protocols and traffic handling, Proceedings of the IPDPS 2002, 15-19 April 2002 Page(s): 94 – 99

[8] IEEE Std 802.1D – 2004. Media Access Control (MAC) Bridges.

[9] IST EuQoS project homepage www.euqos.eu

[10] ITU-T Recommendation Y.1541, Network performance objectives for IP-based services, 2002

[11] JANOWSKI R., KRAWIEC P., BURAKOWSKI W., On assuring QoS in Ethernet access network, Proc. of The Third International Conference on Networking and Services ICNS 2007. June 19-25, 2007 - Athens, Greece.

[12] JANOWSKI R., Performance limits of priority queues, Proc. of HET-NETs'05 The 3rd International working conference on Performance modeling and evaluation of heterogeneous networks, Ilkley, U.K., July, 2005

[13] LIAN F.-L., TU Y.-CH., LI CH.-W., Ethernet Switch Controller Design for Real-Time Control Applications, Proceedings of the 2004 IEEE International Conference, 2-4 Sept. 2004 Page(s):464 - 467 Vol.1

[14] NS2: www.isi.edu/nsnam/ns

[15] PEDREIRAS P., LEITE R., ALMEIDA L., Characterizing the Real-Time Behavior of Prioritized Switched-Ethernet, In 2nd RTLIA, Porto, Portugal, June 2003.

[16] ROBERTS J., MOCCI U., VIRTAMO J. (eds.), Broadband network teletraffic: Performance evaluation and design of broadband multiservice networks, Final report COST 242, LNCS1155, Springer 1996.

[17] TARASIUK H., JANOWSKI R., BURAKOWSKI W., Admissible traffic load of real time class of service for inter-domain peers, Proceedings of ICNS&ICAS'2005 Conference, Papeete, Thaiti, October 2005, IEEE press.

[18] VARADARAJAN S., Experiences with EtheReal: a fault-tolerant real-time Ethernet switch, Proc. of 8th IEEE International Conference on Emerging Technologies and Factory Automation, vol. 1, 2001, pp. 183-194

Wide-sense nonblocking networks, Multicast, Multirate, Multi-$\log_2 N$ switching networks

Wojciech KABACIŃSKI[1], Tomasz WICHARY[2]

WIDE-SENSE NON-BLOCKING LOG2(*N*; *M*; *P*) MULTIRATE SWITCHING NETWORKS WITH MULTICAST CONNECTIONS WITH DISCRETE BANDWIDTH

Abstract: In this paper we are going to investigate the non-blocking operation of multirate log2(N; m; p) switching networks with multicast connections. The nonblocking operation of log2(N; 0; p) switching networks for point-to-point connections was firstly considered in [5], [6]. Conditions under which these switching networks are nonblocking for multicast connections were proposed by Tscha and Lea [7] and corrected by the same authors in [8]. They considered a wide-sense nonblocking switching networks and proposed a control algorithm based on the blocking window of a given size. Their results were later generalized by Danilewicz and Kabaciński to blocking windows of any size [9], [10], [11]. Multirate log2(N; 0; p) switching networks for point-to-point connections were considered by Lea in [14] and by Kabaciński and Żal in [15]. Frank K. Hwang and Bey-Chi Lin in [17] made an attempt to extended these connections to multicast log2(N; m; p) networks but encountered some difficulties which were commented by Danilewicz and Kabaciński in [18] where the correct results were given. In this paper we will extend these known results to multirate switching networks with multicast connections and this is the continuation of investigation is given in [16], [17], [18], [19] and [20]. The nonblocking operation at the connection level is considered.

1. INTRODUCTION

In future telecommunication networks large capacity routers will be needed. These routers will use high-capacity and high-speed switching networks based on optical transmission units and optical switching elements. Many papers have been published concerning the space and time domain as well as the wavelength domain (WDM) switching [1], [2], [3], [21], [22].

One of the architecture considered for both high-speed electronic and photonic switching is a switching network composed of log2(N; m; p) stages. However, the main drawback of such architecture is its blocking characteristics. To create nonblocking architecture two methods were proposed: vertical stacking and horizontal cascading. Switching networks obtained by vertically stacking p copies of log2(N; m; p) switching networks are called multi-log2N switching networks. Nonblocking operation of such networks for point-to-point connections were considered in [5], [6]. Horizontal cascading method introduces a greater number of stages between each inlet-outlet pair. These additional stages introduce a greater delay. The major benefit of this method is the lower cost

[1] Faculty of Electronics and Telecommunications, Poznan University of Technology, wojciech.kabacinski@et.put.poznan.pl
[2] Nokia Siemens Networks Sp. z o.o., tomasz.wichary@nsn.com

than in the vertical stacking method. The switching network obtained by the vertical stacking method is called the multi-log2N switching network. It consists of p copies of vertically stacked log2(N; m; p) networks. The benefit of VS method is the same number of stages. Conditions under which these switching networks are nonblocking for multicast connections were proposed by Tscha and Lea [7] and corrected by the same authors in [8]. They considered a wide-sense nonblocking switching networks and proposed the control algorithm based on the blocking window of a given size. Their results were later generalized by Danilewicz and Kabaciński to the blocking window of any size [9], [10], [11]. Multirate multi-log2N switching networks for point-to-point connections were considered by Lea in [14] and by Kabaciński and Żal in [15].

In [16] were given results for log2(N; 0; p) switching networks and discrete bandwidth case, when $1 \le t \le \lfloor n/2 \rfloor$. In [17] were given results for log2(N; 0; p) switching networks and continuous bandwidth case, for $B \in (1-b;\beta]$ and $B \in (1-2b;1/2]$, $b \in (1/4;1/2]$ sub cases. In [18] were given results for log2(N; 0; p) switching networks and discrete bandwidth case, when $\lfloor n/2 \rfloor + 1 \le t \le n-1$. In [19], [20], these results were extended for log2(N; m; p) switching networks and discrete bandwidth case. In this paper nonblocking conditions for multirate multicast connections will be extended for and discrete bandwidth case, when $\lceil (m+n)/2 \rceil < t < n$.

The paper is organized as follows. In Section 2, the model used in this paper is described. In Section 3, non-blocking conditions for log2(N; m; p), when $\lceil (m+n)/2 \rceil < t < n$ for discrete bandwidth case are given and proved. In Section 4 some analysis were given finally.

2. MODEL DESCRIPTION

Multi-log2N switching networks are constructed by vertically stacking p copies of log2(N; m; p) networks. Each of log2(N; m; p) switching networks is called a plane and is composed of 2 x 2 switches arranged in log2(N; m; p) stages. There are some equivalent topologies [6]. The topology of one network can be transform to another network by changing the position of the switches. These networks are self-routing.

Similarly as in [3], [4], and [7] we will use the algorithm based on blocking windows. In this concept a set of all outputs is divided into subsets. Each such subset is called a blocking window (BW). A new multicast connection is divided into subconnections, were outputs requested in each subconnection belonging to one blocking window.

Definition 1. Let the set of inputs $O = \{0, 1, 2, ..., N-1\}$ will be divided into N/K subsets $O_i = \{K \cdot i, K \cdot i + 1, K \cdot i + 2, ..., K \cdot i + (K-1)\}$, where $i \in [0; N/K - 1]$, $K = 2^t$, $t \in [1; n]$. Each of subsets O_i will be called the blocking window (BW).

Definition 2. Let the O_i will be diveded into $K/2^{\lfloor n/2 \rfloor}$ subsets $SBW_k =$ $= \{K \cdot i + 2^{\lfloor n/2 \rfloor} \cdot k, K \cdot i + 2^{\lfloor n/2 \rfloor} \cdot k + 1, K \cdot i + 2^{\lfloor n/2 \rfloor} \cdot k + 2, ..., K \cdot i + 2^{\lfloor n/2 \rfloor} \cdot (k+1) - 1\}$, where $K = 2^t$, $k = 0, 1, ..., K/2^{\lfloor n/2 \rfloor} - 1$, $\lfloor n/2 \rfloor + 1 \le t \le n-1$. Each of SBW_k will be called the subblocking window.

Throughout the paper we will use also following definitions and notations:

Definition 3. Let i, $1 \leq i \leq n+m-t-1$ means number of stage. Connections set from inputs belong to SI_i^m may be realized to $NBW_i = 2^{n-t}$, for $1 \leq i \leq m$ and $2^{n+m-t-i}$, for $m < i \leq n+m-1$ blocking windows.

Definition 4. Let SI_j^m will be set of inputs (exluding input x and SI_i^m for $i < j$, $1 \leq i \leq n-1$, $1 \leq j \leq n$) such as connection from these inputs may intersect with connection $\langle x;Y;\omega \rangle$, through stage j. Inputs belong to $|SI_j^m|$, $|SI_j^m| = 2^{j-1}$ for $1 \leq j \leq n$; 2^n; for $n < j < m+n-1$ will be called accessible inputs from stage j.

In the same way we will define SO_j^m:

Definition 5. Let SO_j^m will be set of outputs (exluding output y and SO_i^m for $i > j$, $1 \leq i \leq n-1$, $1 \leq j \leq n$) such as connection to these outputs may intersect with connection $\langle x;Y;\omega \rangle$, through stage j. Outputs belong to SO_j^m, $|SO_j^m| = 2^{n+m-j-1}$ for $m \leq j \leq n+m-1$; 2^n, for $1 \leq j < m$ will be called accessible outputs from stage j.

Definition of NCP_i (number of connection paths) can be found in [16], [17]. We will use also the bipartite graph representation described in these papers.

Let's $\langle x;Y;\omega \rangle$ be a new connection. This connection will go through vertices x and y in stages 0 and n, respectively, and some vertices in stages from 1 to $n-1$. When circuit switching is used, no paths are allowed to intersect at any vertex in one plane. In the case of multirate switching more connections may be realized through the same vertex provided that the sum of weights of these connections will be less than or equal to 1 (i.e. to the corresponding inter-stage link capacity).

In stage n, connection $\langle x;Y;\omega \rangle$ can not be blocked and we may still realize $R(SO_{n+m}^m;\omega)$ connections of weight b to ouput y_0. In stage $n+m-i$, where $1 \leq i \leq t-j$, connection $\langle x;Y;\omega \rangle$ may be blocked by $\langle k;l;b \rangle$, where $k \in SI_{n+m-i}^m$, $l \in \{y_0\} \cup \bigcup_{j=1}^{i} SO_{n+m-j}^m$. In stage $n+m-i$, these connections may block $\gamma_{SO}(i;\omega)$ alternative paths for a new one. In accessible outputs from stage $n+m-i$ may still left free bandwidth ($R(SO_{n+m-i}^m;\omega)$ connections of weight b). These connections will not block a new plane or alternative path and will be considered in stage $n+m-i-1$.

We will use the following algorithm for setting up a new connection $\langle x;Y;\omega \rangle$.

Algorithm 1:

Step 1. Divide connection $\langle x;Y;\omega \rangle$ to subconnections $\langle x;Y_i;\omega \rangle$ such that Y_i contains all possible outputs of set Y belonging to the same blocking window O_i.

Step 2. Try to set up the subconnection though one of already occupied planes starting from plane used for setting up the last connection.

Step 3. If Step 2 failed set up the subconnection through one of free planes (not occupied planes).

3. NON-BLOCKING CONDITIONS

In the following Theorems we will use following definitions:

$$\alpha(\omega)=\lfloor(1-\omega+b)/b\rfloor, \tag{1}$$

$$R(SI_0^m;\omega)=\lfloor(\beta-\omega)/b\rfloor, \tag{2}$$

$$R(SI_i^m;\omega)=\left(SI_i^m\lfloor\beta/b\rfloor+R(SI_{i-1}^m)\right)\otimes\alpha(\omega), \tag{3}$$

$$R(SO_{n+m}^m;\omega)=\lfloor(\beta-\omega)/b\rfloor, \tag{4}$$

$$R(SO_{n+m-i}^m;\omega)=\left(SO_{n+m-i}^m\lfloor\beta/b\rfloor+R(SO_{n+m-i+1}^m)\right)\otimes\alpha(\omega), \tag{5}$$

$$Q_{n+m-1}(\omega)=R(SO_{n+m-2}^m), \tag{6}$$

$$Q_{n+m-i}(\omega)=\left(2\cdot Q_{n+m-i+1}(\omega)\right)\otimes\alpha(\omega),\ \text{where } 2\le i\le t, \tag{7}$$

$$\gamma_{SI}(i;\omega)=\lfloor(2^{i-1}\lfloor\beta/b\rfloor+R(SI_{i-1}^m;\omega))/\alpha(\omega)\rfloor \tag{8}$$

$$\gamma_{SO}(i;\omega)=\lfloor(2^{i-1}\cdot\lfloor\beta/b\rfloor+R(SO_{n+m-i+1}^m;\omega))/\alpha(\omega)\rfloor, \tag{9}$$

$$C(i;x)=\lfloor k_1(j-2-i+1;j)/(j-2)\cdot 2^{i-1-(n\otimes 2)}\cdot\gamma_{SO}(n-t+(j-2-i+1);\omega)\cdot\alpha(\omega)\rfloor, \tag{10}$$

$$k_1(x;j)=\begin{cases}2^{t-x}\lfloor\beta/b\rfloor-(3-5\cdot[1-(n\otimes 2)])\cdot\gamma_{SO}(n-t+x;\omega)\cdot\alpha(\omega)- \\ \quad -(n\otimes 2)\cdot\left\lfloor\dfrac{2^{n-t+x-1}\lfloor\beta/b\rfloor+\sum_{i=1}^{n-t+x}2^{n-i}\otimes\alpha(\omega)}{\alpha(\omega)}\right\rfloor\cdot\alpha(\omega), & \text{for } x=j-2; \\ k_1(j-2;j)-C(1;x)\cdot(j-2-2\cdot(n\otimes 2))\cdot\gamma_{SO}(n-t+x;\omega)\cdot\alpha(\omega), & \text{for } x=j-3; \\ k_1(j-3;j)- \\ -\sum_{i=1}^{j-2-(n\otimes 2)-x}C(i+1;x)\cdot(j-2-2\cdot(n\otimes 2))\cdot 2^i\cdot\gamma_{SO}(n-t+x;\omega)\cdot\alpha(\omega), & \text{for } x\le j-4,\end{cases} \tag{11}$$

$$k_2(x;j)=\lfloor k_1(x;j)/a_2\rfloor,\quad a_2=\lfloor(2^{n-t+x-1}\lfloor\beta/b\rfloor+R(SO_{t-x+2}^m;\omega))/\alpha(\omega)\rfloor\cdot\alpha(\omega), \tag{12, 13}$$

$$k_3(x;j)=\lfloor 2^{t-j}\lfloor\beta/b\rfloor/a_3\rfloor,\quad a_3=\lfloor 2^{n-t+x}\lfloor\beta/b\rfloor/\alpha(\omega)\rfloor\alpha(\omega), \tag{14, 15}$$

$$k_4(x;j)=\lfloor k_1(x;j)/2^{j-x-(n\otimes 2)}\cdot\gamma_{SO}(t-x+1;\omega)\cdot\alpha(\omega)\rfloor, \tag{16, 17}$$

where,
$a\otimes b$ means the rest of dividing a by b,
$\lceil x\rceil$ means round up to the nearest integer, $\lfloor x\rfloor$ means round down to the nearest integer.

Theorem 1. *The $log_2(N; m; p)$ multirate switching network is nonblocking in the wide sense for $\lceil (m+n)/2 \rceil < t < n$, $m=1$ for discrete bandwidth case when algorithm 1 is used if and only if:*

$$p \geq S(B;j)+1, \tag{18}$$

where

$$S(\omega;j) = p_1^1 + p_1^2 + p_2 + p_3 + 1, \text{ for } j=1, \tag{19}$$

$$S(\omega;j) = p_1^1 + p_1^2 + p_1^3 + p_1^4 + p_2 + p_3 + 1, \text{ for } j=2, \tag{20}$$

$$S(\omega;j) = p_1^1 + p_1^2 + p_1^3 + p_1^4 + p_1^5 + \sum_{x=2}^{j-2} p_1^{j-x+4} + p_1^{j+4} + p_2 + p_3 + 1, \text{ for } j \geq 3. \tag{21}$$

Proof. Let $\langle x;Y;\omega \rangle$ be a new connection. This connection may be a point-to-point connection or to be part of multicast connection. This connection can be blocked by connections of weight greater then $1-\omega$ which passes through any of nodes belonging to the path of the new connection.

Let $y_0 \in SBW_0 \cap Y$. In stage $n+m-i$, $1 \leq i \leq t-j$, a new connection may be blocked by $\langle k;l;b \rangle$, $k \in SI^m_{n+m-i}$, $l \in \{y_0\} \cup \bigcup_{j=1}^{i} SO^m_{n+m-j}$. In the worst state connections set up to SBW_0 these connections may block p_1^1 planes, where

$$p_1^1 = \sum_{i=1}^{t-j} \gamma_{SO}(i;\omega)/2. \tag{22}$$

Let $y_1 \in SBW_1 \cap Y$ for n is even. In stage $n+m-i$, $1 \leq i \leq t-j-1$, a new connection may be blocked by $\langle k;l;b \rangle$, $k \in SI^m_{n+m-i}$, $l \in \{y_0\} \cup \bigcup_{j=1}^{i} SO^m_{n+m-j}$. These connections may block

$$p_1^{2a} = \sum_{i=1}^{t-j-1} \gamma_{SO}(i;\omega)/2 \tag{23}$$

planes. In outputs belog to SBW_0, $\gamma_{SO}(t-j;\omega)\cdot\alpha(\omega)$ connections of weight b are aready set. Hence $|SI^m_{t-j}| \cdot \lfloor \beta/b \rfloor - \gamma_{SO}(t-j;\omega)\cdot\alpha(\omega)$ connections of weight b may be still set and they may block together

$$p_1^2 = \sum_{i=1}^{t-j-1} \gamma_{SO}(i;\omega)/2 + \lfloor (2^{t-j-1} \lfloor \beta/b \rfloor - \gamma_{SO}(t-j;\omega)\cdot\alpha(\omega))/\alpha(\omega) \rfloor \tag{24}$$

planes.

Let $y_1 \in SBW_1 \cap Y \in \varnothing$ for n is odd. In stage $n+m-i$, $1 \leq i \leq t-j$, a new connection may be blocked by $\langle k;l;b \rangle$, $k \in SI^m_{n+m-i}$, $l \in \{y_0\} \cup \bigcup_{j=1}^{i} SO^m_{n+m-j}$. In the worst state connections set up to SBW_0, may block p_1^2 planes, where

$$p_1^2 = \lfloor 2^{t-j-1} \lfloor \beta/b \rfloor / \alpha(\omega) \rfloor. \tag{25}$$

For $t \geq \lfloor n/2 \rfloor + 2$, outputs belongs to SBW_2 may be block by configuration of $\sum_{i=1}^{t-j-1} \gamma_{SO}(i;\omega)$ and by connections from inputs accessible from stage $t-j$.

Let $Y_2 \in SBW_2 \cap Y$, $|Y_2| = 2$. In stage $n+m-i$, $1 \leq i \leq t-j$, a new connection may be blocked by $\langle k;l;b \rangle$, $k \in SI^m_{n+m-i}$, $l \in \{y_0\} \cup \bigcup_{j=1}^{i} SO^m_{n+m-j}$. These connections may block p_1^3 planes, where p_1^3 where

$$p_1^3 = \sum_{i=1}^{t-j-1} \gamma_{SO}(i;\omega) \tag{26}$$

for n is even and by (22) for n is odd.
Let $Y_3 \in SBW_3 \cap Y$, $|Y_3| = 2$. In stage $n+m-i$, $1 \leq i \leq t-j$, a new connection may be blocked by $\langle k;l;b \rangle$, $k \in SI^m_{n+m-i}$, $l \in \{y_0\} \cup \bigcup_{j=1}^{i} SO^m_{n+m-j}$. These connections may block p_1^4 planes where p_1^4 is also given by (26). For n is even we have: In stage $n+m-i$, $1 \leq i \leq t-j-1$, a new connection may be blocked by $\langle k;l;b \rangle$, $k \in SI^m_{n+m-i}$, $l \in \{y_0\} \cup \bigcup_{j=1}^{i} SO^m_{n+m-j}$. These connections may block p_1^{4a} planes, where p_1^{4a} is given by (23). In outputs belog to SBW_0 and SBW_2, $2 \cdot \gamma_{SO}(t-j;\omega) \cdot \alpha(\omega)$ connections of weight b are aready set. Hence $|SI^m_{t-j}| \cdot \lfloor \beta/b \rfloor - 2 \cdot \gamma_{SO}(t-j;\omega) \cdot \alpha(\omega)$ connections of weight b may be still set up to them and they may block

$$p_1^{4b} = \lfloor (2^{t-j-1} \lfloor \beta/b \rfloor - 2 \cdot \gamma_{SO}(t-j;\omega) \cdot \alpha(\omega)) / \alpha(\omega) \rfloor \tag{27}$$

planes. In general connection to outputs belong to SBW_4 may block p_1^4 planes, where

$$p_1^4 = \sum_{i=1}^{t-j-1} \gamma_{SO}(i;\omega)/2 + \lfloor (2^{t-j-1} \lfloor \beta/b \rfloor - 2 \cdot \gamma_{SO}(t-j;\omega) \cdot \alpha(\omega)) / \alpha(\omega) \rfloor. \tag{28}$$

Let's consider two case for connections set to outputs belong to SBW_5, when n is even.

Let $Y_4 \in SBW_5 \cap Y$, $|Y_4| = 3$. For one of outputs belong to Y_4 we will consider following configuration of connections. In stage $n+m-i$, $1 \leq i \leq t-j-1$, a new connection may be blocked by $\langle k;l;b \rangle$, $k \in SI^m_{n+m-i}$, $l \in \{y_0\} \cup \bigcup_{j=1}^{i} SO^m_{n+m-j}$. For rest of ouputs belong to Y_4 we will consider two configuration of connections: so, in stage $n+m-i$, $1 \leq i \leq t-j-2$, a new connection may be blocked by $\langle k;l;b \rangle$, $k \in SI^m_{n+m-i}$, $l \in \{y_0\} \cup \bigcup_{j=1}^{i} SO^m_{n+m-j}$. In general these connections may block p_1^{5a} planes, where

$$p_1^{5a} = 3 \cdot \sum_{i=1}^{t-j-2} \frac{\gamma_{SO}(i;\omega)}{2} + \frac{\gamma_{SO}(t-j-1;\omega)}{2}, \tag{29}$$

Let $Y_4 \in SBW_4 \cap Y$, $|Y_4| = 2$. For this case, $p_1^{5a} = 2 \cdot p_1^{2a}$, where p_1^{2a} is given by (22).

Let consider stage $t-x+1$, $1 \leq x \leq j-2$. In that stage we have $|SI^m_{t-x+1}| = 2^{t-x}$ and $|SO^m_{t-x+1}| = 2^{n-t+x-1}$. $\sum_{i=1}^{n-(t-x)} \gamma_{SO}(i;\omega)$ will take bandwidth in outputs accessible from stage $t-x+1$ and next ones. On outputs belong to SO^m_{t-x+1} may left bandwidth equals to $k_1(x;j)$ connections of weight *b*. Hence we have $k_2(x;j)$ such configurations and maximum $k_3(x;j)$ in one SBW. The mensioned configuration of connection will occupy outpus in $k_4(x;j)$ next SBW.

So, we obtain:

$$p_1^{j-x+4} = k_4(x;j) \cdot 2^{j-x-(n \otimes 2)} \cdot \sum_{i=1}^{n-(t-x)} \gamma_{SO}(i;\omega). \tag{30}$$

For $x = 1$, SBW will be counted up from $2 \cdot k_4(j-2;j) - 1 - (n \otimes 2)$ to $2^j - 1$ and connections to outputs belong to that SBW may block p_1^{j+4} planes where

$$p_1^{j+4} = \left(2^j - 5 - (n \otimes 2)\right) \cdot \sum_{i=1}^{n-(t-x)} \gamma_{SO}(i;\omega), \text{ for } j-2=1,$$
$$p_1^{j+4} = \left(2^j - \sum_{i=2}^{j-2} k_4(i;j) - 5 - (n \otimes 2)\right) \cdot \sum_{i=1}^{n-(t-x)} \gamma_{SO}(i;\omega), \text{ for } j-2>1.$$

At stages from 1 to $t - 2j + (n \otimes 2)$ we will consider multicast connections. At the stage i, $1 \le i \le t$ connections $\langle k;l;b \rangle$, $k \in \{x\} \cup \bigcup_{j=1}^{i} SI_j^m$ may be set up to $|L| \in NBW_i - 1$ blocking windows. The rest of free bandwidth may be used by connections at stage $i+1$. Generally, at stage i, connections $\langle k;l;b \rangle$ may block $\gamma_{SI}(i;\omega) \cdot (2^{n-t-i} - 1)$ alternative paths for a new one. In the worst state they may block p_2 planes, where

$$p_2 = \gamma_{SI}(1;\omega) \cdot 2^{n-t} + \sum_{i=2}^{t-2j+(n \otimes 2)} \gamma_{SI}(i;\omega) \cdot 2^{m+n-t-i} \tag{31}$$

In BW_0 may left free bandwidth. In the worst state, connections to this free bandwidth at outputs belonging to this BW_0, may block p_3, where
for *n* is even

$$p_3 = \begin{cases} A_1, & \text{for } j=1; \\ A_2, & \text{for } j=2 \text{ lub } j-2>1 \text{ or } j-2=1,\, p_1^{5a} > p_1^{5b}; \\ A_3, & \text{for } j-2=1,\, p_1^{5b} \ge p_1^{5a}, \end{cases} \tag{32}$$

for *n* is odd

$$p_3 = \begin{cases} A_1, & \text{dla } j=1; \\ A_2, & \text{dla } j=2; j \ge 3, j-2=1; \\ A_3, & \text{dla } j-2>1, \end{cases} \tag{33}$$

$$A_1 = R(SO_{t-j}^m;\omega) + \left[2^{t-j-1} \lfloor \beta/b \rfloor - \gamma_{SO}(t-j;\omega) \cdot \alpha(\omega)\right] \otimes \alpha(\omega), \text{ for } n \text{ is even;} \tag{34}$$
$$A_1 = R(SO_{t-j}^m;\omega) + \left[2^{t-j-1} \lfloor \beta/b \rfloor\right] \otimes \alpha(\omega), \text{ for } n \text{ is odd;} \tag{35}$$

$$A_2 = A_1 + \left[k_1(x;j) - (k_2(x;j)-1) \cdot \sum_{i=1}^{n-(t-x)} \gamma_{SO}(i;\omega) \cdot \alpha(\omega)\right] \otimes \alpha(\omega) \tag{36}$$

$$A_3 = A_2 + \sum_{x=2}^{j-2} (k_2(x;j)-1) \cdot R(SO_{t-x+1}^m;\omega) +$$
$$+ 6 \cdot R(SO_{t-j-1}^m;\omega) + (2^j - 5) \cdot R(SO_{t-x+1}^m;\omega), \text{ for n is even;} \tag{37}$$

$$A_3 = A_2 + R(SO^m_{t-j};\omega) + R(SO^m_{t-j-1};\omega) + [2^{t-j-1}\lfloor\beta/b\rfloor - 2\cdot\gamma(t-j;\omega)]\otimes\alpha(\omega) +$$
$$+ (2^j - 4)\cdot k_3(x;j)\cdot R(SO^m_{t-x+1};\omega), \text{ for } n \text{ is odd}; \tag{38}$$

In the worst case, these sets of planes (p_1^1, p_1^2, p_1^3, p_1^4, p_1^5, p_1^{j-x+4}, p_1^{j+4}, p_2, and p_3) are disjoint and one additional plane for a new connection $\langle x;Y;\omega\rangle$ is needed. Equations (p_1^1, p_1^2, p_1^3, p_1^4, p_1^5, p_1^{j-x+4}, p_1^{j+4}, p_2, and p_3) must be maximized through all possible values of ω so we obtain following formula:

$$p \geq \max_{b\leq\omega\leq B}\{S(\omega)\} + 1, \tag{39}$$

where $S(\omega)$ is given by (19-21) formulae. After puting $\omega = B$ to formulae (39) we got (18).

□

Theorem 2. *The log₂(N; m; p) multirate switching network is nonblocking in the wide sense for* $\lceil (m+n)/2 \rceil < t < n$, $m \geq 2$ *for discrete bandwidth case when algorithm 1 is used if and only if:*

$$p \geq S(\omega) + 1, \tag{40}$$

where

$$S(\omega) = \lfloor p_1 + p_2 + p_3 + p_4 \rfloor. \tag{41}$$

Proof. Sufficient proof we will prove by showing the worst state of the switching network.

Let $\langle x;Y;\omega\rangle$, $Y \in BW_0$, $|Y| = 2^{t-2}$ be a new connection. In the worst state these connections may be block by connections set to BW_0 and they may block p_1 planes, where

$$p_1 = 2^{t-2}\lfloor\lfloor\beta/b\rfloor + R(SO^m_{n+m};\omega)/\alpha(\omega)\rfloor + \sum_{i=2}^{t} 2^{t-2i}\lfloor 2\cdot Q_{n+m-i+1}(\omega)/\alpha(\omega)\rfloor, \tag{42}$$

$$Q_{n+m-1}(\omega) = R(SO^m_{n+m-1};\omega),\ Q_{n+m-i}(\omega) = 2\cdot Q_{n+m-i+1}(\omega)\otimes\alpha(\omega), \text{ for } 2 \leq i \leq t. \tag{43-44}$$

At stages from 1 to t we will consider multicast connections. At the stage i, $1 \leq i \leq t$ connections $\langle k;l;b\rangle$, $k \in \{x\} \cup \bigcup_{j=1}^{i} SI_j^m$ may be set up to $|L| \in NBW_i - 1$ blocking windows. The rest of free bandwidth may be used by connections at stage $i+1$. Generally, at stage i, connections $\langle k;l;b\rangle$ may block $\gamma_{SI}(i;\omega)\cdot(2^{n-t-i} - 1)$ alternative paths for a new one. In the worst state they may block p_2 planes, where

$$p_2 = \sum_{i=1}^{m}[\gamma_{SI}(i;\omega)/2^i]\cdot(2^{n-t} - 1) + \sum_{i=m+1}^{t}\gamma_{SI}(i;\omega)\cdot(2^{n-t-i} - 2^{-m}). \tag{45}$$

In each BW may left free bandwidth. We may set connections to ouputs belong to the BW through stage $t+1$. These connections may block p_3 planes, where

$$p_3 = 2^t\lfloor\beta/b\rfloor - \lfloor(2^t\lfloor\beta/b\rfloor\otimes\alpha(\omega) - \omega/b)/\alpha(\omega)\rfloor(2^{n-2t-1} - 2^{-m}), \text{ for } t < n-1$$

and 0 for $t = n-1$ (46)

In BW_0 may still left free bandwidth. We may realize $V^0(\omega)$ connections of weight b to outputs belong to BW_0. In BW_i, $i \in \langle 1; 2^{n-t}-1 \rangle, i \in N$ may also left free bandwidth and we may realize $2^t \cdot \lfloor \beta/b \rfloor \otimes \alpha(\omega)$ connections of weight b to outpus belong to them. There connections through stage $t+1$ may block p_4 additional planes, where

$$p_4 = \lfloor (V^0(\omega) + 2^t \lfloor \beta/b \rfloor \otimes \alpha(\omega)) / \alpha(\omega) \rfloor \cdot 2^{-m}, \text{ for } t < n-1$$
and 0 for $t = n-1$, (47)

$$V^0(\omega) = Q_{n+m-t}(\omega),\ V^i(\omega) = (V^{i-1}(\omega) + 2^{i-1} \cdot (2^t \lfloor \beta/b \rfloor \otimes \alpha(\omega))) \otimes \alpha(\omega). \qquad (48\text{-}49)$$

In the worst case, these sets of planes (p_1^1, p_1^2, p_1^3 and p_1^4) are disjoint and one additional plane for a new connection $\langle x; Y; \omega \rangle$ is needed. Equations (p_1^1, p_1^2, p_1^3 and p_1^4) must be maximized through all possible values of ω so we obtain following formula:

$$p \geq \max_{b \leq \omega \leq B} \{S(\omega)\} + 1, \qquad (50)$$

where $S(\omega)$ is given by (41) formula. After puting $\omega = B$ to formulae (50) we got (40).

□

4. CONCLUSIONS

In this paper we extend knows results for multi-log2(N; 0; p) to multi-rate multi-log2(N; m; p) with multicast connections. We proved wide sense non-blocking conditions for discrete bandwidth, when $\lceil (m+n)/2 \rceil < t < n$. Due to editorial requirements we skipped numerical and graphical examples. As a result of analysing numerical outcomes we affirm that when ω increase then the number of planes increases as well. The explanation is that $S(\omega)$ is a nondecreasing function and mostly is expressed in denominator. If ω increases, denominator decreases. The highest values of $S(\omega)$ is given for $\omega = B$. We also analysing of influence size of blocking window by changing t value. We emphasize that the number of planes is inversely proportional of tendency to size of blocking window.

It should be noted that number of planes depend also on sort of a new connection $\langle x; Y; \omega \rangle$ and t and j.

In Theorem 1, Y is set of outputs at least of one output in each SBW, when n is even and by analogy with excepting output in SBW_1, when n is odd. In Therem 2, when $m \geq 2$, we have $|Y|$ equals to 2^{t-2} outputs for maximizing of blocking planes for $\langle x; Y; \omega \rangle$.

REFERENCES

[1] DANILEWICZ G., KABACIŃSKI W., *Struktury s-sekcyjnych pól komutacyjnych zbudowanych z komutatorów λ*, KST 95 – conference materials, Bydgoszcz, September 1995, vol. E, pp. 157-166.

[2] FUJIWARA M., *A coherent photonic wavelength-division switching system for broadband networks*, J. Lightwave Technology, vol. 8, No. 3, 1990, pp. 416-422.

[3] GERSTEL O., *On the future of wavelength routing networks*, IEEE Networks, vol. 10, No. 6, 1996, pp.14-20.

[4] HWANG F.K., JAJSZCZYK A., *Nonblocking multiconnection networks*, IEEE Transactions on Communications, Vol. COM-34, No. 10, 1986, pp. 1038-1041.

[5] KACZMAREK S., *Własności równoległych połączeń pól komutacyjnych*, Przegląd Telekomunikacyjny, vol. LVI, No. 2, 1983, pp. 54-56.

[6] LEA C.-T., *Multi-Log2N networks and their applications in high-speed electronic and photonic switching systems*, IEEE Transactions on Communication, vol. 38, No. 10 October 1990,pp. 1740-1749.

[7] TSCHA Y., LEA K.H., *Non-blocking conditions for multi-log2N multiconnection networks*, IEEE GLOBECOM 1992, pp.1600-1604.

[8] TSCHA Y., LEA K.H., *Yet another result on multi-log2N networks*, IEEE Transactions on Communication, vol. 47, No. 9, September 1999, pp.1600-1604.

[9] DANILEWICZ G., KABACIŃSKI W., *Wide-Sense Non-blocking Multi-Log2N Broadcast Switching Networks*, International Conference on Communications ICC 2000 New Orleans, LA USA, June 2000, pp. 1440-1444.

[10] DANILEWICZ G., KABACIŃSKI W., *Non-blocking multicast multi-log2N switching networks*, First Polish-German Teletraffic Symposium 2000, Dresden, September 2000, pp. 201-210.

[11] DANILEWICZ G., KABACIŃSKI W., *Wide-sense and strict-sense non-blocking operation of multicast multi-log2N switching networks*, IEEE Transactions on Communications, vol. 50, No. 6, June 2002, pp. 1025-1036.

[12] DANILEWICZ G., KABACIŃSKI W., *Comments "Wide-Sense Nonblocking Multicast Log2(N; m; p) Networks*, IEEE Transactions on Communications, vol. 54, No. 6, June 2006, pp. 980-982.

[13] FRANK K., HWANG AND BEY-CHI LIN, *Wide-Sense Nonblocking Multicast Log2(N; m; p) Networks*, IEEE Transactions on Communications, VOL. 51, NO. 10, October 2003.

[14] LEA C.-T., *Multirate Logd(N,e,p) Networks*, IEEE GLOBECOM 1994, pp.319-323.

[15] KABACIŃSKI W., ŻAL M., *Non-blocking operation of multi-Log2N switching networks*, System Science, vol. 25, No. 4, 1999, pp.83-97.

[16] KABACIŃSKI W., WICHARY T., *Warunki nieblokowalności w polach typu multi-log$_2$N z połączeniami rozgłoszeniowymi typu multi-rate dla pasma dyskretnego*, PWT 2004, 9-10 XII

[17] KABACIŃSKI W., WICHARY T., *Multi-log$_2$N Multirate Switching Networks with Multicast Connections,* 10th PSRT, Cracow, Poland, 2003.

[18] KABACIŃSKI W., WICHARY T., *Wide-Sense Non-blocking Multi-log$_2$N Multirate Switching Networks with Multicast Connections*, 12th PSRT, Poznan, Poland, 2005.

[19] KABACIŃSKI W., WICHARY T., *Wide-sense Non-blocking Multirate and Multicast log2(N; m; p) Switching Networks*, Proceedings of 14th Polish Teletraffic Symposium, Zakopane, 2007 pp. 53 – 64.

[20] KABACIŃSKI W., WICHARY T., *Wide-sense Non-blocking Multirate and Multicast log2(N; m; p) Switching Networks*, Theoretical and Applied Informatics, Quarterly, Volume 19, No.3/2007, Gliwice, pp.189-202.

[21] SUZU S., *An experiment on high-speed optical time-division switching*, J. Lightwave Technology, vol. LT-4, No. 7, 1986, pp.894-899.

[22] SUZU S., NAGASHIMA K., *Optimal broadband communications network architecture utilizing wavelength-division switching technology*, Proc. First Topical Meeting on Photonic Switching, Incline Village, Nevada, 1987, pp. 134-137.

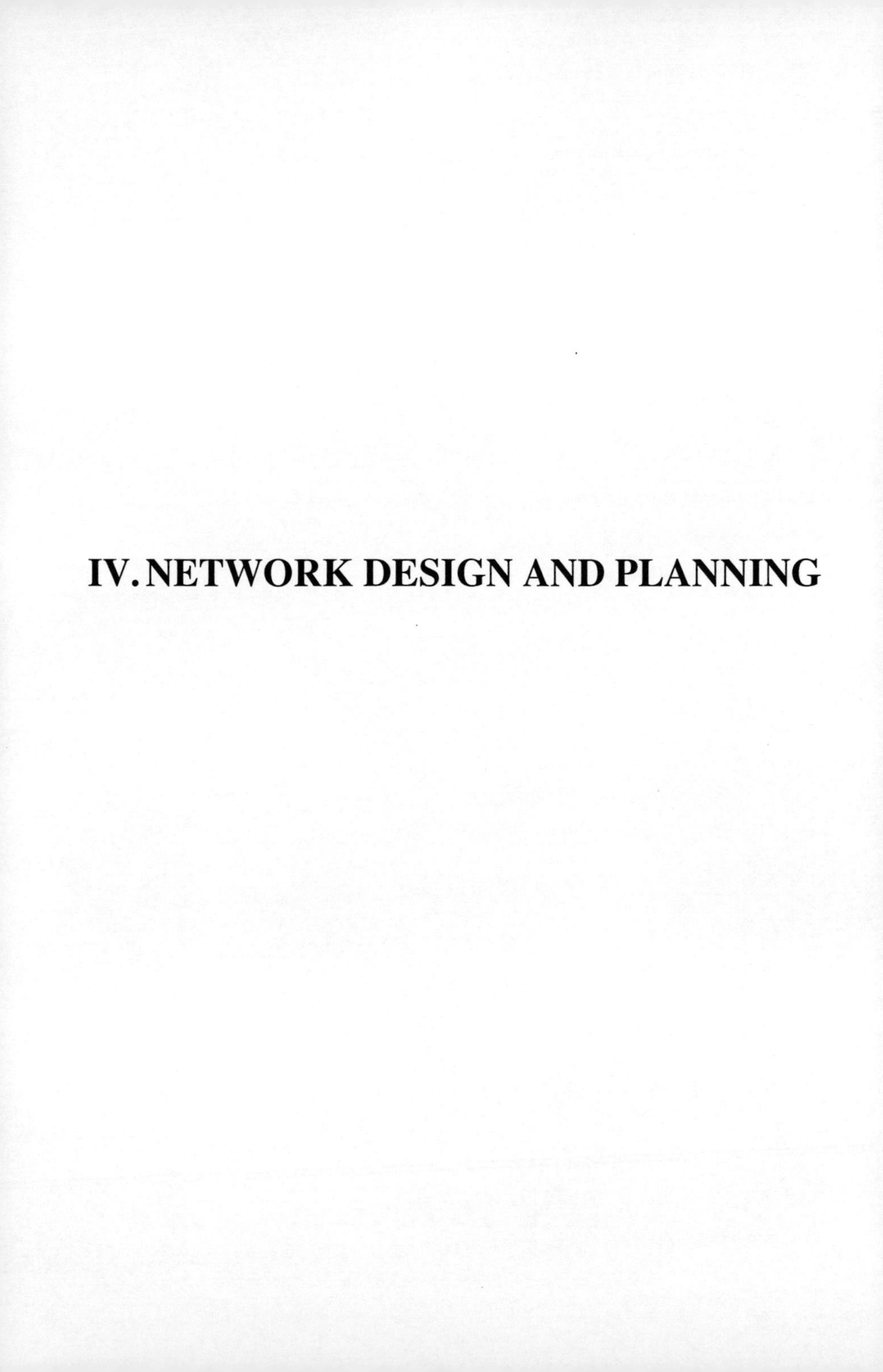

IV. NETWORK DESIGN AND PLANNING

Keywords – Inter-domain routing, distributed optimization

Mariusz MYCEK[1]
Michał PIÓRO[1,2]
Artur TOMASZEWSKI[1]

A SCHEME FOR COOPERATIVE OPTIMIZATION OF FLOWS ON INTER-DOMAIN LINKS

As today's market forces implementation of QoS enabled services spanned over multiple administrative domains, isolated and locally optimized inter-domain routing decisions become increasingly inadequate. Instead, coordinated routing models are required together with joint optimization goals. Available papers and standardization documents focus on description of technical means for deployment of inter-domain transport services giving little (or no) attention to the problem of evaluating effective inter-domain routing patterns. Our paper aims at closing this gap. It presents an iterative distributed process where domains cooperatively determine a (sub)optimal, with respect to a common utility function, flow of inter-domain traffic. If all the cooperating domains adhere to the results of this process, they can reduce their operational costs, speed up operations, and increase profits.

1. INTRODUCTION

Inter-domain routing in IP/MPLS networks results from policies (rules) the operators apply independently within each domain. These rules may determine what inter-domain flows are allowed to enter and to transit a domain, which border routers should handle these flows, which border routers and inter-domain links are to be used to reach particular destinations, etc. If carefully applied, these rules allow an operator to minimize intra-domain resources used for handling of inter-domain traffic, simultaneously fulfilling such traffic engineering goals such as assuring prefixes reachability, inter-domain load balancing, and resilience to inter-domain link outages.

As today's market forces implementation of QoS enabled services spanned over multiple administrative domains, isolated and locally optimized inter-domain routing decisions become increasingly inadequate. Instead, coordinated routing models are required together with joint optimization goals. Available papers and standardization documents focus on description of technical means for deployment of inter-domain transport services giving little (or no) attention to the problem of evaluating effective inter-domain routing patterns. Our paper aims at closing this gap. It presents an iterative distributed process where domains cooperatively determine a (sub)optimal, with respect to a common

[1]Institute of Telecommunications, Warsaw University of Technology, Poland, Nowowiejska 15/19 00-665 Warszawa, e:mail: {mariusz,mpp,artur}@tele.pw.edu.pl
[2]Department of Electrical and Information Technology, Lund University, Sweden

utility function, flow of inter-domain traffic. If all the cooperating domains adhere to the results of this process, they can reduce their operational costs, speed up operations, and increase profits.

Each domain of the Internet acts as Autonomous System and reveals only very limited information about its internal topology and the implemented routing scheme to adjacent domains (by means of exterior gateway protocols (EGP) such as BGP – see [1] and the discussion there). As a result, every domain has a very limited knowledge of the overall network topology what virtually does not allow a domain making optimal (in a global sense) inter-domain routing decisions.

It seems that reaching a (sub)optimal traffic routing in the inherently decentralized Internet environment requires implementation of a distributed process of routing optimization run in the control plane of the cooperating domains. Preliminary results on distributed inter-domain routing optimization can be found for example in [10], [6], [8], [7] and [9]. In [8] a generic multi-domain routing problem (consisting in optimization of bandwidth reservation levels on inter-domain links for traffic flows identified by traffic classes and traffic destinations) is formulated, and its possible decompositions are discussed. In [7] it is shown how to decompose the problem with respect to individual domains using sub-gradient optimization based on Lagrangean relaxation. In [9] it is demonstrated how to resolve an inter-domain routing optimization problem using a distributed process based on sub-gradient optimization combined with recovering of near-optimal bandwidth reservation levels.

The paper is organized as follows. Sections 2 present a formulation and decomposition of the problem (as stated in [8]). Section 3 presents usable ways for aggregation of intra-domain topologies. Section 4 presents results of numerical experiments evaluating effectiveness of the proposed methods in case of simple multi domain networks. Eventually, Section 5 gives a summary.

2. PROBLEM FORMULATION AND DECOMPOSITION

Generally speaking, we consider a problem of maximizing the total utility function for a set of cooperating domains associated with handling of inter-domain traffic. The considered model of the network consists of a directed graph $\mathcal{G} = (\mathcal{V}, \mathcal{E})$ with the set of nodes $\mathcal{V}$ and the set of directed links $\mathcal{E}$ ($\mathcal{E} \subseteq \mathcal{V} \times \mathcal{V}$). For a set of nodes $\mathcal{U} \subseteq \mathcal{V}$ we define the set $\delta^+(\mathcal{U})$ of links outgoing from set $\mathcal{U}$, and the set $\delta^-(\mathcal{U})$ of links incoming to set $\mathcal{U}$. More precisely, $\delta^+(\mathcal{U}) = \{e \in \mathcal{E} : a(e) \in \mathcal{U} \wedge b(e) \notin \mathcal{U}\}$ and $\delta^-(\mathcal{U}) = \{e \in \mathcal{E} : b(e) \in \mathcal{U} \wedge a(e) \notin \mathcal{U}\}$, where $a(e)$ and $b(e)$ denote the originating and terminating node, respectively, of link $e \in \mathcal{E}$. Besides, we shall write $\delta^\pm(v)$ instead of $\delta^\pm(\{v\})$, i.e., when $\mathcal{U} = \{v\}$ is a singleton.

$\mathcal{M}$ is the set of network domains. Each node $v \in \mathcal{V}$ belongs to exactly one domain denoted by $\mathcal{A}(v)$. Hence, set $\mathcal{V}$ is partitioned into subsets $\mathcal{V}_m = \{v \in \mathcal{V} : \mathcal{A}(v) = m\}, m \in \mathcal{M}$. Let, for each domain $m \in \mathcal{M}$, $\mathcal{E}_m^+ = \delta^+(\mathcal{V}_m)$ denotes the set of outgoing inter-domain links of domain m, $m \in \mathcal{M}$, $\mathcal{E}_m^- = \delta^-(\mathcal{V}_m)$ denotes the set of incoming inter-domain links of domain m and $\mathcal{E}_m = \{e \in \mathcal{E} : a(e), b(e) \in \mathcal{V}_m\}$ is the set of *intra-domain* links between the nodes in the same domain m.

The set of all intra-domain links is denoted by $\mathcal{E}_\mathcal{I} = \bigcup_{m \in \mathcal{M}} \mathcal{E}_m$. Further, the set of all *inter-domain* links is denoted by $\mathcal{E}_\mathcal{O}$, where $\mathcal{E}_\mathcal{O} = \{e \in \mathcal{E} : \mathcal{A}(a(e)) \neq \mathcal{A}(b(e))\} = \bigcup_{m \in \mathcal{M}} \delta^+(\mathcal{V}_m) = \bigcup_{m \in \mathcal{M}} \delta^-(\mathcal{V}_m)$. Clearly, the set of intra-domain links is disjoint with the set of inter-domain links. Finally, the capacity of link $e \in \mathcal{E}$ is denoted by c_e and expressed in units of bandwidth, for example in Mb/s.

Set $\mathcal{D}_\mathcal{O}$ represents inter-domain traffic demands between pairs of nodes (not necessarily between all pairs). The originating and terminating node of demand $d \in \mathcal{D}_\mathcal{O}$ is denoted by $s(d)$ and $t(d)$, respectively, and h_d is the traffic volume of demand d, expressed in the same units of bandwidth as capacity of links. Also, $\mathcal{D}_{\mathcal{O}}(s,t) = \{d \in \mathcal{D}_{\mathcal{O}} : s(d) = s \wedge t(d) = t \wedge \mathcal{A}(s) \neq \mathcal{A}(t)\}$ denotes the set of all inter-domain demands from node $s \in \mathcal{V}_m$ to node $t \in \mathcal{V}_n, m \neq n$ (note that there can be more than one demand between a given pair of nodes). The set of all demands originating in domain m is denoted as $\mathcal{D}_{\mathcal{O}m} = \{d \in \mathcal{D}_{\mathcal{O}} : s(d) \in \mathcal{V}_m\}$. The sets $\mathcal{D}_{\mathcal{O}m} = \{d \in \mathcal{D}_{\mathcal{O}} : s(d) \in \mathcal{V}_m\}$, $m \in \mathcal{M}$, define a partition of $\mathcal{D}_{\mathcal{O}}$.

Let x_{et} denote a variable specifying the amount of aggregated bandwidth (called *flow* in the sequel) reserved on intra-domain link $e \in \mathcal{E}_{\mathcal{I}}$ for the traffic destined for (a remote) node $t \in \mathcal{V}$. Then, for each inter-domain link $e \in \mathcal{E}_{\mathcal{O}}$ we introduce two flow variables: r^+_{et} and r^-_{et}. Variable r^+_{et} (respectively, r^-_{et}) denotes the amount of bandwidth reserved for traffic carried on e and destined for t that is reserved by domain $\mathcal{A}(a(e))$ (respectively, $\mathcal{A}(b(e))$) at which link e originates (respectively, terminates). Then for each domain $m \in \mathcal{M}$ we introduce the following flow vectors:

- $\boldsymbol{x}_m = (x_{et} : e \in \mathcal{E}_m, t \in \mathcal{V})$
- $\boldsymbol{r}^+_m = (r^+_{et} : e \in \mathcal{E}^+_m, t \in \mathcal{V})$
- $\boldsymbol{r}^-_m = (r^-_{et} : e \in \mathcal{E}^-_m, t \in \mathcal{V})$
- $\boldsymbol{X}_m = (\boldsymbol{x}_m, \boldsymbol{r}^+_m, \boldsymbol{r}^-_m)$.

The basic conditions that have to be fulfilled in each domain $m \in \mathcal{M}$ are flow conservation constraints

$$\sum_{e \in \delta^+(v) \cap \mathcal{E}_m} x_{et} + \sum_{e \in \delta^+(v) \cap \mathcal{E}^+_m} r^+_{et} \tag{1a}$$

$$- \sum_{e \in \delta^-(v) \cap \mathcal{E}_m} x_{et} - \sum_{e \in \delta^-(v) \cap \mathcal{E}^-_m} r^-_{et} \tag{1b}$$

$$= \sum_{d \in \mathcal{D}_{\mathcal{O}}(v,t)} h_d, \quad t \in \mathcal{V}, v \in \mathcal{V}_m \setminus \{t\} \tag{1c}$$

and capacity constraints

$$\sum_{t \in \mathcal{V}} x_{et} \leq c_e, \qquad e \in \mathcal{E}_m \tag{1d}$$

$$\sum_{t \in \mathcal{V}} r^+_{et} \leq c_e, \qquad e \in \mathcal{E}^+_m \tag{1e}$$

$$\sum_{t \in \mathcal{V}} r^-_{et} \leq c_e, \qquad e \in \mathcal{E}^-_m. \tag{1f}$$

Let $\boldsymbol{\mathcal{X}}_m$ ($m \in \mathcal{M}$) denote the set of all vectors $\boldsymbol{X}_m$ satisfying constraints (1) and, possibly, certain extra domain-specific conditions. Such extra constraints can for example be implied by requirements for the weight-based shortest-path intra-domain routing (see Chapter 7 in [5]) or by QoS-type conditions.

The problem optimizing the total network utility can be stated as follows:

$$\max \ F = \sum_{m \in \mathcal{M}} U_m(\boldsymbol{r}^-_m, \boldsymbol{r}^+_m) \tag{2a}$$

$$\text{s.t.} \quad \boldsymbol{X}_m \subset \boldsymbol{\mathcal{X}}_m, \qquad m \in \mathcal{M} \tag{2b}$$

$$r^+_{et} \leq r^-_{et}, \qquad e \in \mathcal{E}_{\mathcal{O}},\ t \in \mathcal{V}. \tag{2c}$$

The network utility is defined as a sum of objective functions of individual domains. If the objective function of a domain is to maximize the sum of volumes of inter-domain demands that are generated by the domain (and that are successfully handled by the network), the common objective becomes:

$$\max\ F = \sum_{m \in \mathcal{M}} \sum_{d \in \mathcal{D}_{\mathcal{O}m}} z_d h_d \tag{3a}$$

where $z_d, d \in \mathcal{D}_{\mathcal{O}}$ denotes a variable specifying the percentage of volume h_d actually handled in the network, i.e., $z_d h_d$ is the carried traffic of demand d (we assumed $z_d \geq 1, d \in \mathcal{D}_{\mathcal{O}}$ which means, that at least volume h_d of each inter-domain demand must be handled). Certainly, objective functions different from (3) can also be considered.

Let $\boldsymbol{\lambda} = (\lambda_{et} : e \in \mathcal{E}_{\mathcal{O}}, t \in \mathcal{V})$ be a vector of (non-negative) multipliers associated with constraints (2c). As shown in [8], the Lagrangean function $L(\boldsymbol{\lambda}; \boldsymbol{X}), \boldsymbol{\lambda} \geq \mathbf{0}, \boldsymbol{X} = (\boldsymbol{X}_m : m \in \mathcal{M}) \in \boldsymbol{\mathcal{X}} = \bigotimes_{m \in \mathcal{M}} \boldsymbol{\mathcal{X}}_m$ associated with problem (2) is of the following decomposed form:

$$L(\boldsymbol{\lambda}; \boldsymbol{X}) = \sum_{m \in \mathcal{M}} L_m(\boldsymbol{\lambda}_m; \boldsymbol{X}_m). \tag{4}$$

In (4), $\boldsymbol{\lambda}_m = (\lambda_{et} : e \in \mathcal{E}_m^- \cup \mathcal{E}_m^+, t \in \mathcal{V})$ is the sub-vector of $\boldsymbol{\lambda}$ composed of the values λ_{et} for all inter-domain links e originating or terminating in domain $m \in \mathcal{M}$, and $L_m(\boldsymbol{\lambda}_m; \boldsymbol{X}_m)$ denotes the partial Lagrangean corresponding to domain $m \in \mathcal{M}$ equal to

$$\sum_{m \in \mathcal{M}} U_m(\boldsymbol{r}_m^-, \boldsymbol{r}_m^+) + \sum_{t \in \mathcal{V}} \Big(\sum_{e \in \mathcal{E}_m^-} \lambda_{et} r_{et}^- - \sum_{e \in \mathcal{E}_m^+} \lambda_{et} r_{et}^+ \Big), \tag{5}$$

where $\boldsymbol{\lambda}_m \geq \mathbf{0}$ and $\boldsymbol{X}_m \in \boldsymbol{\mathcal{X}}_m$.

The problem dual to (2) (see for example [3]) becomes as follows:

$$w^* = \min_{\boldsymbol{\lambda} \geq \mathbf{0}} w(\boldsymbol{\lambda}). \tag{6}$$

The (non-empty) set of optimal solutions of problem ((6)) will be denoted by $\boldsymbol{\Lambda}^*$. The dual function w is defined as $w(\boldsymbol{\lambda}) = \sum_{m \in \mathcal{M}} w_m(\boldsymbol{\lambda}_m)$ and is computed through resolving separate subproblems:

$$w_m(\boldsymbol{\lambda}_m) = \max_{\boldsymbol{X}_m \in \boldsymbol{\mathcal{X}}_m} L_m(\boldsymbol{\lambda}_m; \boldsymbol{X}_m),\ m \in \mathcal{M}. \tag{7}$$

For any $\boldsymbol{\lambda} \geq 0$, $\boldsymbol{X}(\boldsymbol{\lambda}) \in \boldsymbol{\mathcal{X}}$ will denote the so called *maximizer* of the Lagrangean function (4), i.e., any optimal solution of the Lagrange problem:

$$\boldsymbol{X}(\boldsymbol{\lambda}) = \arg\max_{\boldsymbol{X} \in \boldsymbol{\mathcal{X}}} L(\boldsymbol{\lambda}; \boldsymbol{X}). \tag{8}$$

Any maximizer $\boldsymbol{X}(\boldsymbol{\lambda}) = (\boldsymbol{X}_m(\boldsymbol{\lambda}_m) : m \in \mathcal{M})$ is computed through solving independent subproblems (7):

$$\boldsymbol{X}_m(\boldsymbol{\lambda}_m) = \arg\max_{\boldsymbol{X}_m \in \boldsymbol{\mathcal{X}}_m} L_m(\boldsymbol{\lambda}_m; \boldsymbol{X}_m),\ m \in \mathcal{M}. \tag{9}$$

In the sequel the quantity $\nabla w(\boldsymbol{\lambda})$ will denote a subgradient of the dual function w at point $\boldsymbol{\lambda}$. Subgradients are obtained as a by-product of the (distributed) computation of the values of $w(\boldsymbol{\lambda})$: if $\boldsymbol{X}(\boldsymbol{\lambda})$ is a maximizer of the Lagrangean function (4) for a given $\boldsymbol{\lambda}$, then the corresponding subgradient $\nabla w(\boldsymbol{\lambda})$ is as follows ([3]):

$$\nabla w(\boldsymbol{\lambda}) = (x(\boldsymbol{\lambda})_{et}^- - x(\boldsymbol{\lambda})_{et}^+ : e \in \mathcal{E}_{\mathcal{O}}, t \in \mathcal{V}). \tag{10}$$

3. IMPLEMENTATION ISSUES

In our previous papers (cf. [8], [7]) we evaluated performance of two different methods of resolution of the dual problem (6). They show that a professional "Conic Bundle" algorithm implemented in the library [2] significantly outperforms our implementation of the subgradient algorithm (complemented with a mechanism for recovering a primal solution) both in terms of computational efficiency and quality of the produced results. The only advantage of our approach seemed to be its relative simplicity which stimulated hopes to construct a completely distributed (with no coordinating centre at all) mechanism for optimization of inter-domain routing. Still, according to our best knowledge, such a solution is hardly (if not completely impossible) to be achieved. In the first place, the lack of coordinating centre negatively influences on the subgradient optimization process – it forces to use rather trivial methods of selecting sizes for consecutive subgradient algorithm steps (e.g. constant step size), what it turn decreases convergence speed of the algorithm and hampers (or even makes impossible) the effective recovering a feasible primal solution. (It is well known that an optimal solution to a Lagrangean dual does not in general directly determine any optimal (or even feasible) primal solution (see [3], [4]). In particular this is the case for the dual (6) and the primal (2), even if the objective function is strictly concave, as shown in [8], and contrary to what is stated in [6].) In the second place, no coordinating centre means that no one knows a global solution obtained and no one can assure that this solution is feasible what, for operators of cooperating domains, is unlikely to be agreed on.

In this paper we restricted our investigation to the case where the coordinating centre is present. The main responsibility of the centre is to iteratively solve (by running "Conic Bundle" algorithm) the dual problem (6), to disseminate the dual solution into collaborating domains and to collect from these domains solutions of particular lagrangean subproblems (7). The centre is also responsible for evaluation of the feasibility and the quality of a recovered primal solution and for checking termination criteria.

Results from our previous work show, that for networks of realistic size, the subgradient process requires many (thousands, say) iterations to converge. Since in every iteration the task of maximizing the partial Lagrangean function (5) for each collaborating domain has to be performed, the overall computational burden can become excessive even if computation of the Lagrangeans is distributed over the domains. Hence, to keep the decomposition approach effective, the complexity of intra-domain topology as well as the complexity of intra-domain routing have to be significantly reduced in order to make the Lagrangean subproblems as small and simple as possible. The question is whether and what type of topology aggregation can be applied in practice.

An operational practice is that domain operators choose not to mix intra-domain traffic with inter-domain traffic through providing (for example by means of MPLS tunnels) virtual transit links between border routers of a domain. To engineer these virtual links it is basically important for an operator to know the requirements imposed on the domain by the inter-domain traffic, i.e., the amount of traffic directed to a particular destination (network prefix) for each inter-domain link outgoing from or incoming to the domain, and the amount of inter-domain traffic (per each QoS class) that is to be transited over the domain between any pair of the domain border routers. For determining such link traffic estimates within the proposed routing optimization process it is sufficient to use certain aggregated (simplified) domain topology representation as discussed below.

Three models for the domain topology aggregation are considered. The first topology aggregation model (we will refer to it as Domain Connectivity model (*DC*)) reduces a domain to just one node. In effect, the resulting graph of the whole inter-domain network represents only the domains and their connectivity, but neglects the intra-domain structure. Hence, a node of this graph represents a single network domain while a link represents all the original links connecting border routers of a particular domain pair. Applying our routing optimization process to such a simplified topology results in calculation, for each link of the reduced network graph (i.e, for each pair of collaborating domains), a vector of reservation levels for traffic directed to every node of the graph (every domain of the original network).

Certainly, the *DC* aggregation drastically reduces the complexity of the domain representation and therefore considerably shortens the computation time required for the Lagrangean sub-problems resolution. Still, from the operator's viewpoint, *DC* has at least two fundamental drawbacks. First, loads of individual inter-domain links cannot be specified because these links have no individual representation in the aggregated graph. Hence the bandwidth reservation levels applied by the two domains at the end nodes of each particular inter-domain link may (and in practice, will) be inconsistent. Second, there is no information specifying through which border router the inter-domain traffic enters (leaves) the domain. This is because in each domain more than one border router can have an inter-domain link to a border-router of a particular neighbor domain. Thus, these two features make exact dimensioning of virtual transit links difficult.

The two major drawbacks of the *DC* aggregation are eliminated in the second and in the third models (referred to as Full Matrix model (*FM*) and Star model (*Star*) respectively) in which each border router is explicitly represented in the aggregated graph, and so is each original inter-domain link. This results in precise representation of the inter-domain network topology. The intra-domain topology is represented in a simplified way. In the case of *FM* aggregation nodes representing border routers of a domain are fully connected within a complete sub-graph (observe that for representation of traffic sources and destinations inside a domain, introduction of additional nodes and links may be necessary). In the case of *Star* aggregation, border routers are connected through spoke connections and the artificial hub node. Each link of such a domain sub-graph represents a virtual (or part of such) transit link through the domain corresponding, in a particular implementation scenario, to an MPLS LSP through the domain. In a slightly more complex model a set of parallel links can be introduced between each pair of the border routers, e.g., one link per each QoS class. Despite the simplicity of the models, carefully chosen capacities of those virtual links and additional constraints imposed on the reservation levels on those links can reflect operator's inter-domain traffic handling policies. It should be noted, that despite some advantages of *FM* aggregation over *Star* aggregation (each intra-domain transit link has an explicit representation in *FM* model) the potential drawback of the *FM* approach is the limited reduction of the domain topology (from the partial Lagrangeans computation viewpoint).

It should be noted that the considered aggregations models are only the exemplary ones – provided consensus exists on use of explicit representation of inter-domain links and of explicit representation of domain border nodes – all the operators of collaborating domains are free to adopt any model of aggregation of intra-domain topology (their selection has only local meaning).

4. NUMERICAL EXPERIMENTS

Our numerical experiments investigate how the model of aggregation of intra-domain topology (cf. Section 3) influences on the convergence time of the inter-domain routing optimization process (described in Section 2) and how quality of the produced results depends on time of the computations. The reported computations were performed on an Intel class computer with a 2.0 GHz Core 2 Duo processor and 1 GB of RAM, using CPLEX 10.1. The subgradient optimization was solved using a professional "Conic Bundle" algorithm implemented in the library [2].

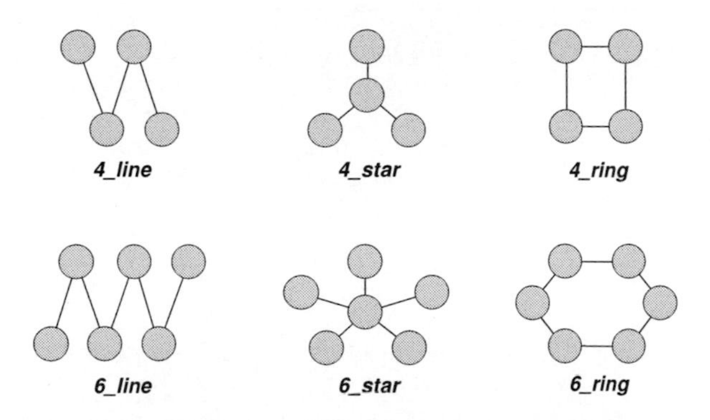

Figure 1: Domain connectivity graphs of considered networks

	complexity parameters		computational cost					
	#inter-domain links	#domain border nodes	*DC*		*FM*		*Star*	
			I	$T[s]$	I	$T[s]$	I	$T[s]$
4_line	12	2/4/2.75	18	0.4	172	156	102	11
4_ring	16	3/4/3.5	18	0.5	198	106	98	20
4_star	12	2/4/2.5	33	0.5	133	98	80	7.8
6_line	20	2/3/2.33	17	0.5	62	161	60	20
6_ring	24	3/3/3	14	0.2	167	622	141	70
6_star	20	2/3/2.16	21	0.3	86	138	149	42
4_test	20	2/4/3	17	0.4	363	488	284	61
7_test	44	3/6/4.42	76	3.4	-	-	361	1140

Table 1: Complexity of considered networks and related computation costs

We analyzed six simple networks whose domain connectivity graphs are presented in Figure 1 together with two more complex networks *4_test* and *7_test* with four and seven domains respectively. General parameters characterizing complexity of the considered networks are presented on the left side of Table 1. The first column of the table denotes the total number of (unidirectional) inter-domain links and the second column (using the notation *A/B/C*) shows the minimal (*A*), the maximal (*B*) and the average (*C*) number of border nodes in every domain.

The right part of Table 1 shows number of iterations (column *I*) and the computations time (column *T*) the subgradient algorithm required to achieve assumed accuracy ($\beta = 1.05$) in dependency on the applied model of intra-domain topology aggregation.

One can observe (cf. Table 1) that for all the considered networks the optimization process takes reasonable time to converge (with exception for *FM* aggregation of *7_test* network for which the algorithm did not achieve assumed accuracy within limit of thirty minutes of computation time). Application of *DC* aggregation (which drastically reduces complexity of the network – cf. Section 3)

	DC			*FM*			*Star*		
	P	N	V	P	N	V	P	N	V
4_line	.99	0	0	.97	0	0	.97	0	0
4_ring	.99	0	0	.99	0	0	.99	0	0
4_star	.96	0	0	.98	0	0	.98	0	0
6_line	.97	0.06	0.03	.97	0.12	0.02	.97	0.07	0.02
6_ring	.95	0.15	0.02	.95	0	0	.95	0	0
6_star	.97	0	0	.97	0	0	.97	0	0
4_test	.91	0	0	.88	0	0	.88	0	0
7_test	.98	0.04	0.02	-	-	-	.84	0	0

Table 2: Quality of the recovered primal solutions for particular aggregations

results in the shortest (of order of seconds) convergence times, *Star* aggregation is somewhere in the middle, and *FM* aggregation requires the longest time to converge (two up to ten times longer than that is required in the case of *Star* aggregation.) The simple observation is that increasing complexity of intra-domain topologies results in prolonging of the required computation time. It is worth to note that despite the big difference in computation times required for *Star* and *FM* aggregations, associated numbers of iterations for both these cases are similar.

Table 2 shows selected quality parameters of the recovered primal solutions obtained for particular aggregation models. The accuracy of the final solution is denoted by $\beta = \frac{\tilde{w}}{F^*}$, where F^* denotes the (known, computed in a centralized way) optimal primal objective, and $\tilde{w}$ denotes the value of the final dual solution (which is always an upper bound for F^*). For all the computations reported in Table 2 value $\beta = 1.05$ was assumed. Let $\mathcal{D}_\mathcal{O}$ be the set of all inter-domain demands (i.e., $d \in \mathcal{D}_\mathcal{O}$ if, and only, if $s(d)$ and $t(d)$ belong to different domains), and let $\overline{h} = \frac{\sum_{d \in \mathcal{D}_\mathcal{O}} h_d}{|\mathcal{D}_\mathcal{O}|}$ be the average volume of an inter-domain demand. Further, let $\mathcal{Q} = \{(e,t) : e \in \mathcal{E}_\mathcal{O}, t \in \mathcal{V}, r^+_{et} > 0\}$ be the set of all pairs (e,t) with non zero inter-domain reservation level r^+_{et}, and let $\mathcal{R} = \{(e,t) \in \mathcal{Q} : r^+_{et} > r^-_{et} + \varepsilon\overline{h}\}$ be the set of all pairs (e,t) with primal infeasible reservation levels r^+_{et} and r^-_{et}. The quantity ε is a small positive constant used to take care of numerical inaccuracies (and, potentially, for intrinsic uncertainty of demand volumes measurements). In our experiments, we assumed $\varepsilon = 0.01$. The quality of a solution is described by parameters P, N and V. P is equal to the objective value of the recovered primal solution divided by the true optimal primal objective value F^*, $N = \frac{|\mathcal{R}|}{|\mathcal{Q}|}$ denotes the fraction of pairs (e,t) that violate constraint ((2)c), and $V = \frac{\sum_{(e,t)\in\mathcal{R}} (r^+_{et} - r^-_{et})}{\overline{h}|\mathcal{R}|}$ is the relative mean value of the violation. The quality of the recovered primal solutions (cf. Table 2) is very good – value of the objective function for the recovered primal solution is always close to optimal and there are practically no violated inter-domain reservation levels (in few cases where they are violated, the scale of this violation is negligible).

A careful reader can observe, that in our algorithm we took advantage of known (pre-computed) optimal value of the global primal problem (2). That knowledge allowed us to evaluate the quality of recovered primal solution in hand and to set a criterion for stopping the optimization process. In a real distributed environment that value is hardly to be known and another termination criteria must be found. Let F^* denote the optimal solution of global problem (2), w^i denote the value of objective of the dual problem (6) in iteration i and F^i_r denote value of the objective function for a recovered primal solution in iteration i. Figure 2 presents how relative value of a solution of the dual problem

(6) ($\beta^i = \frac{w^i}{F^*}$) and relative value of the recovered primal solution ($P^i = \frac{F_r^i}{F^*}$) performs as the the execution of the algorithm progresses (an application of *Star* aggregation to *7_test* network has been assumed). One can observe that β^i and P^i are approaching the value 1 from the opposite sides and that ratio $\frac{P^i}{\beta^i} = \frac{F_r^i}{w^i}$ could be used to set the required stopping criterion.

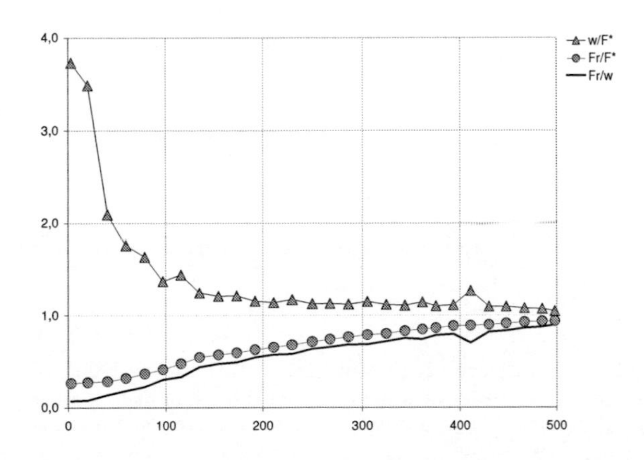

Figure 2: Relative dual and relative recovered primal values during the optimization process

The questions arise how to determine such value of ratio $\frac{F_r^i}{w^i}$ that guarantees satisfactory good results and if value of that ratio could be used as a sole criterion for termination of the optimization process. Let N^i and V^i denote temporary (after iteration i) values of quality parameters N and V (these parameters are defined earlier at the begining of the present section). Graphs presented at Figure 3 shows how values of these quality parameters (N^i,V^i) together with the value of ratio $\frac{F_r^i}{w^i}$ performs during the optimization process – these graphs present results for application of *Star* aggregation to *7_test* network (on the left side) and to *6_star* network (on the right side). One could observe that they differs – the left graph shows the optimization process which starts with a relatively week recovered primal solution which the solution is being gradually improved during consecutive iterations. The initial recovered primal solution is strongly infeasible (a big number of inter-domain reservation level consistency constraints (1f) are violated and the scale of this violation is significant) but the scale of the infeasibility falls rapidly as the optimization progresses. One can conclude, that for values of ratio $\frac{F_r^i}{w^i} \geq 0.5$ the scale of infeasibility becomes negligible and further iterations (if performed) result only in a better (closer to optimum) primal solution. At the right graph already the initial recovered primal solution is almost-optimal and remains at that level during the whole optimization process. The scale of infeasibility decreases slower than in the previous case. The important observation is that value of ratio $\frac{F_r^i}{w^i}$ does not increase smoothly (please observe the peek about 20th iteration) what the phenomenon may misled the optimization process causing its immature termination.

The general conclusion is that value of ratio $\frac{F_r^i}{w^i}$ is a meaningful optimality parameter still, to reliably decide on termination, also the scale of the infeasibility (as indicated by parameters N^i and V^i) must be simultaneously controlled.

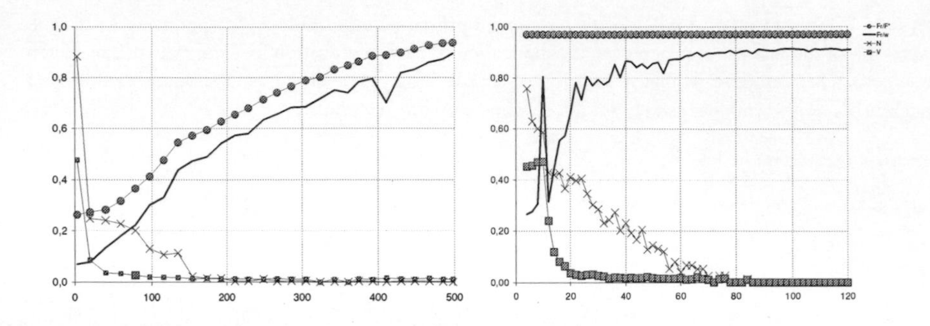

Figure 3: Quality parameters of recovered primal solution during the optimization process

5. CONCLUSIONS

Because of a highly distributed nature of the Internet, and because the operators do not disclose sensitive information concerning their domains, any process of inter-domain traffic routing optimization must be distributed across the collaborating domains. Despite some interesting initial trials (see [6], [10] for the two-domain case; [8] for a general approach), an effective, distributed optimization processes for inter-domain routing optimization is yet to be found.

In this paper we have presented a modified version of introduced in [8] and [7] global multi-domain routing design problem (2) consisting in optimization of bandwidth reservation levels on inter-domain links for traffic flows identified by traffic classes and traffic destinations and we have shown how to decompose the problem with respect to individual domains using Lagrangean relaxation.

However, for networks of realistic size, the subgradient optimization process, which if the method of choice for resolution of the dual problem, requires many iterations to converge. Hence, to keep the proposed decomposition approach effective, the complexity of intra-domain topology as well as the complexity of intra-domain routing have to be significantly reduced in order to make the Lagrangean subproblems as small and simple as possible. In this paper we proposed and evaluated three types of topology aggregation models. We examined how the adopted model of intra-domain topology aggregation influences the convergence time of the considered inter-domain routing optimization process, and for each network aggregation model how the convergence time depends on the total number of domains.

In this paper we presented that the ratio $\frac{F_r^i}{w^i}$ is a meaningful parameter describing level of optimality of a current recovered primal solution. Still we also pointed out the case, where value of that ratio does not increase monotonically as the optimization process progresses and concluded, that to reliably decide on termination of the optimization process the continuous monitoring of scale of infeasibility of current recovered primal solution is also necessary.

Acknowledgements
The research presented in this paper has been funded by "Optimization Models for NGI Core Network" (Polish Ministry of Science and Higher Education, grant N N517 397334) and EURO-NF (FP7 Network of Excellence). M. Pióro has been also supported by "Modeling and Design of Core Internet Networks" (Swedish Research Council, grant 621-2006-5509).

REFERENCES

[1] FEAMSTER N., BORKENHAGEN J., REXFORD J. *Guidlines for interdomain traffic engineering*. ACM SIGCOM Computer Communications Review, vol. 33, no. 5, pp. 19–30, 2003.

[2] HELMBERG C. *Conic bundle 0.1*. Tech. rep., 2005. Fakultät für Mathematik, Technische Universität Chemnitz.

[3] LASDON L. *Optimization Theory for Large Systems*. MacMillan, 1970.

[4] MINOUX M. *Mathematical Programming: Theory and Algorithms*. John Wiley & Sons, 1986.

[5] PIÓRO M., MEDHI D. *Routing, Flow, and Capacity Design in Communication and Computer Networks*. Morgan Kaufman, 2004.

[6] SHRIMALI G., AKELLA A., MUTAPCIC A. *Cooperative inter-domain traffic engineering using Nash bargaining and decomposition. Proceedings of the IEEE Conference on Computer Communication INFOCOM'2007*. 2007.

[7] TOMASZEWSKI A., PIÓRO M., DZIDA M., MYCEK M., ZAGOŻDŻON M. *A subgradient optimization approach to inter-domain routing in IP/MPLS networks. Proceedings of the IFIP Networking'2007, Atlanta*. 2007.

[8] TOMASZEWSKI A., PIÓRO M., DZIDA M., MYCEK M., ZAGOŻDŻON M. *Towards distributed inter-domain routing optimization for IP/MPLS networks*. Tech. rep., 2007. Institute of Telecommunications, Warsaw University of Technology, Poland.

[9] TOMASZEWSKI A., PIÓRO M., MYCEK M. *A distributed scheme for inter-domain routing optimization. Proceedings of the 6th International Workshop on Design and Reliable Communication Networks DRCN'2007, La Rochelle*. 2007.

[10] WINNICK J., JAMIN S., REXFORD J. *Trafffic engineering between neighboring domains*. Tech. rep., 2002. (http://www.cs.princeton.edu/~jrex/publications.html).

Keywords – Network design, network resilience, computational complexity, multi-commodity flows

Mateusz ŻOTKIEWICZ[1,2]
Michał PIÓRO[1,3]
Artur TOMASZEWSKI[1]

COMPLEXITY OF RESILIENT NETWORK OPTIMIZATION

Path restoration is one of the basic mechanisms for securing telecommunication networks against failures. In the paper we discuss the complexity of certain variants of a multi-commodity flow network optimization problem related to state-independent path restoration mechanisms with no stub-release. We demonstrate that most variants of the considered problem are $\mathcal{NP}$-complete. Depending on the variant, we show how either the partition problem or the problem of finding an arc-disjoint pair of paths that connect two specific pairs of nodes can be reduced to the considered problem. We also demonstrate that at the same time the considered problem is difficult to approximate. The complexity results of the paper are important as they can help to devise proper algorithms for resilient network design tools. All the derived results are novel.

1. INTRODUCTION

Path restoration (PR) is one of the basic mechanisms for securing telecommunication networks against failures. Although PR can be complex in terms of implementation, it is among the most efficient mechanisms as far as the extra link capacity required for flow protection is concerned. The idea of the mechanism is as follows. The traffic demand between a pair of nodes is realized using several different network paths to route the connections (i.e., path flows, either non-bifurcated or bifurcated) in the normal (failure-free) state. When a failure occurs, the affected primary connections are restored using backup connections (again, by means of either non-bifurcated or bifurcated path flows) along the surviving routes. In any failure state the total capacity of the temporarily established backup flows and the primary flows that survive is equal to the requested volume of the traffic demand.

In the case considered in this paper, the backup flows are established using the protection (backup) pool of links' capacity (i.e., protection capacity). The protection link capacity is separated from the working link capacity used by primary flows that traverse a given link. In effect, the working and protection link capacity form two separate pools of resources, so the backup flows do not use the working capacity released on links as a by-product of a failure (it is said that backup flows do not exploit the phenomenon of *stub-release*). It is important that backup capacity *is shared* between

[1]Institute of Telecommunications, Warsaw University of Technology, Nowowiejska 15/19 00-665 Warszawa, Poland, e:mail: {mzotkiew,mpp,artur}@tele.pw.edu.pl
[2]TELECOM & Management SudParis, France
[3]Department of Electrical and Information Technology, Lund University, Sweden

restoration flows established in different failure states, so the considered case *is not* the so called hot-standby. Finally, we assume that primary flows have to be restored in exactly the same way each time they fail—such a strategy is usually referred to as state-independent restoration.

Another possibility, not considered in this paper, is when backup flows are allowed to use also the working capacity that is released by the failed primary connections on surviving links, so that the backup flows can exploit the phenomenon of stub-release. This in fact leads to the usage of links' capacity as a common pool of resources.

In this paper, we discuss the complexity of various variants of the multi-commodity flow network optimization problem related to the above described state-independent PR mechanism with no stub-release. In particular, we prove that in most cases the problem is $\mathcal{NP}$ complete. Depending on the variant, we show how either the partition problem or the problem of finding an arc-disjoint pair of paths that connect two specific pairs of nodes can be reduced to the considered problem. Besides, we discuss the complexity of approximation schemes that solve the various variants of the problem and show that approximation is also difficult. All the results derived in this paper are original. It is worth noticing that the complexity results for multi-commodity flow optimization problems related to design of resilient networks are of great theoretical interest as they help to develop proper algorithms used in network design tools. Such complexity issues have been considered by several authors, see [2, 5, 8–11, 13, 14].

Network optimization problems considered in this paper stem from the so called single backup restoration design problem in which each traffic demand is routed on exactly one primary path, and when this path fails, it is restored on one fixed backup path that is supposed to protect its primary counterpart in all failure situations that affect the primary path. The objective is to minimize the total cost induced by the (working) link capacity used by all primary flows and the (protection) capacity used by the backup paths. Note that the backup capacity can be shared. In effect, we have to find one primary/backup path-pair for each demand so that the resulting link capacities are optimal. The obtained optimization problem turns out to be in general $\mathcal{NP}$-complete (as shown in Section 4), and can be modeled as a mixed-integer programming (MIP) (see [10, Chapter 9]).

The problem setting characterized above can be modified in several valid ways, first of all, by admitting bifurcation of both primary and backup flows. This leads to a linear relaxation of the MIP formulation of the single backup restoration design problem. Like other similar bifurcated resilient design problems (see [9] and [10, Chapter 9]), this linear relaxation can be easily formulated as a linear program (LP) using the link-path formulation (see [1, 10]). However, such link-path formulations require predefined lists of candidate paths. Therefore, the resulting LP formulations are not compact, as they need a flow variable associated with each candidate path. This leads to an exponential number of variables, if all (elementary) paths of the network graph are to be taken into account in the problem. To overcome the non-compactness, column (path) generation can be used (see [1, 10]). However, it turns out that the related path generation (pricing) problems are, as observed for example in [10, 14], difficult for several cases of the path generation problem, and have recently been proved to be $\mathcal{NP}$-complete, see [2, 5, 8, 11] (a survey of path generation issues in resilient network design can be found in a recent survey [9]). $\mathcal{NP}$-completeness of path generation for the LP version of our basic problem is demonstrated in [11]. This, however, is not a surprise since, as we show in Section 5, the problem itself is $\mathcal{NP}$-complete.

The paper is organized as follows. In Section 2 we present a formal model of the considered problem. It is followed by Section 3 that acts as an overview of the complexity of different variants of the problem. In Sections 4 and 5 we present two proofs of the $\mathcal{NP}$-completeness of specific variants of the problem. Then, in Section 6 we discuss the complexity of approximation schemes. The paper ends with a summary in Section 7.

2. PROBLEM FORMULATION

The considered network is modeled by a directed graph $\mathcal{G}(\mathcal{V},\mathcal{E})$ composed of set $\mathcal{V}$ of nodes and set $\mathcal{E}$ of links. Set $\mathcal{D}$ represents directed end-to-end demands. The number of demands, $|\mathcal{D}|$, will be denoted by D. The source and target of demand $d \in \mathcal{D}$ are denoted by s_d and t_d, respectively, and assumed to be different from each other. The volume of demand $d \in \mathcal{D}$ is given by h_d ($h_d > 0$). The demand volumes are realized by means of path flows assigned to (directed) paths form s_d to t_d. The cost of realizing one unit of demand on link $e \in \mathcal{E}$ is denoted by ξ_e. Each link $e \in \mathcal{E}$ has its *primary* (basic) capacity, denoted by y'_e, used for realizing flows in the normal operating state (i.e., *primary flows*), and the protection capacity y''_e used for restoration of failed primary flows by means of *backup flows*. Both y'_e and y''_e ($e \in \mathcal{E}$) will be variables subject to optimization.

The family of all failure states (called a *failure scenario*) is denoted by $\mathcal{S}$ where each *failure state* (also called a failure situation) $s \in \mathcal{S}$ is identified by the set of failing links, so $s \subset \mathcal{E}$. Family $\mathcal{S}$ is assumed to include the failure-less state $\mathcal{O}$ (formally equal to the empty subset of the set $\mathcal{E}$) in which all links are operational (state $\mathcal{O}$ is sometimes called the *normal state*). It is assumed that links fail totally. Set $\mathcal{S}_e = \{s \in \mathcal{S} \ : \ e \notin s\}$ will denote the set of all states $s \in \mathcal{S}$ in which link $e \in \mathcal{E}$ is available.

The set of all candidate paths that can be used for carrying flows is denoted by $\mathcal{P} = \bigcup_{d\in\mathcal{D}} \mathcal{P}_d$, where $\mathcal{P}_d$ is the set of candidate paths for demand $d \in \mathcal{D}$. We assume that the paths are elementary (do not contain loops) so, for each demand $d \in \mathcal{D}$, set $\mathcal{P}_d$ is a subset of the set of all elementary paths from s_d to t_d, and each path $p \in \mathcal{P}$ can be identified with the set of the links it traverses, so that $p \subseteq \mathcal{E}$. Further, $\mathcal{S}_p = \{s \in \mathcal{S} \ : \ p \cap s = \emptyset\}$, ($\mathcal{S}_p \subseteq \mathcal{S}$) denotes the set of all states $s \in \mathcal{S}$ in which path $p \in \mathcal{P}$ is available, and $\bar{\mathcal{S}}_p = \mathcal{S} \setminus \mathcal{S}_p$ is the set of all states $s \in \mathcal{S}$ where path $p \in \mathcal{P}$ fails, i.e., $\bar{\mathcal{S}}_p = \{s \in \mathcal{S} \ : \ p \cap s \neq \emptyset\}$.

For the considered path restoration mechanism we will use the following notation. For a given candidate path $p \in \mathcal{P}_d$ assigned to demand $d \in \mathcal{D}$, the set of all candidate backup paths that can be used for protecting path p is denoted by $\mathcal{Q}_p$. Certainly, $\mathcal{Q}_p \subseteq \mathcal{P}_d$ and for all $s \in \mathcal{S}$, $(p \cap s \neq \emptyset) \Rightarrow (q \cap s = \emptyset)$, i.e., paths p and q never fail simultaneously (and therefore are called *failure-disjoint*). In this context, a path $p \in \mathcal{P}$ is called the *primary path* and all paths from $\mathcal{Q}_p$ are called its *backup paths*. The set of all candidate failure-disjoint primary/backup path-pairs $r = (p,q)$ for demand $d \in \mathcal{D}$ will be denoted by $\mathcal{T}_d$, i.e., $\mathcal{T}_d = \{r = (p,q) \ : \ p \in \mathcal{P}_d, q \in \mathcal{Q}_p\}$. For each link $e \in \mathcal{E}$, and demand $d \in \mathcal{D}$, the set of all pairs $r = (p,q) \in \mathcal{T}_d$ such that $e \in p$ will be denoted by $\mathcal{R}'_{ed}$. Also, for each link $e \in \mathcal{E}$, demand $d \in \mathcal{D}$, and in each failure state $s \in \mathcal{S}_e$, the set of all pairs $r = (p,q) \in \mathcal{T}_d$ such that $e \in q$ and $s \in \bar{\mathcal{S}}_p$ will be denoted by $\mathcal{R}''_{eds}$. Finally, the flow of demand $d \in \mathcal{D}$ allocated to pair $r = (p,q) \in \mathcal{T}_d$ will be denoted by variable x_{dr}.

The generic form of the basic optimization problem considered in this paper is denoted by $\mathbb{P}$ and is as follows.

$$\text{minimize } F(y) = \sum_{e \in \mathcal{E}} \xi_e (y'_e + y''_e) \tag{1a}$$

$$\sum_{r \in \mathcal{T}_d} x_{dr} = 1 \qquad d \in \mathcal{D} \tag{1b}$$

$$\sum_{d \in \mathcal{D}} \sum_{r \in \mathcal{R}'_{ed}} h_d x_{dr} \leq y'_e \qquad e \in \mathcal{E} \tag{1c}$$

$$\sum_{d \in \mathcal{D}} \sum_{r \in \mathcal{R}''_{eds}} h_d x_{dr} \leq y''_e \qquad e \in \mathcal{E},\ s \in \mathcal{S}_e \setminus \{\mathcal{O}\} \tag{1d}$$

$$y'_e, y''_e \geq 0 \qquad e \in \mathcal{E} \tag{1e}$$

$$x_{dr} \geq 0 \qquad d \in \mathcal{D}, r \in \mathcal{T}_d \tag{1f}$$

Problem $\mathbb{P}$ consists in minimizing the cost of primary and protection capacity installed on the links subject to a number of constraints. Constraint (1b) assures that all demand volumes are realized (x_{dr} is a fraction of h_d allocated to pair $r = (p, q) \in \mathcal{T}_d$) while constraint (1c) states that in the normal state the load of link $e \in \mathcal{E}$ cannot exceed its primary capacity. The flow summation on the left-hand side of constraint (1d) is taken over all path-pairs whose backup path contains the considered link e and whose primary path fails in the considered failure state s. Clearly, problem (1) assumes that the pool of protection capacity $y''_e, e \in \mathcal{E}$ is shared by the demands in different situations.

In the balance of this paper we will examine $\mathcal{NP}$-completeness of different variants of problem (1). The variants are distinguished according to the following four criteria:

- **Criterion 1.** (a) Non-bifurcated flow (IP): $x_{dr} \in \{0, 1\}, d \in \mathcal{D}, r \in \mathcal{T}_d$. (b) Bifurcated flow (LP): $x_{dr} \in \mathbb{R}^+, d \in \mathcal{D}, r \in \mathcal{T}_d$.

- **Criterion 2.** (a) Predefined lists of path-pairs: $\mathcal{T}_d$ are given in advance; the lists are limited in length polynomially with the number of nodes. (b) All possible elementary path-pairs in $\mathcal{T}_d$.

- **Criterion 3.** (a) Single link failure scenario: $|s| = 1, s \in \mathcal{S} \setminus \mathcal{O}$. (b) Arbitrary link failure scenario (failures of more than one link at a time are admitted): $|s| \geq 1, s \in \mathcal{S} \setminus \mathcal{O}$. We assume that the number of failures is polynomially bounded by the number of network nodes.

- **Criterion 4.** (a) Single demand: $D = 1$. (b) Multiple demands: $D \geq 1$.

3. OVERVIEW OF THE COMPLEXITY RESULTS

As indicated in criterion 1, by IP we denote the version of problem $\mathbb{P}$ that is characterized by binary flow variables (non-bifurcated flows), and by LP—its bifurcated counterpart (linear relaxation of IP) with continuous flows.

We first observe that the variant specified by criterion 1b (LP) and criterion 2a (predefined path-pairs lists polynomially bounded in length) is always polynomial, no matter what criteria 3 and 4 are. This is because in this case formulation (1) becomes a compact linear programming problem, and as such can be solved in time which is polynomial with the size of the problem (see [7]). On

Table 1: Complexity overview ($\mathcal{NPC}$: $\mathcal{NP}$-complete, $\mathcal{P}$: polynomial)

		single failures	multiple failures
$D = 1$	LP	$\mathcal{NPC}$ [Section 5]	$\mathcal{NPC}$ [5]
	IP	$\mathcal{P}$ [12]	$\mathcal{NPC}$ [5]
$D > 1$	LP	$\mathcal{NPC}$ [because $D = 1$ is $\mathcal{NPC}$]	$\mathcal{NPC}$ [because $D = 1$ is $\mathcal{NPC}$]
	IP	$\mathcal{NPC}$ [Section 4, see also Section 6]	$\mathcal{NPC}$ [because $D = 1$ is $\mathcal{NPC}$]

the other hand, when LP with all possible (elementary) path-pairs is considered (variant 1b, 2b) then formulation (1) becomes non-compact as the number of path-pairs is exponential with the size of the graph. As we will demonstrate in Section 5, this variant is $\mathcal{NP}$-complete, no matter what criteria 3 and 4 are.

The second observation concerns variants assuming criterion 1a (IP). For the IP case the complexity is already determined by 2a, no matter what variants of criterions 3 and 4 are considered. When variant 2a is $\mathcal{NP}$-complete, variant 2b is also (automatically) $\mathcal{NP}$-complete. The case when variant 1a can be solved in polynomial time is a bit more tricky. This is in fact variant 1a, 3a, 4a, namely: $|s| = 1$ for all $s \in \mathcal{S} \setminus \mathcal{O}$, $\mathcal{D} = \{d\}$ and $x_{dr} \in \{0, 1\}, r \in \mathcal{T}_d$. Consider criterion 2a, i.e., that $\mathcal{T}_d$ are given in advance. Then the optimal solution can be found in polynomial time by examining all pairs in $\mathcal{T}_d$, and selecting the shortest one. Now let us consider the second variant (2b). In this case the problem is just to find a shortest pair of disjoint paths in the network graph, and allocate the whole demand volume h_d to this pair. This can be efficiently solved using Suurballe's algorithm [12]. Hence, both variants 2a and 2b can be solved in polynomial time.

Thus, although according to criteria 1-4 there are 16 variants of problem (1), because of the above remarks, this number can be reduced to 8. These 8 variants are summarized in Table 1, where criterion 2b is assumed for LP, and criterion 2a is assumed for IP.

When criterion 1b is assumed instead of 1a so the flow bifurcation is admitted (vector $x = (x_{dr} : d \in \mathcal{D}, r = (p, q) \in \mathcal{T}_d)$ is continuous), the demand can be routed on many different primary/backup path-pairs simultaneously. How different the IP and the LP versions of the problem are can be seen in Figure 1 depicting the network consisting of two nodes s and t, $N + 1$ links $1, 2, \ldots, N + 1$, and one demand d between s and t with $h_d = 1$. Suppose that $\xi_n = 1, n = 1, 2, \ldots, N + 1$. Clearly, any optimal solution of the IP version consists (by definition) of just one primary/backup path-pair and must cost 2 units. On the contrary, an optimal solution of the LP version uses N different primary paths protected by one common backup path. Such a solution is for example obtained by assigning flow $x_{dr} = \frac{1}{N}$ to each of of the N path-pairs $(p_n, q), n = 1, 2, \ldots, N$ where $p_n = \{n\}$ and $q = \{N + 1\}$. Hence, $y'_n = \frac{1}{N}, y''_n = 0, n = 1, 2, \ldots, N$ and $y''_{N+1} = \frac{1}{N}, y'_{N+1} = 0$, and the backup capacity of link $N + 1$ is shared among the different backup paths in different failure states (each of these failure states affects exactly one link). The cost of this solution is $1 + \frac{1}{N}$, i.e., almost 50% cheaper than the non-bifurcated solution.

In Section 5 we prove that the 1b, 3a (IP, single failures) variant of problem (1) is $\mathcal{NP}$-complete. Moreover, in Section 6 we prove that any approximation of the problem better than a $\frac{4}{3}$-approximation is also $\mathcal{NP}$-complete.

Let us now proceed to the case with multiple demands and with the single failure scenario (3a, 4b). The $\mathcal{NP}$-completeness of the LP case (variant 1b, 3a, 4b) is automatically implied by the $\mathcal{NP}$-completeness of the case with only one demand (criteria 1b, 3a, 4a), since when something is difficult for the single demand case it has to be difficult also for multiple demands. As far as the IP version

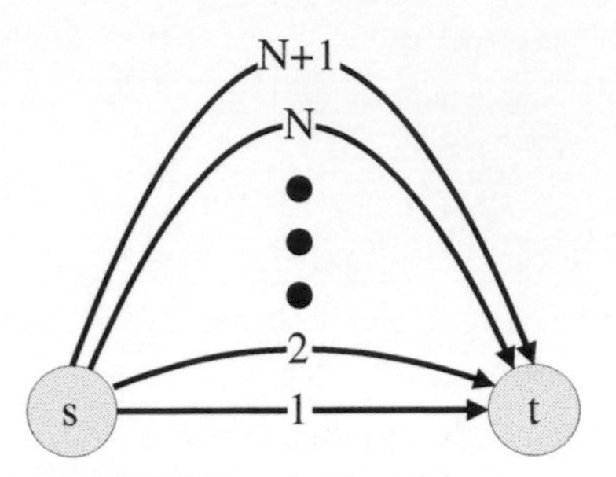

Figure 1: Network of two nodes and N+1 links

is concerned (1a, 3a, 4b), we cannot follow the same way of reasoning, because the single demand case can be solved in polynomial time. In fact, the version admitting multiple demands is $\mathcal{NP}$-complete. In order to prove this we can use a simple construction presented in Section 4 and reduce the PARTITION problem to the considered variant of problem $\mathbb{P}$. However, since this construction gives weak results concerning the complexity of the problem's approximations, in Section 6 we extend the proof from Section 5 to cover the IP variant of the problem admitting multiple demands (variant 1b, 3a, 4a).

As far as the multiple failure variant 1b, 3b, 4b is concerned, its complexity can be deduced from [5] where it is proved that finding a pair of failure-disjoint paths when multiple failures are admitted is $\mathcal{NP}$-complete. As each feasible solution to the problem consist of assigning the demand volume to one or more of such pairs, it is clear that finding any of them cannot be completed in polynomial time.

4. $\mathcal{NP}$-COMPLETENESS PROOF(MANY DEMANDS, IP)

Let us consider the variant of problem $\mathbb{P}$ assuming criteria 1a, 2a, 3a, and 4b, i.e., IP, limited lists of path-pairs, single failures, and many demands. In order to prove the $\mathcal{NP}$-completeness of this case, we will demonstrate a reduction of the PARTITION problem to the considered problem. PARTITION was proved to be $\mathcal{NP}$-complete in [4].

Consider a given sequence $\mathcal{H} = (h_1, h_2, \ldots, h_D)$ of positive integer numbers. Denote by $\sum \hat{\mathcal{D}}$ the sum $\sum_{d \in \hat{\mathcal{D}}} h_d$ defined for any subset $\hat{\mathcal{D}}$ of the set of indices $\mathcal{D} = \{1, 2, \ldots, D\}$. Problem PARTITION consists in answering the question whether there is a partition (split) of the set $\mathcal{D}$ into two subsets $\mathcal{D}'$ and $\mathcal{D}''$ ($\mathcal{D}' \cup \mathcal{D}'' = \mathcal{D}, \mathcal{D}' \cap \mathcal{D}'' = \emptyset$), such that $\sum \mathcal{D}'$ is equal to $\sum \mathcal{D}''$. Partition $(\mathcal{D}', \mathcal{D}'')$ will be called a *valid partition.*

Consider an instance $F_{\mathcal{H}}$ of PARTITION for a given sequence $\mathcal{H}$. The corresponding instance $P_{\mathcal{H}}$ of problem (1) is specified by means of a network that consists of only two nodes, s and t, connected by three parallel links $e = 1, 2, 3$ (see Figure 2). The unit capacity costs are all equal to 1, $\xi_e = 1$ for $e = 1, 2, 3$. The set of demands is defined as $\mathcal{D} = \{1, 2, \ldots, D\}$ and each demand $d \in \mathcal{D}$ has its source in node s and its sink in node t, its volume is given by h_d. Also, the (single) failure of each of the three links is included into the considered failure scenario, so the scenario contains the normal state $\mathcal{O}$ and the three failure states corresponding to the three links. Let $H = \sum \mathcal{D}$.

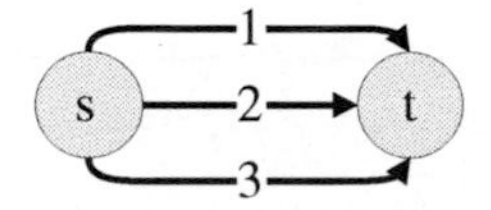

Figure 2: Simple network

Proposition 4.1 *The optimal objective of $P_{\mathcal{H}}$ is equal to $\frac{3}{2}H$, if and only if, $F_{\mathcal{H}}$ forms a valid partition.*

Proof. We first prove that the optimal objective F^* of $P_{\mathcal{H}}$ is not less than $\frac{3}{2}H$. Indeed, it must be that $y'_1 + y'_2 + y'_3 \geq H$ (because the total demand volume to be realized from s to t is equal to H, and that $y''_2 + y''_3 \geq y'_1$, $y''_1 + y''_3 \geq y'_2$ and $y''_1 + y''_2 \geq y'_3$ (because when link e fails then its primary capacity y'_e must be restored on the protection capacity of the two remaining links). Summing up the last three inequalities and using the first inequality we get $2 \cdot (y''_1 + y''_2 + y''_3) \geq y'_1 + y'_2 + y'_3 \geq H$, and hence $y''_1 + y''_2 + y''_3 \geq \frac{1}{2}H$. Thus, $F^* \geq \frac{3}{2}H$.

Next, we show that a valid partition $(\mathcal{D}', \mathcal{D}'')$ yields an optimal solution of $P_{\mathcal{H}}$. Such a solution is obtained by assigning the path-pair $r' = (p', q)$ to all demands from $\mathcal{D}'$, and the path-pair $r'' = (p'', q)$ to all demands from $\mathcal{D}''$ where the primary path p' is composed of link $e = 1$, the primary path p'' is composed of link $e = 2$, and the common backup path q is composed of link $e = 3$. Then $y'_1 = y'_2 = y''_3 = \frac{1}{2}, y'_3 = y''_1 = y''_2 = 0$, and the resulting objective $F(y)$ is equal to $\frac{3}{2}H$, thus optimal.

Finally, we prove that any optimal solution of $P_{\mathcal{H}}$ defines a valid partition of $\mathcal{D}$. Let $y'_e, y''_e, e \in \mathcal{E}$ be such an optimal solution. Introducing the notation $y_e = y'_e + y''_e, e \in \mathcal{E}$ we can write that $y_1 + y_2 + y_3 = \frac{3}{2}H$ (because the solution is optimal and hence $F(y) = \frac{3}{2}H$) and $y_2 + y_3 \geq H$, $y_1 + y_3 \geq H$ and $y_1 + y_2 \geq H$ (because when link e fails then the total demand volume must be carried on the remaining links). The only solution $y = (y_1, y_2, y_3)$ fulfilling these conditions is $y_1 = y_2 = y_3 = \frac{1}{2}H$. Now consider any fixed link e. Jointly, the primary capacity y'_e and the backup capacity y''_e of this link carry exactly $\frac{1}{2}\sum_{d \in \mathcal{D}} h_d$ of flow. Define $\mathcal{D}' = \{d \in \mathcal{D} : \exists\, r = (p, q) \in \mathcal{T}_d, x_{dr} = 1 \wedge (e \in p \vee e \in q)\}$ and $\mathcal{D}'' = \mathcal{D} \setminus \mathcal{D}'$. Then, clearly, $\sum \mathcal{D}' = \frac{1}{2}H$ and hence $(\mathcal{D}', \mathcal{D}'')$ is a valid partition of $\mathcal{D}$. ■

5. $\mathcal{NP}$-COMPLETENESS PROOF (ONE DEMAND, LP)

Now let us consider the linear relaxation of problem $\mathbb{P}$ (i.e., problem (1)) admitting all possible path-pairs, with the full single link failure scenario (i.e., all links can fail, one at a time), and with only one demand ($D = 1$). Below, we prove that this variant (i.e., variant 1b, 2b, 3a, 4a) is also $\mathcal{NP}$-complete by reducing the 2DIV-PATHS problem to it. 2DIV-PATHS was proved to be $\mathcal{NP}$-complete in [3].

Consider a directed graph and two pairs of its vertices (s_1, t_1) and (s_2, t_2). Assume that all four vertices are different. Problem 2DIV-PATHS consists in answering the question whether there exist two arc-disjoint directed paths, one from s_1 to t_1, and the second from s_2 to t_2, in a given network. We note that the problem cannot be solved using the polynomial Suurballe's algorithm [12], since the algorithm cannot assure that in the two resulting paths s_1 will be connected to t_1 and not to t_2.

Consider an instance $F_{\mathcal{G}}$ of 2DIV-PATHS for a given graph $\mathcal{G} = (\mathcal{V}, \mathcal{E})$ and $s_1, s_2, t_1, t_2 \in \mathcal{V}$. The corresponding instance of problem (1) is denoted by $P_{\mathcal{G}}$ and is modeled by means of a directed network $\mathcal{G}' = (\mathcal{V}', \mathcal{E}')$ depicted in Figure 3. The network consists of $8 + |\mathcal{V}|$ nodes and $13 + |\mathcal{E}|$ links. The original graph $\mathcal{G}$ forms a subgraph of network $\mathcal{G}'$ and is depicted as a "cloud" in the figure.

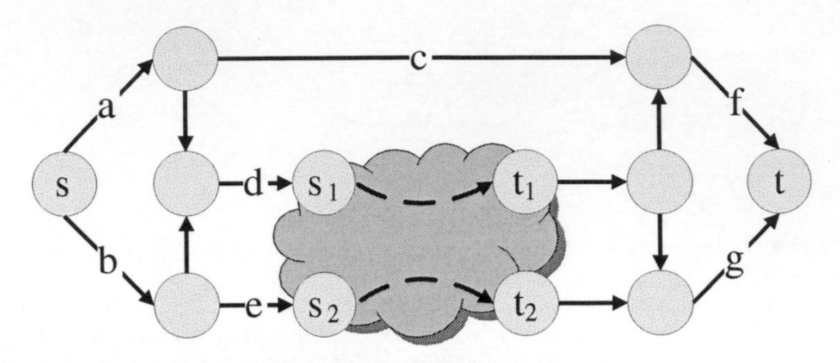

Figure 3: Network $\mathcal{N}$ resulting from graph $\mathcal{G}$

In the following proof we directly refer only to 7 specific links denoted by a, b, c, d, e, f, g and to 6 specific nodes denoted by s, t, s_1, s_2, t_1, t_2. Most of the unit capacity costs ξ_e are set to 0, and only $\xi_c = \xi_d = \xi_e = 1$. There is only one demand d from node s to node t, and $h_d = 1$.

Proposition 5.1 *The optimal objective of $P_{\mathcal{G}}$ is less or equal to $\frac{3}{2}$, if and only if, the answer to $F_{\mathcal{G}}$ is YES.*

Proof. Consider $F_{\mathcal{G}}$ and suppose that the two disjoint paths from the question exist. We denote these paths by $s_1 \to t_1$ and $s_2 \to t_2$. We shall construct a solution to $P_{\mathcal{G}}$ with the objective function value equal to $\frac{3}{2}$. To do this we route the demand using two primary/backup path-pairs. Each of these pairs carries flow $x = \frac{1}{2}$. The first primary flow traverses links a, c and f, while its corresponding backup flow is routed through links b, d, path $s_1 \to t_1$ and link g. The second primary flow travels through links b, e, path $s_2 \to t_2$ and link g; its corresponding backup flow passes links a, d, path $s_1 \to t_1$, and link f. Notice that each primary/backup flow pair is arc-disjoint. Moreover, the two primary paths are also arc-disjoint, and both backup flows use link d. Therefore they can share the backup capacity y''_d on it. Thus, we have to reserve $\frac{1}{2}$ of primary capacity on links c and e, and $\frac{1}{2}$ of backup capacity on link d. It means that the objective of the corresponding solution of $P_{\mathcal{G}}$ is equal to $\frac{3}{2}$. Observe that this solution would not be feasible, if $s_1 \to t_1$ and $s_2 \to t_2$ were not arc-disjoint.

Let us now assume that $F_{\mathcal{G}}$ has no solution, i.e., it is impossible to route two flows, the first from s_1 to t_1 and the second from s_2 to t_2, using arc-disjoint paths. Then, several cases can occur, including the existence of two other arc-disjoint paths: one from s_1 to t_2, and one and from s_2 to t_1. Below we prove that in all such cases the objective of $P_{\mathcal{G}}$ cannot be smaller than 2.

We define three sets of links, namely $\mathcal{A} = \{a, b\}$, $\mathcal{B} = \{c, d, e\}$ and $\mathcal{C} = \{f, g\}$. Notice that each of these sets cuts the network, so removing any of them makes that a path between nodes s and t does not exist. Moreover, the links in the sets are directed in the way making it impossible to traverse more than one link from any of these sets using only one path. Therefore, each path from $\mathcal{P}_d$ has to pass exactly one link from each of the sets $\mathcal{A}, \mathcal{B}, \mathcal{C}$.

We divide (all possible) paths in $\mathcal{P}_d$ into groups with respect to the links they traverse. In effect, we distinguish $|\mathcal{A}| \cdot |\mathcal{B}| \cdot |\mathcal{C}| = 12$ different groups; out of them only 6 are not empty. For instance, there is no path that traverses links a, c and g. The available groups are shown in Table 2. For each of the groups we calculate all possible groups of backup paths (certainly, a primary path and its backup path have to be arc-disjoint). The possible backup path groups are also shown in the table. Each

Table 2: Possible primary/backup path groups when $F_{\mathcal{G}}$ has no solution

Primary paths	Backup paths
$a - c - f$	$b - d - g$ or $b - e - g$
$a - d - f$	none
$a - d - g$	$b - e - f$
$b - e - f$	$a - d - g$
$b - e - g$	$a - c - f$
$b - d - f$	none
$b - d - g$	$a - c - f$

Table 3: Arc-disjoint primary path group pairs and their possible backup path groups

Primary path 1	Primary path 2	Backup path 1	Backup path 2
$a - c - f$	$b - e - g$	$b - d - g$ or $b - e - g$	$a - c - f$
$a - c - f$	$b - d - g$	$b - d - g$ or $b - e - g$	$a - c - f$
$a - d - g$	$b - e - f$	$b - e - f$	$a - d - g$

group is identified by an appropriate triple. For instance, a group of paths that traverse links a, c and f is denoted by $a - c - f$.

We notice that any simple non-bifurcated solution of $P_{\mathcal{G}}$ that uses just one primary/backup path-pair has the objective function equal to 2, as both the primary and the backup path have to traverse one link from set $\mathcal{B}$, and $\xi_e = 1$ only if $e \in \mathcal{B}$. Moreover, it is impossible to reduce the cost corresponding to the reserved primary capacity. Therefore, in order to improve the solution, we have to find a way to reduce the cost corresponding to the reserved backup capacity. In order to do this, we have to analyze all possible arc-disjoint pairs of primary paths, as backup capacity can be shared only by backup flows that belong to primary/backup path-pairs whose primary flows cannot fail simultaneously. In the considered case (single failures) this means that the primary paths have to be arc-disjoint. All possible arc-disjoint primary path-pairs, together with all possible backup paths protecting them, are shown in Table 3.

It is clearly seen in the table that when $F_{\mathcal{G}}$ has no solution it is impossible to find two arc-disjoint primary/backup path-pairs in such a way that the primary paths of these pairs are also arc-disjoint, and their backup paths traverse the same link $e \in \mathcal{B}$. Therefore, when $F_{\mathcal{G}}$ has no solution (or its solution is not known) the cost of $P_{\mathcal{G}}$ cannot be lower than 2. ■

6. COMPLEXITY OF APPROXIMATION SCHEMES

From the proof of Proposition 5.1 we can draw conclusions concerning complexity of various approximation schemes that solve the 1b, 3a, 4a variant (LP, single failures, one demand) of problem $\mathbb{P}$. When a solution to an instance $F_{\mathcal{G}}$ of 2DIV-PATHS exists (and can be found), the objective of a solution to the corresponding instance $P_{\mathcal{G}}$ is not greater than $\frac{3}{2}$. On the other hand, when a solution to $F_{\mathcal{G}}$ is not known, solutions to $P_{\mathcal{G}}$ cannot use such a path-pair. Therefore, the corresponding objectives must be equal at least to 2. Thus, since we have no way to solve 2DIV-PATHS in polynomial time, any approximation to the considered variant of $\mathbb{P}$ better than a $\frac{4}{3}$-approximation has to be $\mathcal{NP}$-complete. Note that the same reasoning can be applied also to the 1b, 3a, 4b variant (LP, single failures, many

Table 4: Overview of the complexity of approximation schemes

		single failures	multiple failures
$D=1$	LP	$<\frac{4}{3}$-approximation is $\mathcal{NPC}$	any approximation is $\mathcal{NPC}$
	IP	$\mathcal{P}$	any approximation is $\mathcal{NPC}$
$D>1$	LP	$<\frac{4}{3}$-approximation is $\mathcal{NPC}$	any approximation is $\mathcal{NPC}$
	IP	$<\frac{4}{3}$-approximation is $\mathcal{NPC}$	any approximation is $\mathcal{NPC}$

demands).

From the proof of Proposition 4.1 (IP, single failures, many demands) we can also try to draw some conclusions concerning complexity of approximation schemes. Suppose that the answer to the considered instance of the PARTITION problem ($F_{\mathcal{H}}$) is NO. In such a case, the quality of a solution to the corresponding instance $P_{\mathcal{H}}$ depends on how precise we can approximate a solution to $F_{\mathcal{H}}$. Note that PARTITION is equivalent to a special case of the SUBSET-SUM problem, for which Kellerer et al. in [6] presented a fully polynomial approximation scheme that solves it within accuracy ε in time $O(\min\{n\cdot\frac{1}{\varepsilon}, n+\frac{1}{\varepsilon^2}log(\frac{1}{\varepsilon})\})$. Therefore, we cannot conclude from the considerations of Section 4 that problem $\mathbb{P}$ in the considered variant *cannot* be approximated within a given accuracy ε in polynomial time. That is why, we will show how to modify the proof of Proposition 5.1 to cover the version of the problem considered in Section 4.

The modification is simple. Consider set $\mathcal{D}$ of demands instead of just one demand d. Consider a special case when demands from the set $\mathcal{D}$ can be divided into two sets $\mathcal{D}'$ and $\mathcal{D}''$ such that $\sum_{d\in\mathcal{D}'} h_d = \sum_{d\in\mathcal{D}''} h_d$. Without loss of generality we can assume that $\sum_{d\in\mathcal{D}} h_d = 1$. Now we can treat the set $\mathcal{D}$ as one demand that can be split into halves. Note that in the proof of Proposition 5.1 we deal with one demand which is equally split into two flows. Therefore, we can apply the reasoning from Section 5 also to the case discussed in Section 4. That is why, all polynomial approximations to the 1a, 3a, 4b variant (IP, single failures, many demands) of problem $\mathbb{P}$ also cannot be better than a $\frac{4}{3}$-approximation.

For variants 3b (multiple failures) of $\mathbb{P}$, the problem is straightforward, as it was proved in [5] that finding a pair of failure-disjoint paths, when multiple failures occur, is $\mathcal{NP}$-complete. Therefore, finding any feasible solution to the considered problem is $\mathcal{NP}$-complete. That is why, all approximation schemes that give any guarantee concerning the quality of a solution have to be $\mathcal{NP}$-complete. That concerns both LP and IP variants.

The results concerning the complexity of approximation schemes that solve different variants of the problem $\mathbb{P}$ are presented in Table 4.

7. SUMMARY

In the paper we have discussed computational complexity of various variants of the multi-commodity flow network optimization problem $\mathbb{P}$ assuming state-independent flow restoration with no use of stub-release. In particular, we have proved that the most important variants of $\mathbb{P}$ are $\mathcal{NP}$ complete. Depending on the variant, we have shown how PARTITION or 2DIV-PATHS can be reduced to the considered problem. Besides, we have discussed complexity of related approximation schemes and we have shown that they are in general also difficult. The complexity results of the paper are important as they can help to devise proper algorithms for resilient network design tools.

Acknowledgements

The research presented in this paper has been funded by "Optimization Models for NGI Core Network" (Polish Ministry of Science and Higher Education, grant N517 397334) and COST 293 Action GRAAL. M. Pióro has been also supported by "Modeling and Design of Core Internet Networks" (Swedish Research Council, grant 621-2006-5509).

REFERENCES

[1] AHUJA R., MAGNANTI T., ORLIN J. *Network Flows: Theory, Algorithms, and Applications*. Prentice Hall, 1993.

[2] COUDERT D., DATTA P., PERENNES S., RIVANO H., VOGE M.E. *Shared risk resource group: Complexity and approximability issues*. Parallel Processing Letters, 2006. To appear. Preliminary version available as INRIA technical report 5859, 2006.

[3] FORTUNE S., HOPCROFT J., WYLLIE J. *The directed subgraph homeomorphism problem*. Tech. rep., Ithaca, NY, USA, 1978.

[4] GAREY M., JOHNSON D. *Computers and Intractability : A Guide to the Theory of NP-Completeness (Series of Books in the Mathematical Sciences)*. W. H. Freeman, 1979. ISBN 0716710455.

[5] HU J. *Diverse routing in optical mesh networks*. IEEE Trans. Com., vol. 51, no. 3, pp. 489–494, 2003.

[6] KELLERER H., MANSINI R., PFERSCHY U., SPERANZA M. *An efficient fully polynomial approximation scheme for the subset-sum problem*. J. Comput. Syst. Sci., vol. 66, no. 2, pp. 349–370, 2003. ISSN 0022-0000.

[7] KHACHIYAN L. *A polynomial algorithm for linear programming*. Soviet Mathematics Doklady, vol. 20, pp. 191–194, 1979.

[8] MAURRAS J.F., VANIER S. *Network synthesis under survivability constraints*. 4OR, , no. 2, pp. 52–67, 2004.

[9] ORLOWSKI S., PIÓRO M. *On the complexity of column generation in survivable network design*. Tech. rep., Zuse Institut Berlin and Warsaw University of Technology, http://ztit.tele.pw.edu.pl/TR/NDG/occg08.pdf, 2008.

[10] PIÓRO M., MEDHI D. *Routing, Flow, and Capacity Design in Communication and Computer Networks*. Morgan Kaufman, 2004.

[11] STIDSEN T., PETERSEN B., RASMUSSEN K., SPOORENDONK S., ZACHARIASEN M., RAMBACH F., KIESE M. *Optimal routing with single backup path protection. Proceedings INOC, Spa, Belgium*. 2007.

[12] SUURBALLE J. *Disjoint paths in a network*. Networks, vol. 4, pp. 125–145, 1974.

[13] TOMASZEWSKI A., PIÓRO M., ŻOTKIEWICZ M. *On the complexity of resilient network design*, 2008. Submitted to Networks: an International Journal.

[14] WESSÄLY R. *Dimensioning Survivable Capacitated NETworks*. Ph.D. thesis, Technische Universität Berlin, 2000.

Keywords – Multi-Layer Networks, Mathematical Programming, Computational Study

Andreas BLEY[1]
Roman KLÄHNE[1]
Ullrich MENNE[2]
Christian RAACK[1]
Roland WESSÄLY[2]

MULTI-LAYER NETWORK DESIGN
A MODEL-BASED OPTIMIZATION APPROACH

In this paper, we present a model-based optimization approach for the design of multi-layer networks. The proposed framework is based on a series of increasingly abstract models – from a general technical system model to a problem specific mathematical model – which are used in a planning cycle to optimize the multi-layer networks. In a case study we show how central design questions for an IP-over-WDM network architecture can be answered using this approach. Based on reference networks from the German research project EIBONE, we investigate the influence of various planning parameters on the total design cost. This includes a comparison of point-to-point vs. transparent optical layer architectures, different traffic distributions, and the use of PoS vs. Ethernet interfaces.

1. INTRODUCTION

Our economy is increasingly dependent on reliable access to high-quality broadband services at reasonable cost. Network and service providers face the challenge of offering their services at a competitive price while still being profitable. This calls for innovative network planning methods and optimization tools. To this end, we present a so-called *model-based optimization cycle*, which has been developed in the German research project EIBONE – a cooperative project combing the expertise of operators, like Deutsche Telekom, system vendors, like Alcatel-Lucent, Ericsson, and Nokia Siemens Networks, together with small and medium enterprises, universities, and research institutes.

The proposed framework is based on models with increasing abstraction levels – starting from a detailed *system model* including specifications for the hardware, demand, cost, and network, using an *optimization model* which describes the planning requirements of a particular use-case in a technology independent way, and eventually applying and solving *mathematical models* (mixed-integer programs) to perform an end-to-end optimization of a network design. This approach has the competitive advantage that it is based on a network and system description as it could be stored in the

[1]Zuse Institute Berlin (ZIB), Takustr. 7, D-14195 Berlin, e-mail: {bley, klaehne, raack}@zib.de
[2]atesio GmbH, Sophie-Taeuber-Arp-Weg 27, D-12205 Berlin, e-mail: {menne, wessaely}@atesio.de
[3]This work has been supported by the German research project "EIBONE – efficient integrated backbone"

databases of a network operator and that it uses – albeit the high complexity of today's advanced technology described in the system model – sophisticated mathematical solution methods to compute proven low cost network designs. All results are analyzed and compared on system level.

This framework of increasingly abstract models lays the foundation for solving almost any kind of multi-layer network planning problem. It provides the basis for detailed techno-economic studies, which help to better understand the business implications of technical constraints as well as cost and traffic assumptions. For a network operator, such studies are indispensable for sound long-term decisions about the network architecture or the hardware vendor. Our framework also supports EIBONE partners in other planning tasks for multi-layer NGNs, ranging from tactical expansion planning to reconfiguration tasks to improve the link utilization.

To demonstrate the effectiveness of the approach, we show in a case study how central design questions for an IP-over-WDM network architecture can be answered using our framework. Based on reference networks from the German research project EIBONE we investigate the influence of various planning parameters on the total design cost. The questions addressed in the study are:

- How big are the potential savings by using a more complex transparent optical layer – allowing for ROADMs and OXCs with full optical switching capabilities – compared to a much simpler point-to-point WDM network?
- How big is the influence of the traffic distribution on the optimal network layout and cost? Even for the same total traffic demand, demand predictions stemming from network measurements or from population based geographical traffic models may lead to substantial differences.
- How big is the influence of using more intelligent but also more expensive STM16- and STM64-PoS-interfaces instead of 10Gbit-Ethernet interfaces?

This paper is organized as follows. The overall model-based optimization cycle is presented in Section 2. The different models and some of the transformations are explained. Section 3 is dedicated to the computational case study. We describe the data used in the study and report on the computational results. Eventually, Section 4 provides our conclusion.

2. PLANNING CYCLE

The model-based optimization cycle developed within EIBONE with its three abstraction levels is shown in Figure 1. The specification of a **system model** as a vendor independent *XML-format* has been carried out. It incorporates models of the different hardware, technology layers, cost, and traffic models. It is designed to allow an adequate description of today's as well as tomorrow's multi-layer networks. It facilitates both the mere exchange of data as well as benchmarking all possible aspects within a planning process. Within EIBONE, three reference networks (two German and one US scenario) have been encoded using this XML-format. Together with the different cost and traffic models, these reference networks provide the basis for benchmarking different solution approaches and planning scenarios.

Optimization models, as considered in EIBONE, represent specific planning tasks such as architecture decisions, capacity expansion, or (dynamic) reconfiguration problems in multi-layer transport networks. These optimization problems abstract from the very detailed technological view of the system model to a problem specific view focusing on the functionalities, restrictions, and decisions relevant for the given planning task. The Survivable Network Design data library (SNDlib [9]) can be

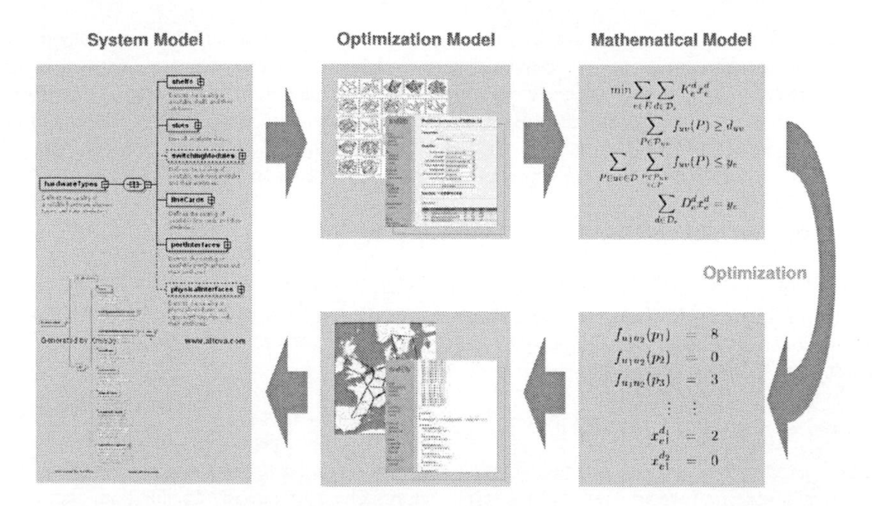

Figure 1: Model-based optimization cycle

considered as a representative of this abstraction level. SNDlib has been compiled out of real-world data from network operators and data stemming from other research projects such as NOBEL [2] and is used by many researchers as a basis for benchmarking their solution approaches by now. The data of SNDlib does not contain information about the particular technological background. It cannot be seen, for instance, whether the links are logical connections between IP routers, whether the link cost includes physical transmission cost, or whether the link capacities are based on ATM, SDH, or Ethernet port capacities. Similarly, we have been developing optimization models which represent the structure of multi-layer networks, including a demand layer, a logical layer and a physical layer. This layering is not coupled with a particular network architecture such as IP-over-OTH, or IP-over-WDM.

Mathematical models and all kinds of algorithmic solution methodologies are used to address the planning problems described by the optimization models. Within EIBONE, two principle approaches have been combined:

- Exact solution methods based on mixed-integer linear programming (MILP) either find a provable optimal solution or a solution within a specified quality guarantee. The running time to close the gap and to prove optimality might increase significantly with the size of the network. Within EIBONE, the efficiency of these methods has been improved through a number of investigations of the underlying fundamental mathematical structure of the problem [7, 10, 11].

- With the second methodology approximate solutions are computed by means of combinatorial heuristics. These can be designed to easily take into account also very special planning requirements and they usually scale better with the size of the network. Heuristics, however, do not provide quality guarantees for the solutions they produce.

Combinations of these two seemingly opposing approaches proved to be very efficient in practice. Heuristics have been integrated into state-of-the-art MILP-solvers to find good quality solutions faster and MILP-based algorithms have been designed to optimally solve critical sub-problems [8].

To close the optimization cycle, transformations are carried out to map the pure mathematical solution to a solution of the considered optimization model and eventually back to a network configuration satisfying the planning task and respecting the hardware and traffic requirements described in the system model. All necessary transformations in the optimization cycle are implemented in Java and fully automated. Network costs are evaluated on system level. The resulting network configuration can be considered to be (close) to optimal with respect to the given cost-model.

3. COMPUTATIONAL STUDY

In this section we present a computational study which is performed using the described model-based optimization cycle. The goal is to give answers to the questions raised in the introduction: the influence of (i) a transparent vs. an opaque optical layer, (ii) the traffic distribution, and (iii) PoS vs. Ethernet interface cards in the IP router.

3.1. NETWORK DATA AND MODELS

We compared two different architectures which differ with respect to cross-connecting channels in the optical layer. Both are an IP-over-WDM architecture. The hardware involved in both alternatives

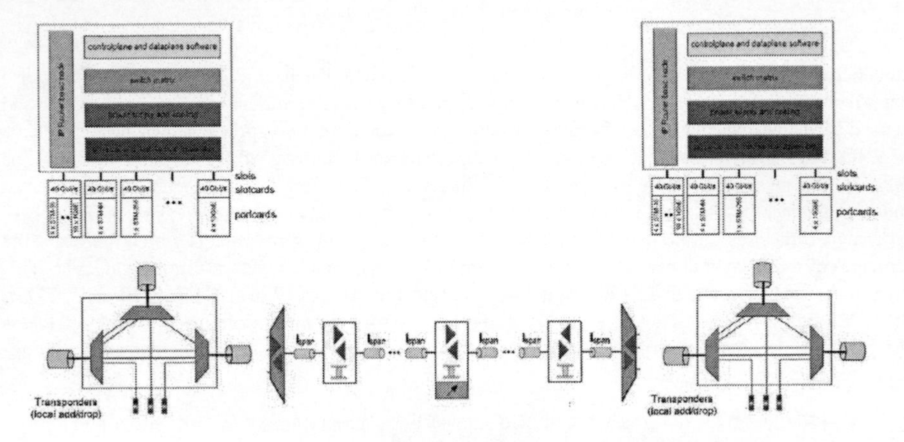

Figure 2: IP layer over WDM layer (Source: NOBEL hardware and cost model [6])

is depicted in Figure 2. In the IP layer, routers are considered having a switching matrix as a central element and slots for slot-cards. The latter host the port-cards which eventually provide capacitated link interfaces. In the WDM layer, the physical link equipment comprises DCFs, DGEs, and OLAs. One WDM-MUX at each end terminates a physical link. For the transparent architecture ROADMs and OXCs can be installed at every node to arbitrarily switch the optical channels, which is not allowed in the point-to-point (opaque) scenario.

Cost values for equipment are based on the NOBEL cost model [6]. The possible capacities at the IP layer nodes range from 640 Gbit/s to 5760 Gbit/s. The cost for the smallest installable router is 16.67, the next larger has a cost of 111.67 which is about 7 times more expensive. The largest one comes at a cost of 315.83. Furthermore, slot-cards, which host the port-cards, are relatively expensive.

It can be seen that the 4xPoS-STM64-SR port-card costs with 18.33 much more than the 4x10GbE-LR port-card with 4.20. The normalizing element in the NOBEL model is the 10G-LH transponder at a cost of 1. The cost of ROADMs and OXCs is about 10, with a degree dependent component. For details, the reader is referred to [6]. The study is performed on:

- the physical (fiber) topology with 50 locations shown in Figure 3, and
- a subset of 17 locations (highlighted in Figure 3) being the traffic sources and the only locations where it is admissible (and mandatory) to install IP router equipment.

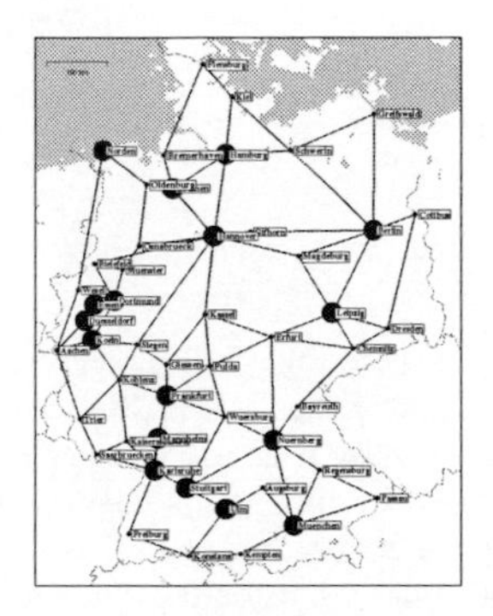

Figure 3: Germany17 over Germany50 reference network

Two types of traffic models are used to create point-to-point traffic matrices, where each of the 17 IP-nodes is a traffic source:

- POP: Dwivedi-Wagner [5], which distributes the traffic classes Voice, Business, and IP dependent on the number of inhabitants, households, and employees. The influence of the distance on the traffic between two locations is highest for Voice and lowest for IP traffic.
- DFN: Measurements of the DFN-Verein [1], which were carried out in 2006 in 5 minute intervals. The measurements over a day, the peak hour over a month, and a year have been used to generate a single point-to-point traffic matrix.

Two traffic matrices have been generated for both models: one with a total traffic volume of 3 Tbit/s (short: 3T) and one with 6 Tbit/s (short: 6T). No survivability constraints have been added for this study. In Figure 4 it can be seen that the traffic is more evenly distributed for Dwivedi-Wagner and that Frankfurt, in particular, is the dominating traffic source for the DFN measurements. It is important to note that the DFN traffic demands reach the limits of the considered router technology, due to its high concentration around Frankfurt. Already for a total traffic volume of 3T it is necessary to install a router with a capacity of 3840 Gbit/s in Frankfurt. There is no feasible network configuration for a total traffic volume of 6T with the DFN traffic distribution, because no router could handle the traffic volumes at the node Frankfurt.

All data (hardware, cost, traffic) has been specified within the system-model. Two optimization models have been defined (one for each considered architecture) that describe abstract two-layer network design problems with discrete capacities for links and nodes of both layers. By introducing

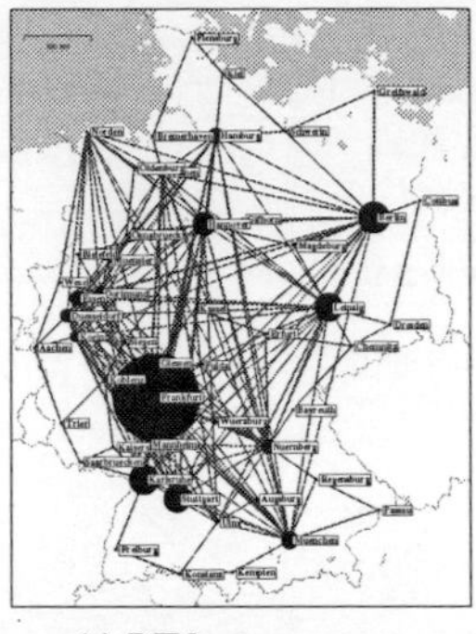

(a) DFN measurements

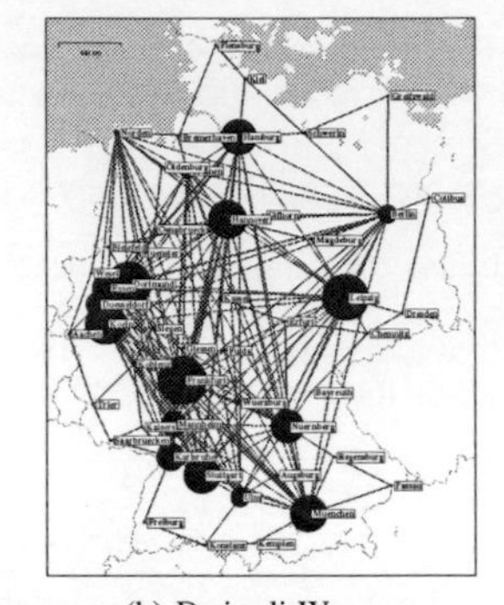

(b) Dwivedi-Wagner

Figure 4: Source traffic distribution

discrete variables and linear inequalities describing all problem constraints in terms of these variables the optimization models have been formulated as MILPs. See Appendix A for a detailed description of the mathematical models. The MILPs have been solved using the general-purpose solver SCIP 1.0 [3, 4]. To accelerate its performance and to improve solutions and lower bounds several problem-specific preprocessing and cutting plane procedures as well as heuristics have been applied in addition. For theoretical and implementational details on these auxiliary methodologies and plug-ins the reader is referred to [7, 8, 10].

All computations have been carried out on an Intel Core 2 Quad 2.66GHz machine with 4 GB main memory. The total running time for a single optimization has been restricted to 60 minutes. It is noticeable that the average gap (proven quality of the result as the relative difference between the solution cost and a mathematically proven lower bound to this cost) has been 2.5 percent, with a maximum of 8.0 percent.

3.2. TRANSPARENT VERSUS OPAQUE OPTICAL LAYER

In the first study, we investigated how big the cost savings are if it is allowed to use full switching capabilities in the optical layer, compared to classical point-to-point WDM links. For this purpose we computed for both architecture alternatives the network configurations for both demand matrices and for the two total traffic volumes 3T and 6T. The resulting costs are shown in Table 1. As expected, the transparent architecture is always more cost-efficient than its point-to-point counterpart: The savings range between 8.31 and 17.53 percent with an average of 12.48 percent.

For the instance POP-3T-ETH we looked into more details. Surprisingly, 76.1 Percent of the savings are WDM savings (WDM-MUX and Medium), only 25% of savings occur in the IP layer. It is also noticeable that in the point-to-point scenario 75 physical and logical links are used with an average of 1.52 IP/WDM hops per IP demand, while for the transparent scenario, 97 logical links and 25 physical links are used with an average of 4.3 WDM hops per IP link and 1.3 IP hops and 6 WDM hops per IP demand. These values show that the reduction of IP hops per IP demand is minimal and, hence, there are almost no savings in the IP layer for the transparent scenario.

Scenario	Point2Point	Transparent	Savings
DFN-3T-ETH	5819.82	5268.80	9.47%
POP-3T-ETH	6838.73	5639.73	17.53%
POP-6T-ETH	11359.24	9879.98	13.02%
DFN-3T-PoS	7316.29	6708.54	8.31%
POP-3T-PoS	8455.67	7110.41	15.91%
POP-6T-PoS	14134.81	12626.63	10.67%
		Average:	12.48%

Table 1: Cost comparison of point-to-point vs. transparent optical layer

3.3. POS- VERSUS ETH-PORTCARDS IN THE IP-LAYER

The NOBEL hardware and cost model allows to install either PoS- or GbE-interface at IP routers. As already mentioned, the 4xPoS-STM64-SR port-card costs with 18.33 much more than the 4x10GbE-LR port-card with 4.20. This raises the following questions: What is the effect of this cost difference on the overall design of the network and how much more expensive is a design using PoS interfaces? The results of allowing PoS-interfaces only (PoS) and GbE-interfaces only (ETH) are shown in Table 2. As expected, there is a significant increase when using the PoS-interfaces. However, we have

Scenario	PoS	ETH	Savings
DFN-P2P-3T	7346.63	5819.82	20.78%
DFN-TRANS-3T	6708.54	5268.80	21.46%
POP-P2P-3T	8455.67	6838.73	19.12%
POP-TRANS-3T	7110.41	5639.73	20.68%
POP-P2P-6T	14157.73	11359.24	19.77%
POP-TRANS-6T	12626.63	9879.98	21.75%
		Average:	20.59%

Table 2: Cost comparison of PoS- and GbE-interfaces

not encountered any big change in the logical or physical topologies. Both architectures use about the same number of logical and physical links with similar capacities. When comparing the total cost of the PoS-interfaces with the total cost of the GbE-interfaces it can be observed that this difference is almost identical to the difference of the total network cost.

3.4. POPULATION BASED VERSUS CENTRALIZED TRAFFIC DEMANDS

In the last part of the study, two very different distributions of the same amount of total traffic have been generated in order to investigate their influence on the architecture and cost of the networks. As described above, the first one is based on measurements in the German national research and education network (DFN) which is operated by the DFN-Verein, and the second one is based on a population model (POP) suggested by Dwivedi-Wagner [5]. The DFN traffic demands have a higher concentration than the traffic demands stemming from the population model. The emanating demand at node Frankfurt is much higher in the DFN scenario than in the population scenario, while the emanating demands at the other nodes are smaller in the DFN scenario than in the population scenario.

Scenario	Population	DFN	Savings
TRANS-3T-ETH	5639.73	5268.80	6.58%
TRANS-3T-SDH	7110.41	6708.54	5.65%
P2P-3T-ETH	6838.73	5819.82	14.90%
P2P-3T-SDH	8455.67	7346.63	13.12%
		Average:	10.06%

Table 3: Cost comparison for different traffic distributions

The total cost values for the different scenarios are shown in Table 3. To handle the DFN demands, one router with capacity 3840 Gbit/s, six with 1280 Gbit/s, and ten with 640 Gbit/s are required in all solutions. For the more evenly distributed population-based traffic demands, no 3840 Gbit/s router but 13 with capacity 1280 Gbit/s and four with 640 Gbit/s are required. These differences in the IP router platform alone yield an additional cost of 453.33 for the population based demands compared to the DFN demands, independent of the architecture of the underlying optical platform. For the transparent network architecture, the cost difference between the networks optimized for the DFN traffic demands and the population based demands is almost equal to the difference in the IP platform, effects of the chosen transmission technology Ethernet or SDH are negligible. For point-to-point networks, the different IP platforms still account for 50% of the total cost difference.

4. CONCLUSIONS

In this paper, we presented a model-based planning cycle in order to optimize multi-layer IP-over-WDM networks using realistic and very detailed system data. Investigating three different scenarios, the main observation is the following: For the realistic NOBEL hardware and cost model, the 12 percent cost difference between the fundamentally different architectures point-to-point WDM-links and transparent optical switches is on average about the same as the cost difference obtained by varying the traffic distribution. The average difference between the networks designed for the DFN measurements and those for the Dwivedi-Wagner model was about 10 percent. Finally, the difference in card cost between PoS- and GbE-interfaces goes directly into the differences between the network cost.

REFERENCES

[1] *DFN-Verein, German national research and education network.* `http://www.dfn.de`.

[2] *IST project NOBEL – phase 2.* `http://www.ist-nobel.org/`.

[3] Achterberg T. *Constraint Integer Programming.* Ph.D. thesis, Technische Universität Berlin, 2007. `http://opus.kobv.de/tuberlin/volltexte/2007/1611/`.

[4] Achterberg T., Berthold T., Heinz S., Koch T., Pfetsch M., Wolter K. *SCIP 1.0–Solving Constraint Integer Programs.* `http://scip.zib.de`, 2008.

[5] Dwivedi A., Wagner R.E. *Traffic model for USA long distance optimal network. Proceedings of the Optical Fiber Communication Conference*, pp. 156–158. 2000.

[6] Huelsermann R., Gunkel M., Meusburger C., Schupke D.A. *Cost modeling and evaluation of capital expenditures in optical multilayer networks*. Journal of Optical Networking, to appear.

[7] Koster A., Orlowski S., Raack C., Baier G., Engel T. *Single-layer Cuts for Multi-layer Network Design Problems*, vol. 44, chap. 1, pp. 1–23. Springer-Verlag, College Park, MD, U.S.A., 2008. Selected proceedings of the 9th INFORMS Telecommunications Conference.

[8] Orlowski S., Koster A., Raack C., Wessäly R. *Two-layer network design by branch-and-cut featuring MIP-based heuristics. Proceedings of the 3rd International Network Optimization Conference (INOC 2007), Spa, Belgium*. 2007.

[9] Orlowski S., Pióro M., Tomaszewski A., Wessäly R. *SNDlib 1.0–Survivable Network Design Library. Proceedings of the 3rd International Network Optimization Conference (INOC 2007), Spa, Belgium*. 2007. `http://sndlib.zib.de`.

[10] Raack C., Koster A., Orlowski S., Wessäly R. *Capacitated network design using general flow-cutset inequalities. Proceedings of the 3rd International Network Optimization Conference (INOC 2007), Spa, Belgium*. 2007.

[11] Raack C., Koster A., Wessäly R. *On the strength of cut-based inequalities for capacitated network design polyhedra*. ZIB Report ZR-07-08, Konrad-Zuse-Zentrum für Informationstechnik Berlin, 2007.

[12] Raghavan S., Stanojević D. *WDM optical design using branch-and-price*, 2007. Working paper, Robert H. Smith School of Business, University of Maryland.

A. THE MATHEMATICAL MODEL

This section provides a brief description of the mixed-integer programming model used for the optimization of the transparent architecture which is an extension of the formulation proposed in [12]. It has the advantage of a very compact description of the flow on the logical layer. This is achieved by aggregating all logical flow variables of a node-pair to a single variable. The model for the point-to-point architecture is similar with minor modifications of the parameters.

The physical network is given by the fiber topology depicted in Figure 3 and represented by an undirected graph $G = (W, E)$. The logical network is defined by a subset V of the nodes W and all node-pairs $V \times V$ (the complete undirected graph spanned by the 17 potential IP router locations). For every node-pair $(i, j) \in V \times V$ a set $P_{(i,j)}$ of admissible (light)paths in the physical network is considered. For our calculations we used the 50 shortest paths in G for every node-pair. Let P be the union of all these paths. Each path $p \in P$ has a set M_p of available capacity modules. These modules correspond to interface cards installable at the end-nodes of the path. A path module m consumes c^m channels in the physical fiber layer and has a total capacity (bit-rate) of C^m. Each module can be installed several times on the corresponding lightpath. A physical link can be equipped with an arbitrary number of fibers each supporting a total of B passing channels. For nodes we introduce the sets N and O of admissible logical and physical node modules, respectively. Only one module can be installed at every logical and physical node. A logical node module n has a maximum switching capacity of C^n. A physical node module o has a restricted number of supported fibers C^o_f, overall channels C^o_c, and add-drop channels C^o_t. Every logical link and node module, physical node module and fiber can be installed at a certain cost.

The given traffic matrix defines a set of demands each having source and target in V and a demand value. By aggregating demands at a common source node a set K of commodities is constructed. This transformation reduces the size of the routing formulation (see for instance [7]) and results in commodities having one source and several target nodes. With every commodity $k \in K$ and every node $i \in V$, a net demand value d_i^k is associated such that $\sum_{i \in V} d_i^k = 0$. The total demand starting or ending at node i, is given by $d_i := \sum_{k \in K} |d_i^k|$.

We introduce the following variables. For every node-pair (i,j) the variables f_{ij}^k and f_{ji}^k describe the flow between i and j in both directions w. r. t. commodity $k \in K$ (aggregating the flow for k on all paths in $P_{(i,j)}$). The value y_p^m counts the number of path modules for module m and path p. For every physical link e the number of provided fibers is given by y_e. The logical node module $n \in N$ is installed at node $i \in V$ if and only if the variable x_i^n is set to 1. Similarily, x_i^o decides whether or not to install the physical node module $o \in O$ at node $i \in W$.

The problem of minimizing the cost of a network satisfying the given demand matrix and the capacity restrictions on both layers can be formulated as the problem of minimizing the cost of all module and fiber configurations satisfying the following set of constraints:

$$\sum_{j \in V \setminus \{i\}} (f_{ij}^k - f_{ji}^k) = d_i^k \qquad \forall i \in V, k \in K \tag{1}$$

$$\sum_{p \in P_{(i,j)}} \sum_{m \in M_p} C^m y_p^m - \sum_{k \in K} (f_{ij}^k + f_{ji}^k) \geq 0 \qquad \forall (i,j) \in V \times V \tag{2}$$

$$\sum_{n \in N} C^n x_i^n - \sum_{p \in \delta_P(i)} \sum_{m \in M_p} C^m y_p^m \geq d_i \qquad \forall i \in V \tag{3}$$

$$\sum_{n \in N} x_i^n \leq 1 \qquad \forall i \in V \tag{4}$$

$$B y_e - \sum_{p \in P:\, e \in p} \sum_{m \in M_p} c^m y_p^m \geq 0 \qquad \forall e \in E \tag{5}$$

$$\sum_{o \in O} C_f^o x_i^o - \sum_{e \in \delta_E(i)} y_e \geq 0 \qquad \forall i \in W \tag{6}$$

$$\sum_{o \in O} C_c^o x_i^o - 2 \sum_{p \in P:\, i \in p} \sum_{m \in M_p} c^m y_p^m \geq 0 \qquad \forall i \in W \tag{7}$$

$$\sum_{o \in O} C_t^o x_i^o - \sum_{p \in \delta_P(i)} \sum_{m \in M_p} c^m y_p^m \geq 0 \qquad \forall i \in W \tag{8}$$

$$\sum_{o \in O} x_i^o \leq 1 \qquad \forall i \in W \tag{9}$$

$$f_{ij}^k, f_{ji}^k \in \mathbb{R}_+,\ y_p^m, y_e \in \mathbb{Z}_+,\ x_i^n, x_i^o \in \{0,1\} \tag{10}$$

The flow conservation equations (1) ensure a feasible routing of the traffic demands. The logical link capacity constraint (2) says that the flow between i and j must not exceed the total capacity installed on all corresponding paths. Inequality (3) guarantees that the logical node capacity is not exceeded. The capacity at a logical node is consumed by the total node demand and the capacity of all terminating channels. The physical link capacity constraint (5) restricts the number of channels that can pass a physical link e. Inequalities (6)-(8) are physical node capacity constraints. Attached fibers, overall channels and terminating channels should not exceed the number of provided physical node ports. With (4) and (9) only one logical and one physical node module can be selected for every node in the logical and physical network, respectively.

V. NETWORK SECURITY

Key words – steganography, information hiding, performance analysis

Wojciech MAZURCZYK* and Józef LUBACZ*

ANALYSIS OF A PROCEDURE FOR INSERTING STEGANOGRAPHIC DATA INTO VoIP CALLS

The paper concerns performance analysis of a steganographic method, dedicated primarily for VoIP, which was recently filed for patenting [1] under the name LACK. The performance of the method depends on the procedure of inserting covert data into the stream of audio packets. After a brief presentation of the LACK method, the paper focuses on analysis of the dependence of the insertion procedure on the probability distribution of VoIP call duration.

1. INTRODUCTION

Communication networks steganography is a method of hiding secret data inside usual data transmitted by users, so that the hidden data cannot be detected (in an ideal case) by scanning the data flow by a third party. A new steganographic method called LACK (Lost Audio PaCKets Steganography) was recently proposed and filed for patenting [1] in Poland. The method is described in some detail in [2].

A detailed review of steganographic methods that may be applied for IP telephony can be found in [2] and [9]. In general, the methods can be classified into the two following groups:

- Steganographic methods that modify packets: network protocol headers or payload fields. Examples of such solutions include modifications of free/redundant headers' fields for IP, UDP or RTP protocols during conversation phase and signalling messages in e.g. SIP[10]. Information hiding which is based on affecting packets' payload usually uses digital audio watermarking algorithms (e.g. DSSS [13] and QIM [14]),
- Steganographic methods that modify packets' time relations. Examples of such solution: affecting sequence order of RTP packets [11] and modifying their inter-packet delay [12].

With respect to the two above groups LACK is a hybrid steganographic method since it modifies both packets' content and their time dependencies.

In general, the LACK method is intended for a broad class of multimedia, real-time applications, but its main foreseen application (at least for now) is VoIP. The proposed method

* Warsaw University of Technology, Faculty of Electronics and Information Technology, Institute of Telecommunications, 15/19 Nowowiejska Str. 00-665 Warsaw, Poland, {W.Mazurczyk, J.Lubacz}@tele.pw.edu.pl

utilises the fact that for usual multimedia communication protocols like RTP (Real-Time Transport Protocol) [3] excessively delayed packets are not used for reconstruction of transmitted data at the receiver (the packets are considered useless and discarded). The main idea of LACK is as follows.

At the transmitter, some selected audio packets are intentionally delayed before transmitting. If the delay of such packets at the receiver is considered excessive, the packets are discarded by a receiver not aware of the steganographic procedure. The payload of the intentionally delayed packets is used to transmit secret information to receivers aware of the procedure so no extra packets are generated. For unaware receivers the hidden data is "invisible".

The effectiveness of LACK depends on many factors such as the details of the communication procedure (in particular the type of codec used, the size of the voice frame, the size of the receiving buffer, etc.) and on the network QoS (packet delay and packet loss probability).

No real-world steganographic method is perfect – whatever the method, the hidden information can be potentially discovered. In general, the more hidden information is inserted into the data stream, the greater the chance that it will be detected, e.g. by scanning the data flow or by some other steganalysis methods. Moreover, the more audio packets are used to send covert data, the greater the deterioration of the quality of the VoIP connection. Thus the procedure of inserting hidden data should be carefully chosen and controlled in order to minimize the chance of detecting inserted data and to avoid excessive deterioration of the QoS.

To avoid excessive deterioration of the QoS lost packet ratio must be kept below certain accepted level. This level depends on the speech codec used. For example, according to [8], maximum loss tolerance is 1% for G.723.1, 2% for G.729A, 3% for G.711 codecs. If special mechanism to deal with lost packets at the receiver is utilized, like the PLC (Packet Loss Concealment) [15], acceptable level of lost packets e.g. for the G.711 codec increases from 3% to 5%. Thus this value provides us with upper limit for transmission rate.

In general, the amount of steganographic data using LACK depends on the acceptable level of packet loss. For example, for the G.711 speech codec with data rate 64 kbit/s and data frame size of 20 ms, if the packet loss probability introduced for LACK purposes is 0.5%, then under condition that packet losses do not exceed acceptable level, the theoretical hidden communication rate is 320 b/s.

In the present paper we shall focus on the hidden data insertion rate *IR* [bits/s]. Obviously, *IR* depends on the amount of hidden data to be sent and on the call duration. In principle, the call duration may be adjusted to the amount of hidden data to be sent. This however could cause that the distribution of calls applying LACK differs from the call duration distribution of LACK-less calls, and as a consequence make LACK vulnerable to statistical steganalysis based on call duration. Thus rather than adjusting the call duration to the amount of hidden data to be sent, it is preferable to adjust the hidden data insertion rate *IR* to LACK-less calls duration distribution. This, in turn, requires making *IR* dependent on that distribution. Obviously, this is not an important question in case of sporadic LACK use; it becomes important in case of a predefined group of frequent LACK users. In the present paper we focus on such a case.

Moreover, in the presented analysis we consider the dependence of *IR* of a particular call on the elapsed time of that call, i.e. we consider *IR* that can (potentially) be made time dependent, adjusted to the foreseen residual call duration. As shown in our analysis, such time-dependent *IR* procedure potentially allows for decreasing the *IR* during the call duration, compared to the *IR* at call initiation time. In effect, the negative influence of LACK on QoS can be decreased. Such an effect can be achieved for call duration distributions with coefficient of variation greater than 1; available experimental data concerning VoIP call duration distributions seem to indicate that this is a realistic assumption for real-life VoIP calls.

It should be emphasised that the LACK procedure introduced in this paper can be utilized by decent LACK users who use their own VoIP calls to exchange covert data, but also by intruders who are able to covertly send data using third party VoIP calls (e.g. in effect of earlier successful attacks by using trojans or worms or by distribution of modified version of a popular VoIP software). This is a usual tradeoff requiring consideration in a broader steganography context which is beyond the scope of this paper.

2. THE VoIP CALL DURATION DISTRIBUTION

For PSTN the call duration probability distribution was well known based on extensive experimental research. For many decades an exponential distribution was assumed a good enough approximation for engineering purposes. VoIP is a relatively new service and thus only few reliable experimental data is available, so in many research papers concerning IP voice traffic (e.g. [4], [5], [6]) the exponential call duration is still assumed. Current experiments prove however that this assumption is far from being realistic.

Birke et al. [7] captured real VoIP traffic traces (about 150 000 calls) from FastWeb, an Italian telecom operator. The obtained call duration probability distribution is reproduced in Fig. 1 with a solid line. To illustrate qualitatively the degree in which the experimental results differ from the exponential distribution and some other chosen distributions (hyperexponential and log-normal) these were drawn with broken lines in Fig.1. As can be seen, the differences are considerable and no straightforward approximation of the experimental data with standard distributions is available. In particular, the exponential distribution is far from being realistic.

The experimental data yield average call duration $E(D) = 117.31$ and standard deviation $\sigma(D) = 278.74$, thus the coefficient of variation $C_v = \sigma(D)/E(D) = 2.37$ (for exponential distribution $C_v = 1$).

To achieve an analytic approximation of the experimental data a combination of some standard distributions can be used, for example:

$$f_D(x) = \begin{cases} \dfrac{1}{1.55x\sqrt{2\pi}} e^{-\frac{(\ln(x)-3.8)^2}{4.805}} & for \quad 0 \le x < 27.5 \\ 0.000114e^{-0.00114x} + 0.027252e^{-0.03028x} & for \quad 66.5 < x \le 27.5 \\ \dfrac{1}{1.55x\sqrt{2\pi}} e^{-\frac{(\ln(x)-3.8)^2}{4.805}} & for \quad 66.5 \le x \le 455 \end{cases} \qquad (1)$$

The above analytic approximation is quite complex and of little practical use for our purposes, i.e. for establishing the dependence of the insertion rate *IR* on some simple enough characterization of the call duration distribution. Our guess was that this can be achieved through characterizing a considerably wide range of call duration distribution types with the coefficient of variation C_v and then expressing the *IR* through C_v.

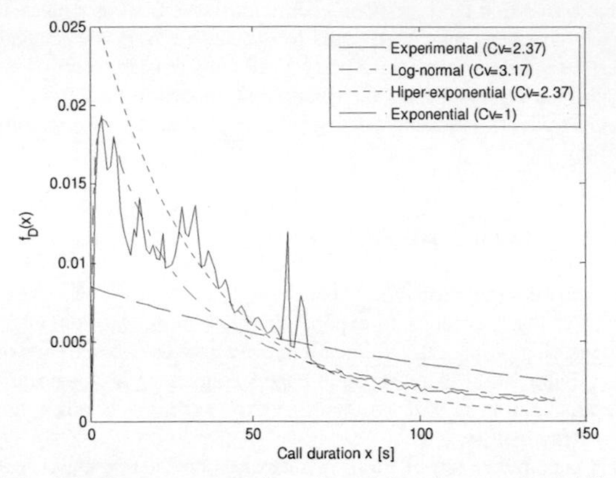

Fig. 1. VoIP call duration – comparison of experimental data with selected probability distributions

A reasonably wide range of call distribution types can be achieved and effectively analyzed with the 2-parameter Weibull distribution, with appropriately chosen parameters: the shape parameter $k > 0$ and the scale parameter $\lambda > 0$. The complementary cumulative probability distribution function ($\overline{F}_D$) and probability density function f_D are as follows:

$$\overline{F}_D(x;k,\lambda) = e^{-\left(\frac{x}{\lambda}\right)^k}$$
$$f_D(x;k,\lambda) = \frac{k}{\lambda}\left(\frac{x}{\lambda}\right)^{k-1} e^{-\left(\frac{x}{\lambda}\right)^k} \quad (2)$$

The λ parameter was set so to achieve the above experimental average call duration time $E(D) = 117.31$ and the k parameter was varied so to obtain a wide range of C_v values. In Table 1 the analyzed values are summarized.

Weibull parameters	k=3.4, λ=130.57	k=2, λ=132.37	k=1.2, λ=124.71	k=1, λ=117.31	k=0.8, λ=103.54	k=0.6, λ=77.97	k=0.5, λ=58.65	k=0.4, λ=35.3
C_V	0.32	0.52	0.84	1	1.26	1.76	2.23	3.14

Table 1. Chosen Weibull distribution parameters k and λ and corresponding C_v values.

In Fig. 2 the Weibull probability distribution is depicted for the parameters from Tab. 1 to illustrate the resulting wide range of distribution shapes. Note by the way that for $k = 1$ the Weibull distribution equals the exponential distribution ($C_v = 1$), for $k = 2$ it becomes the Rayleigh distribution ($C_v = 0.52$) and for $k = 3.4$ it resembles the normal distribution ($C_v=0.32$).

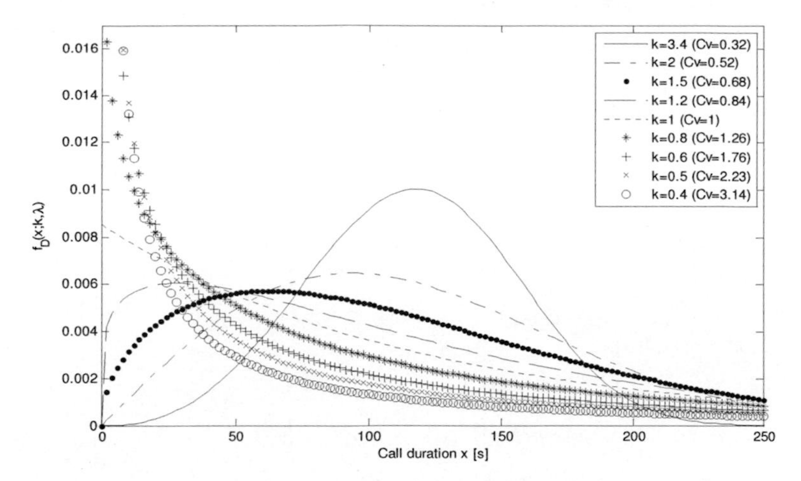

Fig. 2. Weibull distribution for various k, λ and C_V

3. TIME DEPENDENT INSERTION RATE

For an arbitrary instant of a call the average residual call duration is well know to be equal

$$E(R) = \frac{E(D^2)}{2E(D)} \tag{3}$$

or equivalently

$$E(R) = \frac{C_V^2 + 1}{2} E(D) \tag{4}$$

Suppose that at the beginning of a call the insertion rate is set to $IR = S/E(D)$, where S is amount of steganographic data to be sent covertly. As mentioned in section 1, if $C_v > 1$, and thus $E(R) > E(D)$, which seems to be the case for VoIP calls as indicated above, then beginning from some arbitrary instant of the call we may decrease the insertion rate to $IR = S/E(R)$, which is beneficial from the point of view of QoS and resistance to detection of the hidden data.

The above indicates that it reasonable to make the insertion rate dependent on the elapsed time of a call. It is nevertheless not practical to use the classical definition of residual call duration since it involves an arbitrary time instant and not the current call duration. We are rather interested in the expected call duration on condition it has already lasted t units of time:

$$E(D \mid D>t)=\frac{1}{P(D>t)}\int_{t}^{\infty} x f_D(x)dx = t+\frac{1}{\overline{F}_D(t)}\int_{t}^{\infty}\overline{F}_D(x)dx \tag{5}$$

thus

$$t \leq E(D \mid D>t) \leq \frac{E(D)}{\overline{F}_D(t)} \tag{6}$$

For the Weibull distributions considered in the previous section

$$E(D \mid D>t)=t+e^{\left(\frac{t}{\lambda}\right)^k}\int_{t}^{\infty} e^{-\left(\frac{x}{\lambda}\right)^k}dx \tag{7}$$

and

$$t \leq E(D \mid D>t) \leq e^{\left(\frac{x}{\lambda}\right)^k}\lambda\Gamma\left(1+\frac{1}{k}\right) \tag{8}$$

For the parameters from Tab.1 we obtain results shown in Fig.3. The figure illustrates also the $E(D|D>t)$ function for the experimental data presented in the previous section.

The curves from Fig. 3 may be approximated with good accuracy as follows:

$$E(D \mid D>t) \approx 1.32C_v + t\sqrt{C_v} + 0.59 \tag{9}$$

Using this simple approximation we may establish a time-dependent insertion rate we were looking for. Suppose that the amount of remaining steganographic data to be sent at time t is $S_R(t)$. Then the insertion rate at time t may be expressed as

$$IR(t)=\frac{S_R(t)}{E(D \mid D>t)} \tag{10}$$

Finally, we may modify the above $IR(t)$ with a correction factor $CF<1$ to reflect the fact that the LACK procedure may decrease to some extent the QoS of the speech transmission and also to take account of the required robustness to steganalysis: $IR^*(t) = CF \cdot IR(t)$. This however is a very simplified solution of the problem.

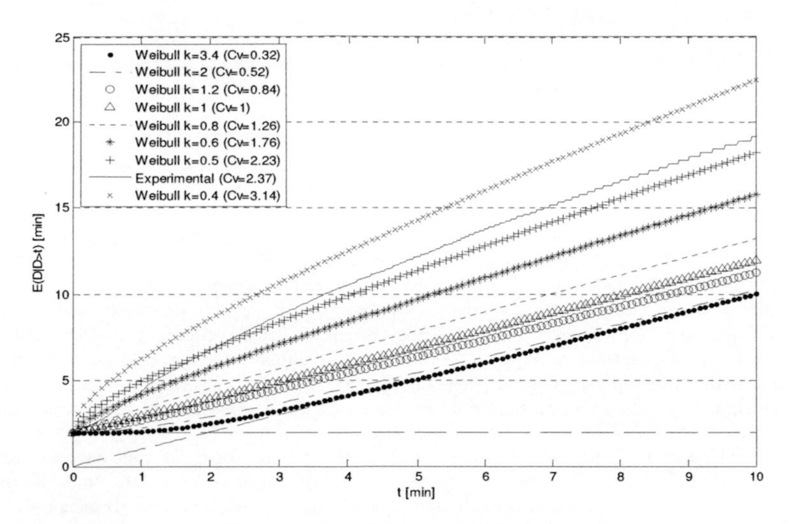

Fig. 3. *E(D|D>t)* for different Weibull distributions

Based on results presented in Fig. 3 and equation 10, for chosen Weibull distributions, the *IR(t)* values are as presented in Fig. 4:

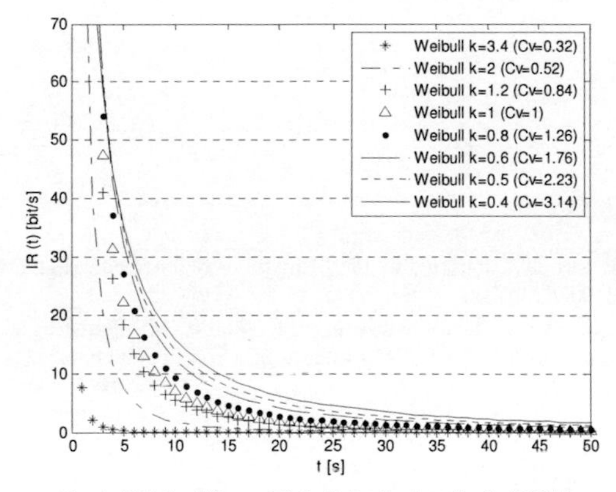

Fig. 4. *IR(t)* for different Weibull distributions for *S=1000* bits

As can be seen, call duration distributions with higher C_v yield higher *IR* values. In effect distributions with higher C_v allow to transmit more steganographic data.

4. CONCLUSIONS AND FUTURE WORK

The LACK steganographic method is a new idea which requires detailed performance evaluation. This paper is only an initial step in this direction. We have focused only on one aspect, namely the dependence of the procedure for inserting hidden data on the call duration probability distribution. It was shown that the insertion rate may be effectively made dependent on the current call duration time, and that this dependence can be expressed with good accuracy through the coefficient of variation of the call duration probability distribution. The derived formulae are simple and can be straightforwardly implemented. The effectiveness of the resulting procedure will depend on the accuracy of the estimated mean call duration and the coefficient of variation of the call duration.

The proposed procedure was made as simple as possible. A more sophisticated version of the procedure would require more detailed information about the call duration probability distribution, which might be too demanding considering the current limited experience with VoIP traffic. Nevertheless a theoretical research seems worthwhile. The authors have analyzed the problem of expressing the insertion rate function $IR(t)$ through a $P(D>x|D>t)$ distribution instead of the $E(D|D>t)$ function which was considered in the present paper. The results will be presented in a future paper.

Another task is to take into account the dependence of the $IR(t)$ function on constraints implied by QoS requirements and by the required resistance of LACK to steganalysis. In the present paper we have practically not considered this problem apart from introducing a correction factor into the $IR(t)$ which is clearly only an indication of the problem.

ACKNOWLEDGMENTS

- This research was partially supported by the Ministry of Science and Higher Education, Poland (grant no. 3968/B/T02/2008/34).
- The authors would like to thank R. Birke, M. Mellia, M. Petracca and D. Rossi from Politecnico di Torino for sharing details of their VoIP experimental data.

REFERENCES

[1] MAZURCZYK W., SZCZYPIORSKI K., *Sposób steganograficznego ukrywania i przesyłania danych w sieci telekomunikacyjnej,* Patent Application no. P-384940, 15 April 2008

[2] MAZURCZYK W., SZCZYPIORSKI K., *Steganography of VoIP Streams,* URL: http://www.arxiv.org/abs/0805.2938

[3] SCHULZRINNE H., CASNER S., FREDERICK R., JACOBSON V., *RTP: A Transport Protocol for Real-Time Applications,* IETF, RFC 3550, July 2003

[4] CHOI Y., LEE J., KIM T.G., LEE K.H, *Efficient QoS Scheme for Voice Traffic in Converged LAN,* Proceedings of International Symposium on Performance Evaluation of Computer and Telecommunication Systems (SPECTS'03), July 20-24, 2003, Montreal, Canada.

[5] MILOUCHEVA I., NASSRI A., ANZALONI A., *Automated Analysis of Network QoS Parameters for Voice over IP Applications,* D41 – 2nd Inter-Domain Performance and Simulation Workshop (IPS 2004).

[6] BARTOLI M., et al., *Deliverable 19: Evaluation of Inter-Domain QoS Modelling, Simulation and Optimization,* INTERMON-IST-2001-34123
URL: http://www.ist-intermon.org/overview/im-wp5-v100-unibe-d19-pf.pdf

[7] BIRKE R., MELLIA M., PETRACCA M., ROSSI D., *Understanding VoIP from Backbone Measurements,* 26th IEEE International Conference on Computer Communications (INFOCOM 2007), 6-12 May 2007, pp. 2027-35, ISBN 1-4244-1047-9

[8] NA S., YOO S., *Allowable Propagation Delay for VoIP Calls of Acceptable Quality,* In Proc. of First International Workshop, AISA 2002, Seoul, Korea, August 1-2, 2002, LNCS, Springer Berlin / Heidelberg, Volume 2402/2002, pp. 469-480, 2002

[9] MAZURCZYK W., SZCZYPIORSKI K., *Covert Channels in SIP for VoIP signalling,* In: Hamid Jahankhani, Kenneth Revett, and Dominic Palmer-Brown (Eds.): ICGeS 2008 - Communications in Computer and Information Science (CCIS) 12, Springer Verlag Berlin Heidelberg, Proc. of 4th International Conference on Global E-security 2008, London, United Kingdom, pp. 65-72, June 2008

[10] HANDLEY M., SCHULZRINNE H., ROSENBERG J., *SIP: Session Initiation Protocol,* IETF RFC 3261, June 2002

[11] BERK V., GIANI A., CYBENKO G., *Detection of Covert Channel Encoding in Network Packet Delays,* Tech. Rep. TR2005-536, Department of Computer Science, Dartmouth College, Nov. 2005
URL: http://www.ists.dartmouth.edu/library/149.pdf

[12] VENKATRAMAN B. R., NEWMAN-WOLFE R. E., *Capacity Estimation and Auditability of Network Covert Channels,* Proc. IEEE Symp. Security and Privacy, May 1995, pp. 186–98.

[13] COX I., KILIAN J., LEIGHTON F., SHAMOON T., *Secure spread spectrum watermarking for multimedia,* IEEE Transactions on Image Processing 6(12): 1997, pp. 1673–1687.

[14] CHEN B., WORNELL G. W., *Quantization index modulation: A class of provably good methods for digital watermarking and information embedding,* IEEE Trans. Info. Theory, vol. 47, no. 4, pp. 1423–1443, May 2001.

[15] ITU-T Recommendation, G.*711: Pulse code modulation (PCM) of voice frequencies,* November 1988.

Key words — Communication system traffic, traffic modeling, anonymous peer-to-peer networks.

Igor MARGASIŃSKI*, Michał PIÓRO*,†

ON TRAFFIC PERFORMANCE MEASURES FOR ANONYMOUS P2P NETWORKS

The modeling of anonymous systems is mostly aimed at the evaluation of the level of provided anonymity. However, the utility of anonymity solutions depends not only on their security, but also on necessary traffic overheads. In this paper, we have proposed an empirical, traffic performance model for anonymous P2P networks. We based our research on a well known information entropy measurement model for an estimate of secure configuration of anonymous forwarding path lengths. Then, we proposed a simulation-based methodology for an evaluation of latency and dynamics of the selected configurations. As an example we used the classical system called CROWDS. We evaluate traffic performance of CROWDS and show that the empirical analysis of the system has been analogous to results obtained analytically for representative boundary conditions.

1. INTRODUCTION

The anonymous P2P networks are the vital domain among the ever increasing variety of anonymous communications solutions. The P2P network architecture is extremely promising for an implementation of effective anonymous techniques since the distributed P2Ps eliminate the presence of trusted third party—privacy trustee. Concerning anonymity, we know a lot about robustness of anonymity systems and particular P2P solutions (e.g., [2], [3], [8], [14]). A mature information theoretic model for the evaluation of anonymity systems ([4], [10]) already is in use. Still, while the research in the filed of anonymous systems measuring is mainly focused on the evaluation of the level of provided anonymity, the modeling of traffic performance of anonymization methods is in our opinion neglected. Yet, the utility of a particular anonymous system results not only from its robustness against attackers and traffic analysis but also its necessary traffic overheads. Roughly speaking, the first question is: "how much anonymity do I get?" and the second, on which we focused our attention in this paper, is: "how much does it cost?" A research in this field was recently introduced in [13] and [5]. These works however, are devoted to the specific system (TOR [7]) and the presented measures in [5] are based on the gathered system's statistics. Still, we have to learn more about the traffic performance's "costs" and its modeling for anonymous networks. In [6] we introduced and used a model for the traffic performance analysis of the new anonymous P2P system (called P2PRIV) and to compare the system with the CROWDS [9]. This paper presents a detailed description of this model.

In 2002, two papers authored by Diaz *et al.* [4] and Serjantov *et al.* [10] simultaneously and independently introduced a new methodology for anonymity measurement based on Shannon's

* Institute of Telecommunications, Warsaw University of Technology
† Department of Electro and Information Technology, Lund University

information theory. The information entropy proposed by Shannon [11] describes the uncertainty associated with a random variable. It can be applied to anonymity quantification by assignment of probability of being an initiator of a specified action in the system to its particular users. Certainly, the sum of all these probabilities should equal 1. Then, based on the information provided by a system (shown by the system to an adversary) it is possible to measure the uncertainty of finding a real initiator. Let X be a discrete random variable, and

$$p_i = \Pr(X = i) \tag{1}$$

where i corresponds to the number of a particular subject/node/user of the analyzed system. For each user from the set of all system users N, the adversary can assign probability of being the initiator p_i. Then the entropy H will by described by

$$H = -\sum_{i=1}^{N} p_i \log_2(p_i) \ . \tag{2}$$

In Equation (2), a base-2 logarithm was used. Therefore the unit of the expressed entropy is a *bit*. This measure discloses a number of bits required for an adversary to explicitly point out the initiator. The adaptation of information theory seems to be a proper method for anonymity quantification as the uncertainty of the observer increases proportionally with H. The anonymity corresponds intuitively to a blending into the crowd of other similar subjects. Entropy measurement, as a quantification of a disorder of a structure of a system, gives a description of this phenomenon and an analytical instrument to measure how particular subjects of the system are distinguishable among the whole population of subjects. To enable the comparison between heterogeneous anonymity systems, a normalization of entropy was proposed by Diaz *et al.* [4]. Let $H_{\max}$ be the maximum entropy for a current number of system users. Entropy reaches the maximum value when all possible users are equiprobable

$$H_{\max} = \log_2(N) \ , \tag{3}$$

where N is the number of users. Then the normalized entropy – degree of anonymity is

$$d = \frac{H}{H_{\max}} = -\frac{\sum_{i=1}^{N} p_i \log_2(p_i)}{\log_2(N)} \ , \tag{4}$$

where p_i is the probability assigned to i subject, describing how likely this subject is perceived by the adversary as the initiator. This metric (4) describes the uncertainty of the system observer (the adversary) in finding the initiator of a specific action (for example sending a request for a specific content) and takes values from [0,1]. The minimum degree of anonymity depends on the purpose of anonymous system. However, in [4] acceptable normalized entropy was restricted to values higher or equaled to $d_{\min} = 0.8$.

In this paper we revised this model of anonymity systems in order to achieve the P2P environment usability and to reflect the practical capabilities of a P2P adversary. In effect, we provide methodology allowing finding a secure configuration of anonymous P2Ps (Section 2). In Section 3 we provide an empirical model of anonymous P2P traffic, capable of evaluating the traffic cost of selected secure configurations of anonymous P2P network. We focus our research on latency and dynamics measures. Latency constitutes a crucial factor for anonymous P2P traffic performance, since the basic common mechanism used to achieve network anonymization is traffic forwarding by a set of middleman nodes. Additionally, we concentrate our research on dynamics measures. Robustness against dynamically changing conditions, such as peers/content migration and traffic bursts introduced by publication of new data, demonstrates the suitability of particular networks anonymization techniques to P2P overlays. Finally, we conclude the paper and discuss a future work in Section 4.

2. ESTIMATION OF SECURE FORWARDING PATH LENGTHS

We will apply the entropy measurement model to quantify anonymity of the CROWDS system. The CROWDS can be used as an example of an attacked system, because it combines anonymity and performance with simplicity and reputability ([12], [15]). The adversary, who foists colluding nodes to the network, can assign probabilities of being the initiator to particular network nodes. Based on [9] and [4] the probability assigned to a predecessor of the first colluding node from the forwarding path is

$$p_{c+1} = 1 - p_f \frac{N-C-1}{N} . \tag{5}$$

The rest of nodes will have assigned equal probabilities since the adversary has no additional information about them. All colluding nodes should not be considered,

$$p_i = \frac{p_f}{N} . \tag{6}$$

According to (2), the entropy of the system will be described by

$$H_{paCROWDS} = \frac{N - p_f(N-C-1)}{N} \log_2 \left(\frac{N}{N - p_f(N-C-1)} \right) + \frac{p_f}{N}(N-C-1)\log_2 \left(\frac{N}{p_f} \right) . \tag{7}$$

The CROWDS maximum entropy is reached when all honest nodes are equiprobably recognized by the adversary as the initiator

$$H_{\max CROWDS} = \log_2(N-C) , \tag{8}$$

then the normalized entropy equals

$$d_{paCROWDS} = \frac{H_{paCROWDS}}{H_{\max CROWDS}} \tag{9}$$

$$d_{paCROWDS} = \frac{\left(N - p_f(N-C-1)\right)\log_2 \left(\frac{N}{N - p_f(N-C-1)} \right) + p_f(N-C-1)\log_2 \left(\frac{N}{p_f} \right)}{N \log_2 (N-C)} . \tag{10}$$

2.1. HOW LONG FORWARDING PATHS CAN DEGRADATE ANONYMITY

In the model proposed by [4] and [10] it is assumed that the adversary has yet colluding nodes among network nodes, which actively anonymize specified request (for example nodes from the forwarding random walk path of CROWDS system). Practically, the scenario may be different, and what is more, the probability that the adversary can find this group of nodes (referred to as an "active set") also determines the quality of the system anonymization.

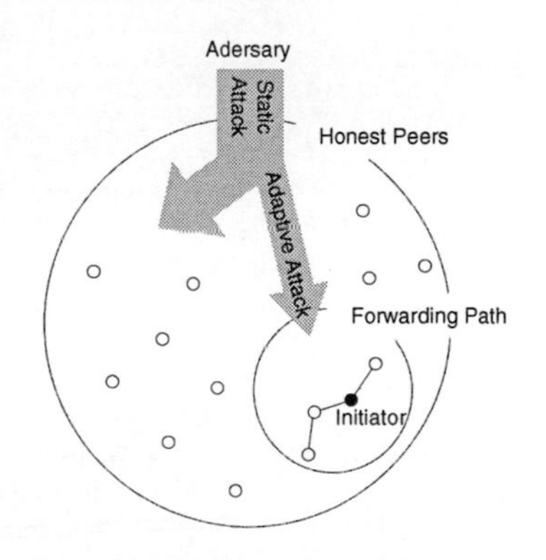

Fig. 1. Range of observation for static and the adaptive attacks.

The scenario described above should be called an adaptive attack, because it is assumed that the adversary is capable of adopting an area of its observation to the scope of activity of system users. It is important to also consider a more general case, where the adversary cannot be certain of a successive collaboration of proper active set. This uncertainlity should be quantified as well. For example, in the CROWDS system an increase of p_f parameter can easily increase the system active sets. However, if the active set is too numerous, even for large networks, collaboration of active nodes is simple – highly probable. Notice that when the adversary has no collaborating nodes among the active set then all the nodes of the system are equalprobable and the uncertainty of the adversary is maximized. The need for considering the impact of the observer uncertainty in finding proper active nodes was noticed in [4] and a weight mean formula was proposed to compute final d. This seems to be an intuitive attitude. However, the same results can be achieved by using a conditional entropy formula. In [1] a conditional entropy was proposed to describe the generalized scenario of the observation. The conditional entropy describes the entropy of a random variable X under condition of elimination of the entropy of other random variable Y. The conditional entropy expresses then the uncertainty associated with one aspect when the other aspect is certain.

This attack will be referred to as a static attack, as in this scenario the adversary "injects" colluding nodes in a static manner and cannot dynamically predict (adapt, like in previous scenario referred to as adaptive attack) which random nodes will actively anonymize the specified request. Let us return to the CROWDS example. A probability that none of the collaborating nodes can become a member of the random walk forwarding path is

$$p_r = \frac{N-C}{N}(1-p_f)\sum_{i=0}^{\infty}\left(\frac{N-C}{N}p_f\right)^i = 1-\frac{C}{N-p_f(N-C)}, \tag{11}$$

then entropy for passive-static attacks equals

$$H_{psCROWDS} = -\frac{C}{N-p_f(N-C)}\frac{N-p_f(N-C-1)}{N}\log_2\left(\frac{N-p_f(N-C-1)}{N}\right)+ \left(1-\frac{C}{N-p_f(N-C)}\right)p_f\frac{N-C-1}{N}\log_2\left(\frac{p_f}{N}\left(1-\frac{C}{N-p_f(N-C)}\right)\right), \tag{12}$$

and the normalized entropy is

$$d_{psCROWDS} = \frac{H_{psCROWDS}}{H_{\max CROWDS}} \tag{13}$$

$$d_{psCROWDS} = -\frac{C}{N-p_f(N-C)}\frac{N-p_f(N-C-1)}{N\log_2(N-C)}\log_2\left(\frac{N-p_f(N-C-1)}{N}\right)+ \left(1-\frac{C}{N-p_f(N-C)}\right)p_f\frac{N-C-1}{N\log_2(N-C)}\log_2\left(\frac{p_f}{N}\left(1-\frac{C}{N-p_f(N-C)}\right)\right). \tag{14}$$

We will analyze how p_f configuration impacts the entropy of the CROWDS system for both attack scenarios. It is important to remember that p_f value directly affects the forwarding path length – the number of network nodes actively involved in the anonymization process. Figure 2a and Figure 2b show the entropy of CROWDS in the full spectrum of available p_f configuration. We use maximum entropy $H_{\max CROWDS}$ (8) as a reference. First we analyzed three variants of network collaboration level (i) $C = 10\%$ – scenario usually considered in the state of the art; (ii) $C = 5\%$ – more realistic collaboration level for large and public access overlays; and (iii) $C = 20\%$ – scenario for small overlays. The next results, Figure 2c and Figure 3d, present the CROWDS entropy as a function of the number of collaborating nodes C, finally including the global collaboration.

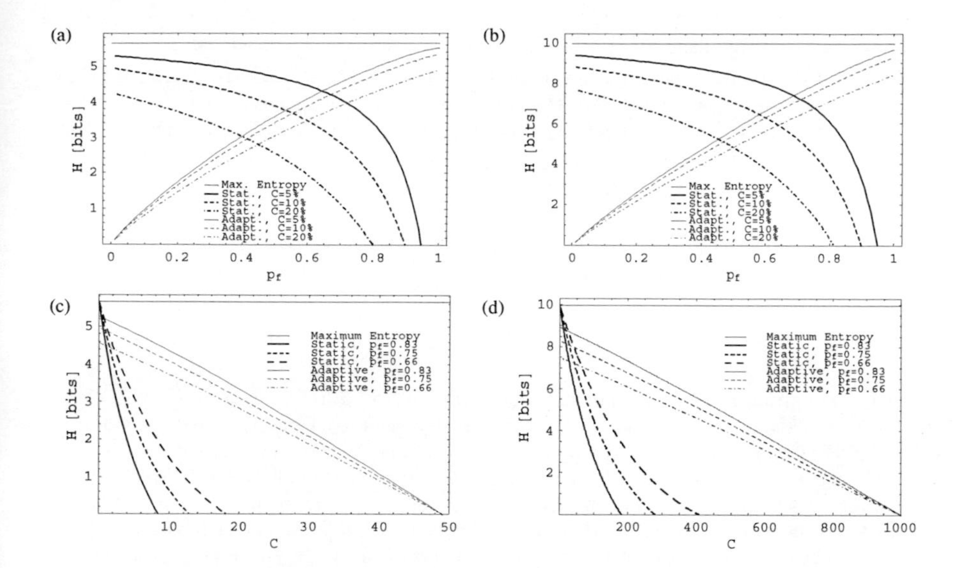

Fig. 2. Entropy of CROWDS, static and adaptive attacks; $N = 50$ (a), (b); $N = 100$ (c), (d).

The results for the two considered attack scenarios are substantially opposite. In the adaptive scenario low entropy, close to zero, is obtained for low p_f values, and high, close to maximum, entropy is achieved for large p_f. In the static scenario, the dependency is quite different and the best results are achieved for the lowest p_f values. As p_f grows, the entropy grows logarithmically smaller. In a small network this decrease (static attack) of entropy is slightly faster in contrast to the adaptive scenario where, in the small network, the decrease of the entropy is slower than for large overlays. This analysis shows that a set of nodes actively involved in the anonymization process should not be too numerous. Longer cascades not only impose larger traffic overheads, but can also make it easier for the adversary to become a member of the forwarding path. Especially in small networks the security of particular systems can be effectively compromised. In small networks, nodes from a forwarding path constitute a significant part of all network nodes. It should be reminded that the analyzed CROWDS system does not include mixing or asymmetric encryptions techniques for traffic analysis protection.

The results show that the secure configuration of forwarding path lengths should be adjusted to the size of the CROWDS overlay network. Taking into account large overlays, more typical for public P2P networks and a collaboration level C lower or equaled to 5%, the p_f configuration of CROWDS should be no lower than roughly 0.6 and no higher than about 0.8. Lower values than 0.6 expose the originator of a particular request against the adaptive adversary. Values higher than 0.8 compromise him to the static attacker. Then the mean forwarding path length of network random walk P equals

$$P = \sum_{i=2}^{\infty} i {p_f}^{i-2}(1-p_f) = \frac{p_f - 2}{p_f - 1} \; . \tag{15}$$

We can emphasize minimum and maximum secure mean path lengths of CROWDS:

$$P_{\min CROWDS} = 4 \; (\, p_f = 0.66\,) \, , \; P_{\max CROWDS} = 6 \; (\, p_f = 0.83\,) \, . \tag{16}$$

Forwarding paths shorter than $P_{\min CROWDS}$ cannot provide sufficient "crowd" of nodes, which actively anonymize the initiator. If the adversary is yet among this set of nodes there should be additional 2 other honest nodes. On the other hand, the forwarding paths longer than 6 nodes ($P_{\max CROWDS}$) becomes too easy to enter, because the size of the "crowd" provided by nodes passively anonymizing the active set becomes insufficient. Table 1 contains the summary of the results represented as a degree of anonymity obtained for promising values of p_f parameter.

Table 1. Degree of anonymity for CROWDS, static and adaptive attacks.

	$C = 5\%$			$C = 10\%$			$C = 20\%$		
pf	**0.66**	**0.75**	**0.83**	**0.66**	**0.75**	**0.83**	**0.66**	**0.75**	**0.83**
Static attack	0.69	0.63	0.58	0.51	0.41	0.32	0.25	0.11	0.0099
Adaptive attack	0.69	0.78	0.82	0.66	0.74	0.79	0.59	0.67	0.71

Our analysis confirmed that the optimum configuration for a realistic level of CROWDS collaboration is about $p_f = 0.75$ (recommended by the system authors). Concerning the collaboration level of $C = 5\%$, the degree of anonymity for CROWDS become close to $d_{\min}$.

As one can expect, the entropy largely depends on the number of colluding nodes. What is more, we can observe a significant impact of static observation on the anonymity of the CROWDS system. CROWDS entropy is significantly lower for static attacks than for adaptive scenarios. Still, both static and adaptive variants of attacks are vital to the anonymity analysis as they correspond to different aspects of the system's anonymity. The static scenario shows more realistic capabilities of the adversary and constitutes a critical point of view on the expansion of the system active sets. However, a more pessimistic attack – adaptive observation – is possible. Even though this scenario happens comparatively rarely, it is important to analyze its consequences.

3. AN EMPIRICAL MODEL OF ANONYMOUS P2P TRAFFIC

For the purpose of the complicated dynamic conditions analysis, we have created a peer-to-peer traffic simulation environment. Each peer of the simulated network retrieves the same algorithm suitable to a simulated protocol (for example Jondo of CROWDS). The simulator traces tasks for each symmetric peer independently. The peers can collect a specified content and can randomly leave the overlay – depriving other users of the content copies. Figure 3 shows an outline of the simulator architecture.

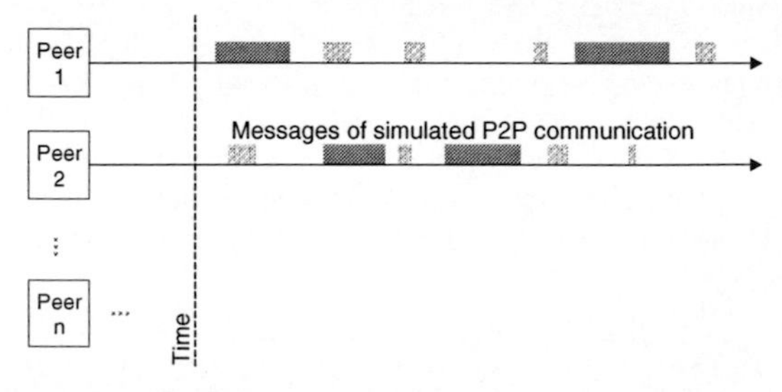

Fig. 3. Architecture of P2P traffic simulator.

The simulator allows setting-up of the following parameters of an overlay traffic:

- mean request arrival rate (λ) – intensity of requests for a content;
- mean download time between two neighbor peers (μ^{-1}) – resultant of a link throughput and an amount of sent data;
- mean request arrival rate for a specified content (λ_{Dc}) – intensity of requests for a specified (for example newly published) content;
- mean migration rate (λ_{Dm}) – intensity of users' leaving and arriving in the overlay network ("churn").

We use Poisson distribution to model a request arrival process. We simulated the CROWDS random walk algorithm to verify the simulation model.

3.1. LATENCY

Let average link throughput between peers be $B = 512$ kb/s and average file size of the shared content $V = 32$ MB. To analyze systems latency quoted as a mean download time we have computed series of simulation with 30 realizations each starting from the maximum request arrival rate per each node

$$\lambda_{max} = \frac{\mu_{min}}{P} = 0.0005\left[s^{-1}\right]. \tag{17}$$

where P is a mean random walk path length (15) and a μ_{min}^{-1} denotes download time between two directly connected nodes (referred in the rest of the work as FTP for simplification),

$$\mu_{min} = \frac{B}{V} = 0.002\left[s^{-1}\right]. \tag{18}$$

Figure 4 shows 95% confidence intervals and 25% to 75% quantiles (marked as boxes) surrounding the mean values of DT for the CROWDS system as the function of parameter λ^{-1}. In the analysis we have assumed CROWDS configuration which is accurate with recommendation of the CROWDS authors and with our analysis: $p_f = 0.75$. It means that the mean number of overlay nodes forwarding a single request equals 5 (15).

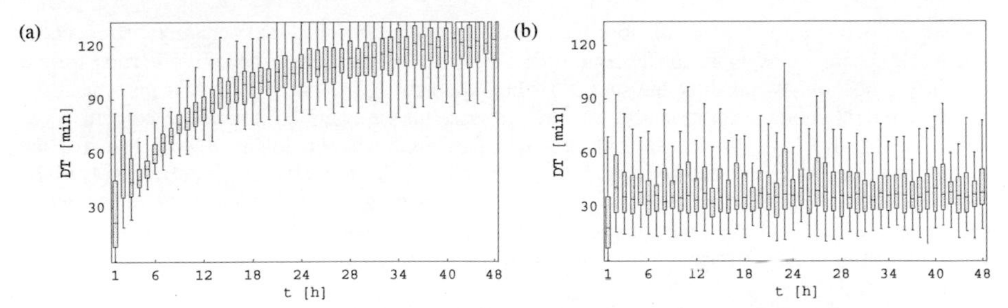

Fig. 4. Mean download time for CROWDS random walk in a period of two days after start of network operation, $N = 100$; maximum request arrival rate (a) and low request arrival rate (b).

Our first remark is that the simulated CROWDS random walk overlay works on the brink of stability for a analytical maximum request arrival rate. Simulations results for a lower request arrival rate showed the stable operation of the overlay.

Secondly, for a low arrival rate we can observe download time much above half an hour. The mean DT of a single file, measured in a period of two days of the system operation, equals 38.82 minutes. The analytically obtained mean time of a file transfer between neighbor nodes equals $\mu_{\min}^{-1} = 8.33$ minutes. In the CROWDS random walk each file is sent through a cascade of nodes. In the configuration of $p_f = 0.75$ the mean number of links in the cascade between source and destination peers equals 4. The simulation results show that DT for the CROWDS random walk is about 4.6 times longer than a FTP DT for a single link.

We repeated our simulation for $P_{\min CROWDS}$ and $P_{\max CROWDS}$ forwarding path lengths. Figure 5 shows results of CROWDS latency in a range of request arrival rate starting from the maximum traffic intensity. We can observe that as P grows, DT grows linearly higher. Addition of one node to the forwarding path results in an addition of $\mu_{\min}$ value to the final DT.

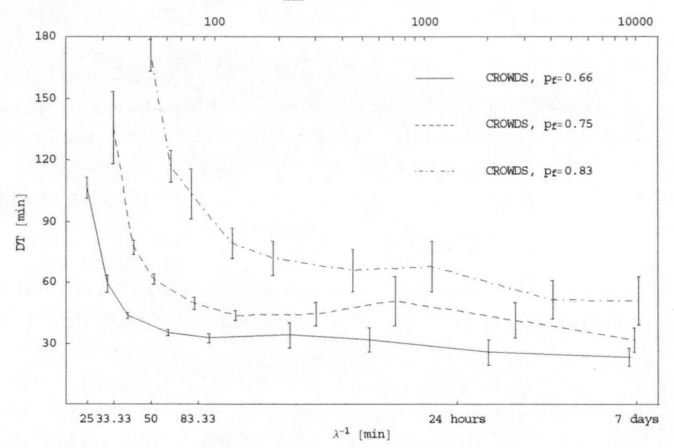

Fig. 5. Latency for CROWDS.

3.2. DYNAMICS

Next we will apply the simulation model to consider the mean DT characteristics under dynamically changing network traffic conditions. We will analyze system behavior starting form a new file publication. We analyze the scenario where the new and popular content is just shared by one of the overlay nodes. Additionally, we take into account the common practice of some users to connect to the overlay only for the purpose of a particular content download and to leave the network just after its successful delivery. Let D be the part of all requests which corresponds to the new file. We will take into account "selfish" users' behavior where simultaneously D percent of copies leaves the overlay network for each request. In this manner we have joined two parameters of the simulation which describe the popularity of a selected content and the migration of overlay users (λ_{Dc} and λ_{Dm}).

Figure 6 shows 95% confidence intervals and 25% to 75% quantiles (marked as boxes) surrounding the mean values of DT. We simulated the overlay under dynamically changing traffic conditions D . The first presented results (Figure 6a) were obtained for dynamics $D = 20\%$. We can observe that mean download time is slightly longer after a new content publications and returns to an initial level after about 5 hours. The further increase of dynamics to the value of $D = 30\%$ (presented on Figure 6b) caused an instability of the system – DT increases in the analyzed period of 36 hours. However, this scenario is highly pessimistic as about ⅓ of all user requests in the overlay network are directed towards the same, new content.

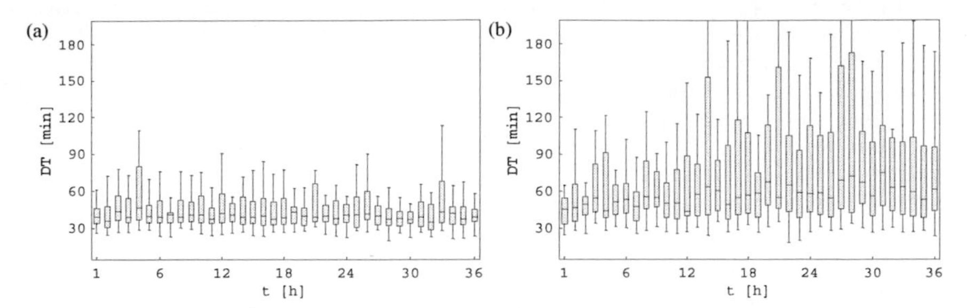

Fig. 6. Reaction of CROWDS to the new content publication, $N = 100$, low request arrival rate, (a) $D = 20\%$, (b) $D = 30\%$.

4. CONCLUSIONS

In this paper, we have proposed an empirical model for an evaluation of traffic performance of anonymous P2P networks. We used an information entropy measurement model ([4], [10]) for estimation of secure configuration of anonymous forwarding path lengths and proposed methodology for evaluation of latency and dynamics of the selected configurations. As an example of an evaluated system we used the classical solution called CROWDS.

The information theoretic model ([4], [10]) of system's anonymity has been revised in order to achieve the P2P environment usability and to reflect the practical capabilities of a P2P adversary. We analyzed the widely described in the state of the art literature, adaptive observation scenario, and also considered more realistic static attacks. Static attacks create awareness of boundless extension of forwarding path lengths. The long paths can impose not only larger traffic overheads, but can also make it easier for the adversary to become a member of set of nodes actively involved in the anonymization process. We have found that both static and adaptive attacks are vital to the anonymity analysis, as they correspond to different aspects of system protection. The static scenario shows more realistic capabilities of the adversary and it exemplifies a critical point of view on the expansion of forwarding paths. However, a more pessimistic attack – an adaptive observation – is possible. This scenario shows the effectiveness of the system's anonymity protection among network nodes actively involved in hiding on an initiator.

We have observed that latency of CROWDS is roughly a multiplication of the direct download time between two neighboring nodes and the forwarding path length. The observation of system's behavior under dynamically changing conditions shows that a stable operation of network random walk from CROWDS can be retained for dynamics lower or equal to 20%.

We have found that the empirical analysis of the network random walk algorithm (taken from CROWDS) is analogous to theoretical values, under boundary conditions. The analytical calculations included: available capacity (traffic intensity of the maximum request arrival rate) and the mean download time for a low-loaded network (low request arrival rate). For the analytically obtained maximum request arrival rate, we have observed that the simulated system works on the brink of stability. For a low traffic intensity the mean download time provided by the simulated system is slightly higher than the analytically obtained results that do not deal with delays introduced by network nodes.

Our future work will include a more detailed analysis of the impact imposed by various anonymous techniques on their traffic performance. An interesting goal is the consideration of overlay dynamics with simultaneous migration of many content resources. The traffic performance analysis, presented in this paper, has covered the following two elements: (i) the evaluation of the latency in the stable operation of simulated overlays; and (ii) the measures of the latency after the publication of a new file – we have considered a series of copying and removing of the single content resource. The future analysis should include modeling of anonymous traffic where many new files are copied and removed in the overlay network simultaneously. Additionally, this research can be based on a sociological analysis of users' habits, as this study can bring the simulation model closer to real networks. The next important direction of the future work is carrying out analytical traffic performance models for anonymous systems.

REFERENCES

[1] N. Borisov: *Anonymous Routing in Structured Peer-to-Peer Overlays*. PhD Thesis, UC Berkeley, 2005.

[2] R. Dingledine, M. J. Freedman, and D. Molnar, "The free haven project: Distributed anonymous storage service," in *H. Federrath, editor, Designing Privacy Enhancing Technologies: Workshop on Design Issues in Anonymity and Unobservability*. Springer-Verlag, LNCS 2009, July 2000.

[3] M. J. Freedman and R. Morris, "Tarzan: A peer-to-peer anonymizing network layer," in *9th ACM Conference on Computer and Communications Security*, Washington, DC, November 2002.

[4] C. Diaz, S. Seys, J. Claessens and B. Preneel, "Towards measuring anonymity," in *Roger Dingledine and Paul Syverson, editors, Proceedings of the Privacy Enhancing Technologies Workshop. Springer-Verlag*, LNCS 2482, April 2002.

[5] K. Loesing, W. Sandmann, C. Wilms, and G. Wirtz, "Performance Measurements and Statistics of Tor Hidden Services," in *Proceedings of the 2008 International Symposium on Applications and the Internet (SAINT)*, Turku, Finland, July 2008.

[6] I. Margasiński and M. Pióro, "A Concept of an Anonymous Direct P2P Distribution Overlay System," in *Proceedings of the 22nd IEEE International Conference on Advanced Information Networking and Applications (AINA2008)*, ISSN 1550-445X, ISBN 978-0-7695-3095-6, pp. 590-597, Ginowan, Okinawa, Japan, March 2008.

[7] D. N. Mathewson and P. Syverson, "Tor: The second generation onion router," in *Proceedings of the 13th USENIX Security Symposium*, August 2004.

[8] M. Rennhard and B. Plattner, "Introducing MorphMix: Peer-to-Peer based Anonymous Internet Usage with Collusion Detection," in *Proceedings of the Workshop on Privacy in the Electronic Society* (WPES 2002), Washington, DC, USA, November 2002.

[9] M. K. Reiter, A. D. Rubin, "Crowds: Anonymity for web transactions," *ACM Transactions on Information and System Security*, 1(1), June 1998.

[10] A. Serjantov and G. Danezis, "Towards an information theoretic metric for anonymity," in *Roger Dingledine and Paul Syverson, editors, Proceedings of the Privacy Enhancing Technologies Workshop*, San Diego, CA, April 2002. Springer-Verlag, LNCS 2482.

[11] C. E. Shannon, *A Mathematical Theory Of Communication*, the Bell System Technical Journal, Vol. 27, pp. 379–423 and pp. 623–656, 1948.

[12] V. Shmatikov, "Probabilistic analysis of anonymity," in *Proceedings of the Computer Security Foundations workshop (CSFW-15 2002)*, pages 119-128, Cape Breton, Nova Scotia, Canada, 24-26 June 2002. IEEE Computer Society.

[13] R. Snader and N. Borisov, "A Tune-up for Tor: Improving Security and Performance in the Tor Network," in *Proceedings of the Network and Distributed Security Symposium – NDSS '08*, February 2008.

[14] P. Tabriz and N. Borisov, "Breaking the collusion detection mechanism of MorphMix," in *Proceedings of the Privacy Enhancing Technologies Workshop (PET)*, June 2006.

[15] M. Wright, M. Adler, B. N. Levine, and C. Shields, "An analysis of the degradation of anonymous protocols," in *Proceedings of the Network and Distributed Security Symposium (NDSS'02)*, San Diego, California, 6-8 February 2002.

VI. CONGESTION AVOIDANCE AND CONGESTION CONTROL

congestion control, optimal control, connection-oriented communication networks

Przemysław IGNACIUK*, Andrzej BARTOSZEWICZ**

LQ OPTIMAL AND NONLINEAR CONGESTION CONTROL IN MULTI-SOURCE CONNECTION-ORIENTED COMMUNICATION NETWORKS

In this paper control theoretic approach is applied to design a new, optimal flow controller for multi-source connection-oriented communication networks. The networks are modeled as discrete time, n-th order systems. On the basis of the system state space description, a feedback control law is derived by minimizing linear quadratic (LQ) cost functional. Then, a nonlinearity is introduced into the basic controller operation, which allows to limit transmission rates growing beyond the network transfer capabilities. Closed-loop system stability of the obtained solution is demonstrated, and conditions for no data loss and full bottleneck link bandwidth utilization in the network are presented, and strictly proved.

1. INTRODUCTION

The need for fast and efficient transfer of large volumes of data and proliferation of services requiring specific flow treatment stimulates the research on new solutions in the area of congestion control in telecommunication networks, in particular connection-oriented ones [3]. So far various authors proposed the use of heuristic [7], proportional-derivative [8], stochastic [5], neural network [6], "on-off" [1], the Smith-predictor [1, 9] and the sliding-mode [4] based controllers to regulate the flow of data in connection-oriented networks. Since high throughput is of primary concern for fulfilling the traffic demand of the users and ensuring high profits of service providers, a successful control strategy should maximize the bandwidth usage and reduce the packet discards.

In this paper, we apply analytical methods to solve the congestion problem in connection-oriented networks serving multiple flows with different propagation delays. On the contrary to other results presented earlier in literature (e.g. [5]), we find the closed form solution of the linear-quadratic (LQ) optimization problem for a network with arbitrary propagation delays, with explicit consideration of the discrete-time nature of feedback information accessibility in the control process. The optimization procedure concentrates on the analytic solution of matrix Riccati equation for a network modeled as the n-th order dynamic system with delay. As the typical approaches for solving Riccati equations are mainly suitable for numerical implementations and systems with predefined dimensions [2], we propose an analytic method to obtain the controller parameters for the considered system. Moreover, since in real networks, the accepted transmission rate is subject to

*, ** Institute of Automatic Control, Technical University of Łódź, 18/22 Stefanowskiego St., 90-924 Łódź, Poland
* pignaciuk@poczta.onet.pl, ** andrzej.bartoszewicz@p.lodz.pl

various limitations, we propose a modified sub-optimal controller with rate saturation. Asymptotic stability of the system with the designed controller implemented, together with full bottleneck link utilization and data loss elimination in the network, is demonstrated. In consequence, the proposed control law allows for the maximum throughput in the communication system.

2. NETWORK MODEL

The connection-oriented network considered in this paper consists of data sources, intermediate nodes and destinations. The sources send data packets at the rate determined by the controller placed at a network node. The packets pass through a series of nodes operating in the store-and-forward mode without the traffic prioritization, to be finally delivered to the destination. However, somewhere on the transmission path a node is encountered, whose output link cannot handle the incoming flow. Consequently, congestion occurs, and packets accumulate in the buffer allocated for that link. We assume that the sources are persistent, and the congestion control problem can be solved only through an appropriate input rate adjustment.

We deal with m data flows, which pass through the bottleneck link. The feedback mechanism for the input rate regulation is provided by means of control units, emitted periodically by each source. These special units travel along the same path as data packets. However, unlike data packets, they are not stored in the queues at the intermediate nodes. Instead, once they appear at the node input link and the feedback information is incorporated, they are immediately transferred at the appropriate output port. As soon as control units reach destination, they are turned back to be retrieved at the origin, and to be used for the transfer speed adjustment, round trip time after they were generated.

The presented scenario is illustrated in Fig. 1. The sources send packets at discrete time instants kT, where T is the discretization period and $k = 0, 1, 2,...,$ in the amounts determined by the controller placed at the bottleneck node. After forward propagation delay T_{fp} packets from source p ($p = 1, 2,..., m$) reach the bottleneck node and are served according to bandwidth availability at the output link $d(kT)$. The remaining data accumulates in the buffer. The packet queue length in the buffer, which at time kT will be denoted as $y(kT)$, and its demand value y_d, are used to calculate the current amount of data $u(kT)$ to be sent by the sources. The m-th share of the total amount, $u(kT)/m$, is recorded as the feedback information in every management unit passing through the node. Once the control units from source p appear at the end system, they are turned back to arrive at their origin backward propagation delay T_{bp} after being processed by the congested node. Since management units are not subject to the queuing delays, round trip time $RTT_p = T_{fp} + T_{bp} = n_pT$, where n_p is a positive integer, remains constant for the duration of the connection. Without the loss of generality we may order the flows in the following way

$$RTT_1 \leq RTT_2 \leq \ldots \leq RTT_{m-1} \leq RTT_m \tag{1}$$

Denoting the number of flows whose round trip time equals jT ($j = 1, 2, \ldots, n_m$) by β_j we have $\sum_{j=1}^{n_m} \beta_j = m$, where some β_j may be equal to zero (including all β_j, for $j < n_1$).

The available bandwidth (the number of packets which may leave the bottleneck node at kT instant) is modeled as an *a priori* unknown, bounded function of time $d(kT)$. If there are packets ready for the transmission in the buffer, then the bandwidth consumed by the sources $h(kT)$ (the number of packets actually leaving the node) will be equal to the available bandwidth. Otherwise,

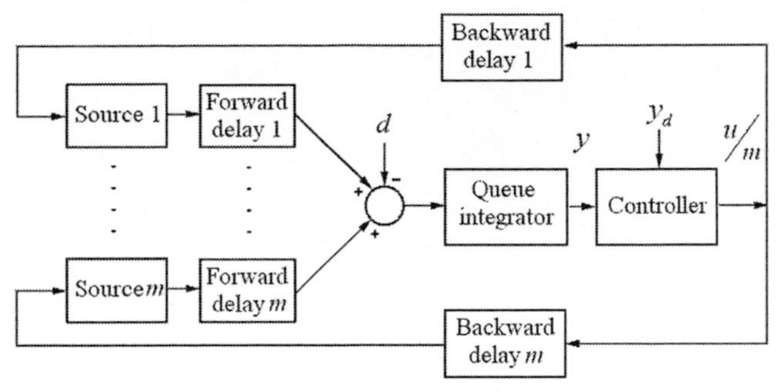

Fig. 1. Network model

the output link is underutilized and the exploited bandwidth matches the data arrival rate at the node. Thus, we may write

$$0 \leq h(kT) \leq d(kT) \leq d_{\max} \tag{2}$$

The rate of change of the queue length depends on the amount of arriving data and on consumed bandwidth h. Assuming that before the connection is established there are no packets in the buffer, the queue length $y(kT \geq 0)$ may be expressed as

$$y(kT) = \sum_{p=1}^{m} \sum_{i=0}^{k-1} \frac{1}{m} u\left(iT - RTT_p\right) - \sum_{i=0}^{k-1} h(iT) = \frac{1}{m} \sum_{j=1}^{n_m} \beta_j \sum_{i=0}^{k-j-1} u\left(iT\right) - \sum_{i=0}^{k-1} h\left(iT\right) \tag{3}$$

The network can also be described in the state space as

$$\begin{aligned} \mathbf{x}\left[\left(k+1\right)T\right] &= \mathbf{A}\mathbf{x}(kT) + \mathbf{b}u(kT) + \mathbf{o}h(kT) \\ y(kT) &= \mathbf{q}^T \mathbf{x}(kT) \end{aligned} \tag{4}$$

where $\mathbf{x}(kT) = [x_1(kT)\ x_2(kT)\ \ldots\ x_n(kT)]^T$ is the state vector with $x_1(kT) = y(kT)$, $\mathbf{A}$ is $n \times n$ state matrix, $\mathbf{b}$, $\mathbf{o}$, and $\mathbf{q}$ are $n \times 1$ vectors

$$\mathbf{A} = \begin{bmatrix} 1 & a_{n-1} & a_{n-2} & \cdots & a_1 \\ 0 & 0 & 1 & \cdots & 0 \\ \vdots & \vdots & \vdots & \ddots & \vdots \\ 0 & 0 & 0 & \cdots & 1 \\ 0 & 0 & 0 & \cdots & 0 \end{bmatrix} \quad \mathbf{b} = \begin{bmatrix} 0 \\ 0 \\ \vdots \\ 0 \\ 1 \end{bmatrix} \quad \mathbf{o} = \begin{bmatrix} -1 \\ 0 \\ \vdots \\ 0 \\ 0 \end{bmatrix} \quad \mathbf{q} = \begin{bmatrix} 1 \\ 0 \\ \vdots \\ 0 \\ 0 \end{bmatrix} \tag{5}$$

and system order $n = n_m + 1$. Since $\sum_{j=1}^{n_m} \beta_j = m$, the elements of the first row of matrix $\mathbf{A}$, $a_j = \beta_j/m$, satisfy $\sum_{j=1}^{n-1} a_j = 1$. We can present system (4) in the alternative form

$$\begin{cases} x_1[(k+1)T] = x_1(kT) + a_{n-1}x_2(kT) + \ldots + a_1 x_n(kT) - h(kT) \\ x_2[(k+1)T] = x_3(kT) \\ \vdots \\ x_{n-1}[(k+1)T] = x_n(kT) \\ x_n[(k+1)T] = u(kT) \end{cases} \tag{6}$$

where the desired system state $\mathbf{x_d} = [x_{d1}\ x_{d2}\ \ldots\ x_{dn}]^T = [x_{d1}\ 0\ 0\ \ldots\ 0]^T$, and $x_{d1} = y_d$ denotes the demand value of the first state variable.

3. PROPOSED CONTROL ALGORITHM

In this section, a discrete-time flow controller for the considered network is designed. We begin with stating the optimization problem. Afterwards, the matrix Riccati equation is solved to determine the controller parameters. Next, a modified control law is defined and its properties are discussed, and strictly proved.

3.1. CONTROLLER DESIGN

The aim of the control action is to bring the system state to the desired value without excessive control effort, or, alternatively speaking, to reduce the closed-loop system error $\mathbf{e}(kT) = \mathbf{x_d} - \mathbf{x}(kT)$ to zero using reasonable data flow rate. Therefore, we seek for a control $u_{opt}(kT)$, which minimizes the quality criterion expressed by the following cost functional

$$J(u) = \sum_{k=0}^{\infty} \left\{ u^2(kT) + w\left[y_d - y(kT)\right]^2 \right\} \tag{7}$$

where w is a positive constant applied to adjust the influence of the controller command and the output variable on the cost functional value. High w implies fast reaction to the changes of networking conditions, yet at the expense of large transmission rates, which may be difficult to be realized in practice. Small w, in turn, reduces excessive flow rates, but lowers the controller responsiveness to the bandwidth fluctuations.

According to [10], for time-invariant discrete-time system (4) the optimal control $u_{opt}(kT)$, minimizing criterion (7), can be presented as

$$u_{opt}(kT) = -\mathbf{g}\mathbf{x}(kT) + r \tag{8}$$

where

$$\begin{aligned} \mathbf{g} &= \mathbf{b}^T\mathbf{K}\left(\mathbf{I_n} + \mathbf{b}\mathbf{b}^T\mathbf{K}\right)^{-1}\mathbf{A} \\ r &= \mathbf{b}^T\left[\mathbf{K}\left(\mathbf{I_n} + \mathbf{b}\mathbf{b}^T\mathbf{K}\right)^{-1}\mathbf{b}\mathbf{b}^T - \mathbf{I_n}\right]\mathbf{k} \\ \mathbf{k} &= -\mathbf{A}^T\left[\mathbf{K}\left(\mathbf{I_n} + \mathbf{b}\mathbf{b}^T\mathbf{K}\right)^{-1}\mathbf{b}\mathbf{b}^T - \mathbf{I_n}\right]\mathbf{k} - w\mathbf{q}y_d \end{aligned} \tag{9}$$

and semipositive, symmetric matrix $\mathbf{K}$ ($\mathbf{K}^T = \mathbf{K} \geq 0$) is determined from the following Riccati equation

$$\mathbf{K} = \mathbf{A}^T\mathbf{K}\left(\mathbf{I}_n + \mathbf{b}\mathbf{b}^T\mathbf{K}\right)^{-1}\mathbf{A} + w\mathbf{q}\mathbf{q}^T \tag{10}$$

Classical approaches for solving (10), as suggested in literature (e.g. in [2]), are mainly suitable for numerical calculations and systems with predefined dimensions. However, in our case analytical solution needs to be found for a system of arbitrary order n. The method proposed in this paper involves iterative (analytical) substitution of $\mathbf{K}$ into the expression on the right hand side of (10) and comparison with its left hand side so that at each step the number of independent variables k_{ij}, where k_{ij} denotes the element in the i-th row and j-th column of $\mathbf{K}$, is reduced. Unfortunately, careful investigation of elements of matrix $\mathbf{K}$ and vector $\mathbf{g}$ obtained for the considered system, and confirmed by numerical computations, revealed a serious drawback of the LQ optimal control, namely, the controller generates negative transmission rates. To eliminate this deficiency and make the scheme applicable for real networks, we introduce the following modification. Notice that each coefficient $a_1, a_2, \ldots, a_{n-1}$ is smaller than or equal to one. Hence, in typical networks the majority of products $a_i a_j \ll 1$, and can be neglected in the calculations of elements of $\mathbf{K}$. As a result, instead of $\mathbf{K}$, we obtain its approximation $\hat{\mathbf{K}}$. In the further part of the paper we will demonstrate that the proposed modification allows to formulate a control law, which indeed guarantees that the established rate never falls below zero.

We begin solution of (10) with the general form of $\hat{\mathbf{K}}$

$$\hat{\mathbf{K}}_0 = \begin{bmatrix} \hat{k}_{11} & \hat{k}_{12} & \ldots & \hat{k}_{1n} \\ \hat{k}_{12} & \hat{k}_{22} & \ldots & \hat{k}_{2n} \\ \vdots & \vdots & \ddots & \vdots \\ \hat{k}_{1n} & \hat{k}_{2n} & \ldots & \hat{k}_{nn} \end{bmatrix} \tag{11}$$

After the first iteration we get

$$\hat{\mathbf{K}}_1 = \begin{bmatrix} \hat{k}_{11} & a_{n-1}\left(\hat{k}_{11} - w\right) & \hat{k}_{13} & \ldots & \hat{k}_{1n} \\ a_{n-1}\left(\hat{k}_{11} - w\right) & a_{n-1}{}^2\left(\hat{k}_{11} - w\right) & \hat{k}_{23} & \ldots & \hat{k}_{2n} \\ \hat{k}_{13} & \hat{k}_{23} & \hat{k}_{33} & \ldots & \hat{k}_{3n} \\ \vdots & \vdots & \vdots & \ddots & \vdots \\ \hat{k}_{1n} & \hat{k}_{2n} & \hat{k}_{3n} & \ldots & \hat{k}_{nn} \end{bmatrix} \tag{12}$$

Now we substitute $\hat{\mathbf{K}}_1$ into the right hand side of (10) and compare with its left hand side, which allows to represent elements $\hat{k}_{i3}$ (i = 1, 2, 3) in terms of $\hat{k}_{11}$. We proceed in this way until all elements of $\hat{\mathbf{K}}$ can be expressed as functions of $\hat{k}_{11}$ and system order n. Final form of $\hat{\mathbf{K}}$ is given as follows (for the sake of clarity we present only the upper part of symmetric matrix $\hat{\mathbf{K}}$)

$$\begin{bmatrix} \hat{k}_{11} & a_{n-1}\left(\hat{k}_{11}-w\right) & \sum_{j=1}^{2} a_{n-j}\left[\hat{k}_{11}-(3-j)w\right] & \cdots & \sum_{j=1}^{n-1} a_{n-j}\left[\hat{k}_{11}-(n-j)w\right] \\ \hat{k}_{12} & a_{n-1}{}^{2}\left(\hat{k}_{11}-w\right) & a_{n-1}\sum_{j=1}^{2} a_{n-j}\left[\hat{k}_{11}-(3-j)w\right] & \cdots & a_{n-1}\sum_{j=1}^{n-1} a_{n-j}\left[\hat{k}_{11}-(n-j)w\right] \\ \hat{k}_{13} & \hat{k}_{23} & \sum_{j=1}^{2} a_{n-j}\cdot\sum_{j=1}^{2} a_{n-j}\left[\hat{k}_{11}-(3-j)w\right] & \cdots & \sum_{j=1}^{2} a_{n-j}\cdot\sum_{j=1}^{n-1} a_{n-j}\left[\hat{k}_{11}-(n-j)w\right] \\ \vdots & \vdots & \vdots & \ddots & \vdots \\ \hat{k}_{1n} & \hat{k}_{2n} & \hat{k}_{3n} & \cdots & \sum_{j=1}^{n-1} a_{n-j}\cdot\sum_{j=1}^{n-1} a_{n-j}\left[\hat{k}_{11}-(n-j)w\right] \end{bmatrix} \tag{13}$$

Substituting $\hat{\mathbf{K}}$ as given by (13) into (10), and comparing the first elements of the resultant matrices, we get

$$\hat{k}_{11} = 1 + w\left(\sum_{j=1}^{n-1} ja_j + 1\right) - \left(\hat{k}_{11} - w\sum_{j=1}^{n-1} ja_j + 1\right)^{-1} \tag{14}$$

Solving for $\hat{k}_{11}$, we arrive at

$$\hat{k}_{11}^{'} = \left[w\left(2\sum_{j=1}^{n-1} ja_j + 1\right) - \sqrt{w(w+4)}\right]/2 \text{ and } \hat{k}_{11}^{"} = \left[w\left(2\sum_{j=1}^{n-1} ja_j + 1\right) + \sqrt{w(w+4)}\right]/2 \tag{15}$$

Only $\hat{k}_{11}^{"}$ guarantees that all the principal minors and the determinant of $\hat{\mathbf{K}}$ are nonnegative, and $\hat{\mathbf{K}}$ is semipositive definite. Therefore, it is the desired solution. Substituting $\hat{\mathbf{K}}$ into the first equation in set (9), we obtain the elements of vector $\hat{\mathbf{g}} = \mathbf{b}^T\hat{\mathbf{K}}\left(\mathbf{I_n} + \mathbf{b}\mathbf{b}^T\hat{\mathbf{K}}\right)^{-1}\mathbf{A}$,

$$\left[1 \quad a_{n-1} \quad (a_{n-1}+a_{n-2}) \quad \ldots \quad (a_{n-1}+a_{n-2}+\ldots+a_1)\right]\alpha \tag{16}$$

where $\alpha = \left(\sqrt{w(w+4)} - w\right)/2$. Applying similar iterative procedure, we determine constant $\hat{r}$ from (9) as $\hat{r} = \alpha y_d$, and the control law

$$u(kT) = \alpha\left[y_d - x_1(kT) - \sum_{j=2}^{n}\left(\sum_{i=1}^{j-1} a_{n-i}\right)x_j(kT)\right] = \alpha\left[y_d - x_1(kT) - \frac{1}{m}\sum_{j=2}^{n}\left(\sum_{i=1}^{j-1}\beta_{n-i}\right)x_j(kT)\right] \tag{17}$$

Alternatively, from (6), we can get the state variables x_j ($j = 2, 3, \ldots, n$) expressed in terms of the control signal generated by the controller at the previous $n-1$ samples as $x_j(kT) = u\left[(k-n+j-1)T\right]$. Substituting this into (17) with $x_1(kT) = y(kT)$, we get

$$u(kT) = \alpha\left\{y_d - y(kT) - \frac{1}{m}\sum_{j=1}^{n-1}\left(\sum_{i=1}^{j}\beta_{n-i}\right)u\left[(k-n+j)T\right]\right\} = \alpha\left\{y_d - y(kT) - \frac{1}{m}\sum_{j=1}^{n_m}\beta_j\sum_{i=k-j}^{k-1}u(iT)\right\} \tag{18}$$

which represents an LQ sub-optimal controller, and completes the design of flow control algorithm for the considered network. The obtained control law can be interpreted in the following way. The current rate assigned for the sources is proportional (with proportionality constant α) to the difference between the target and the current queue length decremented by the amount of data allowed to be transmitted by the sources within their corresponding propagation delays. In the further part of the paper, we will show that the proposed flow control strategy ensures the system stability, eliminates oscillations and data losses, and guarantees the maximum throughput in the considered network.

3.2. STABILITY ANALYSIS

The system is asymptotically stable, if all the roots of the characteristic polynomial of the closed-loop system state matrix $\mathbf{A_c} = [\mathbf{I_n} - \mathbf{b}(\mathbf{c}^T\mathbf{b})^{-1}\mathbf{c}^T]\mathbf{A}$ are located within the unit circle. The roots of the polynomial

$$\det\left(z\mathbf{I_n} - \mathbf{A_c}\right) = z^n + (\alpha - 1)z^{n-1} = z^{n-1}\left[z - (1-\alpha)\right] \tag{19}$$

are located inside the unit circle, if $0 < \alpha < 2$. Since for any w α is always positive and smaller than one, the system is stable, and no oscillations appear at the output.

3.3. MODIFIED CONTROLLER

Real telecommunication networks, due to various limitations, e.g. transmitter rate bounds, link capacity constraints, or processing power deficiency, cannot accommodate flows of arbitrary intensity. Therefore, in this section we propose a new controller, which generates transmission rates not exceeding the upper limit $u_{\max} > 0$. The modified control law is defined as

$$u_{sat}(kT) = \min\left\{\alpha\left[y_d - y(kT) - \frac{1}{m}\sum_{j=1}^{n_m}\beta_j\sum_{i=k-j}^{k-1}u_{sat}(iT)\right], u_{\max}\right\} = \min\left\{u(kT), u_{\max}\right\} \tag{20}$$

Consequently, for any time instant kT the determined transfer rate $u_{sat}(kT) \le u(kT)$ and is limited from above by the maximum value $u_{\max}$.

In the further part of the paper, the properties of the proposed algorithm will be formulated as three theorems. The first theorem states that the transfer speed assigned for the sources is always nonnegative, which is a crucial constraint of any flow control strategy to be implemented in real telecommunication systems. The second proposition defines the memory requirements for the buffers at the bottleneck node which guarantee the loss-free transmission. The third theorem shows how to select the demand queue length in order to obtain full resource usage in the network.

Theorem 1. If controller (20) is applied, then the transmission rate is always nonnegative i.e.

$$\underset{k\ge 0}{\forall}\; u_{sat}(kT) \ge 0 \tag{21}$$

Proof. For $k = 0$ we have $u_{sat}(0) = \alpha y_d \ge 0$. Since by the definition the established transmission rate is positive whenever it is equal to $u_{\max}$, it suffices to show that (21) is satisfied for any $k > 0$ when $u_{sat}(kT) = u(kT)$. Using (3) and (20) we obtain

$$
\begin{aligned}
u_{sat}(kT) = u(kT) &= \alpha\left[y_d - y(kT) - \frac{1}{m}\sum_{j=1}^{n_m}\beta_j\sum_{i=k-j}^{k-1}u_{sat}(iT)\right] \\
&= \alpha\left\{y_d - y\left[(k-1)T\right] - \frac{1}{m}\sum_{j=1}^{n_m}\beta_j u_{sat}\left[(k-j-1)T\right] + h\left[(k-1)T\right] - \frac{1}{m}\sum_{j=1}^{n_m}\beta_j\sum_{i=k-j}^{k-1}u_{sat}(iT)\right\} \\
&= u\left[(k-1)T\right] - \alpha u_{sat}\left[(k-1)T\right] + \alpha h\left[(k-1)T\right]
\end{aligned}
\tag{22}
$$

Since $\alpha \in (0, 1]$, $u_{sat}(kT) \leq u(kT)$ and $h(kT) \geq 0$, $u_{sat}(kT)$ is nonnegative. This ends the proof. □

Theorem 2. If controller (20) is applied to regulate the flow of data in the considered network, then the queue length is always upper-bounded, i.e.

$$
\forall_{k \geq 0} \; y(kT) \leq y_d \tag{23}
$$

Proof. The bottleneck node buffer is empty for any $kT \leq RTT_1$. Hence, it suffices to show that the proposition is satisfied for any $k \geq n_1 + 1$. The transmission rate is nonnegative and $u_{sat}(kT) \leq u(kT)$, hence

$$
\alpha\left[y_d - y(kT) - \frac{1}{m}\sum_{j=1}^{n_m}\beta_j\sum_{i=k-j}^{k-1}u_{sat}(iT)\right] = u(kT) \geq u_{sat}(kT) \geq 0 \tag{24}
$$

which implies $y(kT) \leq y_d - \sum_{j=1}^{n_m}\beta_j / m \sum_{i=k-j}^{k-1} u_{sat}(iT)$. Again, since $u_{sat}(kT) \geq 0$, the queue length $y(kT)$ does not grow above y_d. This completes the proof of Theorem 2. □

Theorem 3. If controller (20) is applied, $u_{max} > d_{max}$, and the demand queue length satisfies

$$
y_d > d_{max}\left(\frac{1}{mT}\sum_{p=1}^{m} RTT_p + \frac{1}{\alpha}\right) \tag{25}
$$

then for any $kT \geq nT + T_{max}$, where $T_{max} = y_d / (u_{max} - d_{max})$, the queue length is strictly positive.

Proof. The theorem assumption implies that we deal with the time instants $kT > nT$, which means that packets from all the sources already contribute to the queue length build-up. Considering signal $u(kT)$ as given by (18), we may distinguish two cases: the situation when $u(kT) < u_{max}$, and the circumstances when $u(kT) \geq u_{max}$.

Case 1: First, we analyze the situation when $u(kT) < u_{max}$. Directly from the definition of function u, we obtain

$$
y(kT) > y_d - u_{max} / \alpha - \frac{1}{m}\sum_{j=1}^{n_m}\beta_j\sum_{i=k-j}^{k-1}u_{sat}(iT) \tag{26}
$$

The maximum rate equals u_{max}. Therefore,

$$
y(kT) > y_d - u_{max} / \alpha - \frac{u_{max}}{m}\sum_{j=1}^{n_m} j\beta_j = y_d - u_{max} / \alpha - \frac{u_{max}}{mT}\sum_{p=1}^{m} RTT_p \tag{27}
$$

Assumption (25) together with the fact that $u_{max} > d_{max}$ guarantee that $y(kT) > 0$.

Case 2: Now, let us study the situation when $u(kT) \geq u_{\max}$. First, we find the last moment $k_0T < kT$ when signal u was smaller than $u_{\max}$. It comes from Theorem 2 that the queue length never exceeds the value of y_d. Furthermore, the packet depletion rate is limited by $d_{\max}$. Thus, the maximum period of time $T_{\max}$, during which the controller may continuously set rate $u_{\max}$ for the sources, can be determined as $T_{\max} = y_d / (u_{\max} - d_{\max})$. This means that instant k_0T indeed exists and $k_0T \geq kT - T_{\max} \geq nT$. With analogy to (26) and (27) we get $y(k_0T) > 0$. Then, the queue length at time instant kT can be expressed as

$$
\begin{aligned}
y(kT) &= y(k_0T) + \frac{1}{m}\sum_{j=1}^{n_m}\beta_j \sum_{i=k_0-j}^{k-j-1} u_{sat}(iT) - \sum_{j=k_0}^{k-1} h(jT) \\
&> y_d - u_{\max}/\alpha - \frac{1}{m}\sum_{j=1}^{n_m}\beta_j \sum_{i=k_0-j}^{k_0-1} u_{sat}(iT) + \frac{1}{m}\sum_{j=1}^{n_m}\beta_j \sum_{i=k_0-j}^{k-j-1} u_{sat}(iT) - \sum_{j=k_0}^{k-1} h(jT) \\
&= y_d - u_{\max}/\alpha - \frac{1}{m}\sum_{j=1}^{n_m}\beta_j \sum_{i=k-j}^{k-1} u_{sat}(iT) + \frac{1}{m}\sum_{j=1}^{n_m}\beta_j \sum_{i=k_0}^{k-1} u_{sat}(iT) - \sum_{j=k_0}^{k-1} h(jT)
\end{aligned}
\tag{28}
$$

In the considered case, for all $k > k_0$ the controller allocates the maximum flow rate value. Hence, using assumption (25), we get

$$
y(kT) > y_d - u_{\max}/\alpha - \frac{u_{\max}}{mT}\sum_{p=1}^{m} RTT_p + u_{\max}(k-k_0) - d_{\max}(k-k_0) > (u_{\max} - d_{\max})(k-k_0) \tag{29}
$$

Since $u_{\max} > d_{\max}$ and $k > k_0$, $y(kT) > 0$. This ends the proof. □

4. SIMULATION RESULTS

We verify the properties of the proposed strategy in a simulation scenario. The model of the network is constructed according to the description given in Section 2. Four connections ($m = 4$) participate in the flow regulation process. They are characterized by the following propagation delays $RTT_1 = 3T$, $RTT_2 = 6T$, $RTT_3 = 6T$, $RTT_4 = 9T$, where $T = 10$ ms represents the discretization period. Hence, the system order $n = n_4 + 1 = 10$, $a_3 = 1/4$, $a_6 = 1/2$ and $a_9 = 1/4$ ($a_1 = a_2 = a_4 = a_5 = a_7 = a_8 = 0$). The maximum available bandwidth $d_{\max}$ is set as 50 packets. The bandwidth actually available for the data transfer at consecutive time instants is shown in Fig. 2 (curve a). We can see from the graph that function d experiences sudden changes of large amplitude, which reflects the most rigorous networking conditions. The maximum rate is adjusted to 55 > 50 packets. Setting $w = 1$ (in this case output error elimination is given no preference over control signal reduction), we obtain gain $\alpha = 0.618$. The demand queue length necessary to guarantee full bandwidth usage is adjusted to 420 > 419 packets according to the assumptions of Theorem 3. As we can see from the plot in Fig. 2 (curve b) the transmission rate generated by the controller is always nonnegative and limited, and quickly converges to stationary values (after the available bandwidth changes) without overshoots and oscillations. The queue length evolution illustrated in Fig. 3 clearly shows that the buffer of size $y_d = 420$ packets is not overflowed, and it is never entirely depleted. This implies that the packet loss is eliminated and the available bandwidth is entirely used by the data traffic.

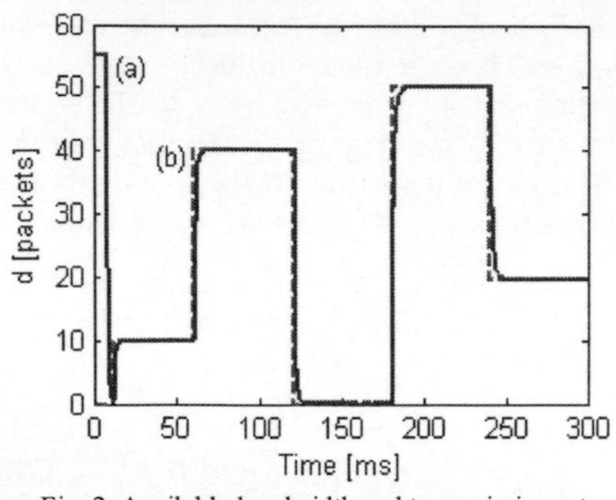

Fig. 2. Available bandwidth and transmission rate

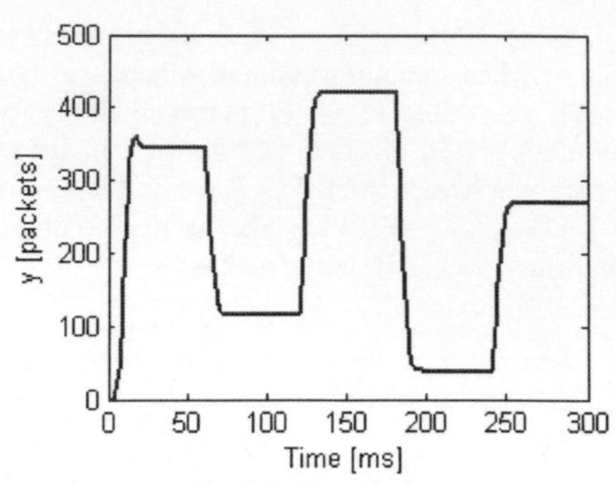

Fig. 3. Buffer occupancy

5. CONCLUSIONS

In this paper, a new method of the flow controller design for connection-oriented networks using control theoretic tools was proposed. The design procedure focused on the minimization of the LQ cost functional and solution of the resultant matrix Riccati equation for the n-th order discrete-time system, representing a multi-source network. For the obtained control law closed-loop system stability was demonstrated, and conditions for data loss elimination and full available bandwidth utilization were formulated, and strictly proved. The rates generated by the controller are always nonnegative and upper-bounded. Moreover, the controller dynamics can be easily tuned by the gain and rate saturation adjustment (within the defined constraints) without the risk instability.

REFERENCES

[1] BARTOSZEWICZ A., *Nonlinear flow control strategies for connection-oriented communication networks*, Proceedings of the IEE – Part D: Control Theory and Applications, Vol. 153 (2006), No. 1, pp. 21-28.

[2] CHU D., LIN W., TAN R., *A numerical method for a generalized algebraic Riccati equation*, SIAM Journal on Control and Optimization, Vol. 45 (2006), No. 4, pp. 1222-1250.

[3] HABIB I., JAJSZCZYK A., AWDUCHE D., *Advances in control and management of connection-oriented networks - guest editorial*, IEEE Communications Magazine, Vol. 44 (2006), No. 12, pp. 58-59.

[4] IGNACIUK P., BARTOSZEWICZ A., *Linear quadratic optimal discrete time sliding mode controller for connection oriented communication networks*, IEEE Transactions on Industrial Electronics, Vol. 55 (2008), No. 11,.

[5] IMER O. C., COMPANS S., BASAR T., SRIKANT R., *Available bit rate congestion control in ATM networks*, IEEE Control Systems Magazine, Vol. 21 (2001), No. 1, pp. 38-56.

[6] JAGANNATHAN S., TALLURI J., *Predictive congestion control of ATM networks: multiple sources/single buffer scenario*, Automatica, Vol. 38 (2002), No. 5, pp. 815-820.

[7] KALYANARAMAN S., JAIN R., FAHMY S., GOYAL R., VANDALORE B., *The ERICA switch algorithm for ABR traffic management in ATM networks*, IEEE/ACM Transactions on Networking, Vol. 8 (2000), No. 1, pp. 87-98.

[8] LENGLIZ I., KAMOUN F., *A rate-based flow control method for ABR service in ATM networks*, Computer Networks, Vol. 34 (2000), No. 1, pp. 129-138.

[9] MASCOLO S., *Smith's principle for congestion control in high-speed data networks*, IEEE Transactions on Automatic Control, Vol. 45 (2000), No. 2, pp. 358-364.

[10] ZABCZYK J., *Remarks on the control of discrete-time distributed parameter systems*, SIAM Journal on Control, Vol. 12 (1974), No. 4, pp. 721-735.

Number of handoffs, call holding time, cell residence time, wireless networks

Wolfgang Bziuk, Said Zaghloul, Admela Jukan [*]

A New Framework for Characterizing the Number of Handoffs in Cellular Networks

The handoff process was widely studied in the literature in the past decade. While significant efforts addressed the mean handoff rate, few attempted to characterize the distribution of the number of handoffs under the assumption that the call and the cell residence times are generally distributed. In this paper, we propose a new analytical framework to obtain the probability generating function of the number of handoffs. Our work covers three generic session behaviours, including new, roaming, and location triggered calls and thus facilitates the calculation of the mean number of handoffs for such session types. When the distributions have rational Laplace transforms, we obtain closed form solutions that are simple-to-compute and hence provide cellular network designers and researchers with deeper insights into the performance of mobility aware wireless protocols. Our numerical examples for the Hyper-Erlang call durations show the practicability of our framework.

1. INTRODUCTION

Numerous research efforts were dedicated to the study of the handoff process in wireless cellular networks in the past decade. The characterization of the handoff process arises as an important problem ([1], [2], [9]) as it directly impacts the performance of signalling protocols, call admission control, and the offered quality of service. As wireless networks evolve, network operators are able to offer different services for home and roaming users. Consequently, the standard research assumptions of exponential call[1] holding time (CHT) and cell residence time (CRT) [3] can lead to intolerable inaccuracies and hence advanced models with more general distributions became a necessity [4]. In this context, distribution models such as hyper-Erlang [1] and SOHYP distributions transforms were utilized to fit field data [2]. These distribution models are desirable for their analytical tractability as they have rational Laplace transforms and hence can be used to study system design parameters such as the channel holding time [5], the call blocking [6], and the handoff probabilitics [7]. In addition, recent work in [8] utilized the concept of the equilibrium renewal process to derive the probability generating function (PGF) for the number of handoffs for an Erlang distributed CHT and general CRT. Non-closed form results were also given for distributions with no rational Laplace transforms, e.g. a lognormal CHT.

In this paper, we present a new framework which uses general distribution models to obtain analytical results for the PGF of the number of handoffs. Our work extends the framework in [5,7]

[*] Institute of Computer and Communication Network Engineering, Hans Sommer-Str. 66, Braunschweig, D-38106, Germany, {bziuk, zaghloul, jukan}@ida.ing.tu-bs.de

[1] The terms call and session are used exchangably in this article.

by deriving the PGF of the number of handoffs allowing a simple evaluation of the moments of the number of handoffs. In this regard, we not only evaluate the PGF of the number of handoffs, but we also consider different session behaviours. While the existing work only addresses the number of handoffs for new calls, i.e., calls generated within cells belonging to the operator under consideration, we specifically study two more scenarios: (i) ongoing calls for roaming users, and (ii) location triggered calls for users entering a service domain. Our work offers the PGF for generally distributed CHT and CRT only by requiring the existence of the Laplace transform of the CHT and the CRT distributions. Furthermore, when the Laplace transforms are rational, we offer closed form solutions. Therefore, our work offers a rather simple way to obtain such distributions by avoiding the nth order differentiation procedure used in [8]. In addition, we illustrate the generality and the practicality of our framework by applying it to the newly proposed call scenarios and provide numerical examples using the Hyper-Erlang distribution. Finally, we show that unlike in previous work, our framework is more comprehensive as it includes the case where the residence time is constant[2]. This case may be of particular interest in highway or train environments.

2. PROBABILITY DISTRIBUTION OF THE NUMBER OF HANDOFFS

2.1. MODEL DESCRIPTION

Researchers have tackled problems involving the characterization of the handoff process in different ways. For instance, Fang [7]'s used the probability that a call incurs at least one more handoffs during its remaining lifetime. However, a more commonly used definition is based on the probability that a call includes a stationary number of handoffs $N(t)=N$ (e.g. see [8] and the references inside). In this paper, we follow the later definition. To do so, let us first define the session model in terms of a set of general random variables. For simplicity and without loss of generality, we assume that calls are always resumed after handoffs, and hence we do not consider the effect of blocking in case of a handoff. The inclusion of blocking is straight forward and can be carried out as in [6]. The call holding time (CHT) S follows the distribution $f_S(t)$ and has a mean of $E\{S\}$. We call the portion of the call holding time while being serviced by a given cellular operator's network, the effective call holding time (ECHT) (S_E). S_E has a general distribution $f_{SE}(t)$, Laplace transform $f_{SE}^*(s)$ and a mean of $E\{S_E\}$. For incoming roaming users, the CHT distribution can be described by the residual of the S (see [9]). The residence time of a user in cell $i, i=1,2,...$ is denoted by R_i and has general distribution $f_{R_i}(t)$, Laplace transform $f_{R_j}^*(s)$, and a mean of $E\{R_i\}$. To simplify the notation let us introduce the sum over the first k residence times as $R(k)$, with density and distribution functions of $f_k(r)$ and $F_k(r)$ respectively.

$$R(k)=\sum_{j=1}^{k} R_j, \quad k\geq 1, \quad R(0)=0 \tag{1}$$

Assuming that the sequence of residence times incurred through a call duration are independent the Laplace transform of the density $f_k(r)$ is,

$$f_k^*(s)=\prod_{j=1}^{k} f_{R_j}^*(s). \tag{2}$$

To simplify the notation it is also assumed that the residence times are identically distributed

[2] Even though this case is easy to evaluate directly, the existing work does not cover it.

(i.e., $R_i = R$ and $E\{R_i\} = E\{R\}$). Now that we have defined the relevant random variables, we proceed to describe three different models for the session behaviour as shown in Fig. 1.

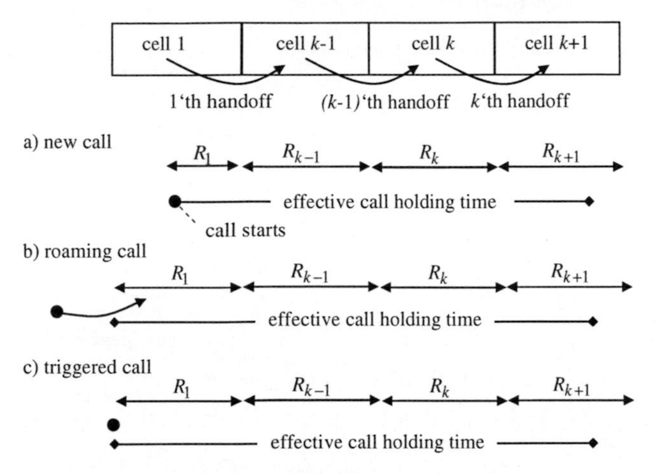

Fig. 1. Different session models and their relationship to the ECHT and CRT

Model 1: In this case, we assume that a user starts his session (i.e., new call) anywhere inside the first cell as shown in Fig.1a). The first residence time incurred by the user is given by the residual $\tilde{R}$ of the cell residence time R, hence we have

$$f_{R_1}^*(s) = f_{\tilde{R}}^*(s) = \frac{1}{sE(R)}\left(1 - f_R^*(s)\right) \tag{3}$$

$$f_k^*(s) = \frac{1}{sE(R)}\left(1 - f_R^*(s)\right)\left(f_R^*(s)\right)^{k-1}, \ k \geq 1. \tag{4}$$

It follows that in this model the effective call holding time $f_{SE}(t)$ is equal to the call holding time,

$$f_{SE}^*(s) = f_S^*(s). \tag{5}$$

Model 2: Here we assume that a user starts his session outside the area of interest, and then roams into the first cell as shown in Fig.1b). In this case, the first residence time is given by cell residence time R, while the effective call holding time is now determined by the residual of the call holding time. Thus, the model is described by the following set of equations:

$$f_{R_1}^*(s) = f_R^*(s) \quad \text{and} \quad f_k^*(s) = \left(f_R^*(s)\right)^k \tag{6}$$

$$f_{SE}^*(s) = f_{\tilde{R}}^*(s) = \frac{1}{sE(S)}\left(1 - f_S^*(s)\right). \tag{7}$$

Model 3: In this case, we assume that a user starts his session just in the moment, when he crosses the boundary of the first cell. This is common for cases where a session is initiated by the wireless system when a user crosses a boundary area (see Fig. 1c). In this case, the first residence time is given by cell residence time, but the ECHT is identical to the call holding time. The model is described by equations (5) - (6).

2.2. HANDOFF PROBABILITY

We will shortly review the work done in [7], to see the general relationship between the handoff probability and the probability distribution of the number of handoffs derived below. Consider Fig. 1a), where a call has already done $k-1$ handoffs. Thus, his effective call holding time must satisfy $R(k-1)=R_1+R_2+...+R_{k-1}\le S_E$. Now let us define $P_H(k)$ as the probability that a call has already incurred k-1 handoffs and that its effective call holding time S_E is large enough to include at least one more handoff, that is $R(k)=R_1+R_2+...+R_k\le S_E$. With a slight modification of the definition given in [7] this probability can be expressed as,

$$P_H(k)=P\{R(k)\le S_E|R(k-1)\le S_E\}=\frac{P\{R(k)\le S_E\}}{P\{R(k-1)\le S_E\}}=\frac{G(k)}{G(k-1)},k\ge 1 \tag{8}$$

where we have set

$$G(k)=P\{R(k)\le S_E\}=P\{S_E\ge R(k)\} \tag{9}$$

and $G(0)=P\{0\le S_E\}=1$. The probability that a call includes no handoffs is given as,

$$P\{S_E<R_1\}=1-P\{R_1\le S_E\}=1-G(1). \tag{10}$$

Assuming that both R and S_E are statistically independent, the probability $G(k)$ can be calculated as,

$$G(k)=P\{R(k)\le S_E\}=\int_{t=0}^{\infty}\int_{r=0}^{t}f_k(r)f_{SE}(t)\,dr\,dt=\int_{t=0}^{\infty}F_k(t)f_{SE}(t)\,dt \tag{11}$$

2.3 HANDOFF- PROBABILITY DISTRIBUTION AND MOMENTS

The goal of this paper is to derive the stationary probability distribution of the stationary number of handoffs N. From Fig. 1a) the probability of making exactly $k\ge 1$ handoffs, (i.e., the session is stopping within cell $k+1$), is given by

$$\begin{aligned}P\{N=k\}&=P_{HO}(k)=P\{R(k)\le S_E<R(k+1)\}=P\{S_E<R(k+1)\}-P\{S_E<R(k)\}\\&=P\{S_E\ge R(k)\}-P\{S_E\ge R(k+1)\}=G(k)-G(k+1)\end{aligned} \tag{12}$$

For $k=0$ we have in accordance with (10)

$$P_{HO}(0)=P\{S_E<R(1)\}=1-G(1) \tag{13}$$

Now that we have characterized the distribution of the number of handoffs, we proceed to evaluate its probability density function. This normally requires to derive (11), however, to avoid the solution of the integral, the method presented in [6] is applied using the inverse Laplace transform of $F_k(t)$. Since the Laplace transform of $F_k(t)$ is given by $f_k^*(s)/s$, we have

$$G(k)=\int_{t=0}^{\infty}F_k(t)f_{SE}(t)\,dt=\int_{t=0}^{\infty}\left(\frac{1}{2\pi j}\int_{\sigma-j\infty}^{\sigma+j\infty}\frac{f_k^*(s)}{s}\cdot e^{st}ds\right)f_{SE}(t)\,dt$$

$$G(k)=\frac{1}{2\pi j}\int_{\sigma-j\infty}^{\sigma+j\infty}\frac{f_k^*(s)}{s}\cdot\left(\int_{t=0}^{\infty}f_{SE}(t)e^{st}dt\right)ds=\frac{1}{2\pi j}\int_{\sigma-j\infty}^{\sigma+j\infty}\frac{f_k^*(s)}{s}\cdot f_{SE}^*(-s)ds, \qquad k\geq 1 \tag{14}$$

Eq.(14) can be viewed as the complex folding of the two functions $f_{SE}^*(s)$ and $f_k^*(s)/s$ evaluated at zero. Since $G(0)=1$, the probability generating function (PGF) for $G(k)$ is derived as,

$$G_Z(z)=\sum_{k=0}^{\infty}G(k)z^k=1+\sum_{k=1}^{\infty}G(k)z^k=1+\hat{G}_Z(z), \tag{15}$$

and combining (2), (14) and the fact that $f_k^*(z)=f_{R_1}^*(z)\left(f_R^*(z)\right)^{k-1}$ we obtain,

$$\hat{G}_Z(z)=\frac{1}{2\pi j}\int_{\sigma-j\infty}^{\sigma+j\infty}\frac{f_{R1}^*(s)f_{SE}^*(-s)}{s}\sum_{k=1}^{\infty}\left(f_R^*(s)\right)^{k-1}z^k ds=\frac{1}{2\pi j}\int_{\sigma-j\infty}^{\sigma+j\infty}\frac{f_{R_1}^*(s)f_{SE}^*(-s)}{s}\frac{z}{1-z\cdot f_R^*(s)}ds. \tag{16}$$

The convergence of the geometric sum requires that $\left|z\cdot f_R^*(s)\right|<1$. Since we want to calculate the moments using (16), the limit $z\to 1$ has to be applied, thus for the moment we have to assume that $\left|f_R^*(s)\right|<1$. From (12) and (16), it is now easy to derive the PGF of the distribution $P_{HO}(k)$ as,

$$\begin{aligned}P_{HO_Z}(z)&=\sum_{k=0}^{\infty}P_{HO}(k)z^k=\sum_{k=0}^{\infty}\left(G(k)-G(k+1)\right)z^k=1-\frac{1-z}{z}\hat{G}_Z(z)\\&=1-\frac{1}{2\pi j}\int_{\sigma-j\infty}^{\sigma+j\infty}\frac{f_{R_1}^*(s)}{s}\frac{1-z}{1-z\cdot f_R^*(s)}f_{SE}^*(-s)ds\end{aligned} \tag{17}$$

Since the behaviour of (17) is not clear as $z\to 1$, an alternative solution can be derived using the transformation $\frac{1-z}{1-z\cdot f_R^*(s)}=1-\frac{z\left(1-f_R^*(s)\right)}{1-z\cdot f_R^*(s)}$, which offers an equivalent result as,

$$P_{HO_Z}(z)=1-\frac{1}{2\pi j}\int_{\sigma-j\infty}^{\sigma+j\infty}\frac{f_{R_1}^*(s)}{s}f_{SE}^*(-s)ds+\frac{1}{2\pi j}\int_{\sigma-j\infty}^{\sigma+j\infty}\frac{f_{R_1}^*(s)}{s}\frac{z\cdot\left(1-f_R^*(s)\right)}{1-z\cdot f_R^*(s)}f_{SE}^*(-s)ds. \tag{18}$$

Let us briefly discuss the properties of (18). From the second integral we observe, that the limit $z\to 1$ is well defined and $P_{HO_Z}(z)$ can be expanded to the whole z-domain. As in [6], the residue theorem can be used to solve the complex contour integrals (17) or (18). Let us assume, that the Laplace-transform $f_{SE}^*(s)$ of the effective call holding time has no branching points and that $f_{SE}^*(s)$, $f_{R_1}^*(s)$ an $f_R^*(s)$ only have isolated poles p_{SE}, p_{R_1} and p_R in the left half of the complex plane with their real parts as $\sigma_{SE}=Re\{p_{SE}\}<0$, $\sigma_{R_1}=Re\{p_{R_1}\}<0$ and $\sigma_R=Re\{p_R\}<0$. Under these assumptions $f_{SE}^*(-s)$ has only isolated poles s_P in the right half of the complex plane, thus for the integration in (17) or (18) we can assume $0<\sigma<-max\{\sigma_{SE}\}$. Now because $Re\{s\}>0$ and since $f_R(t)=0$ for $t<0$, we have $\left|f_R^*(s)\right|\leq 1$ in general. For $|z|<1$ we have $\left|zf_R^*(s)\right|<1$ and the denominator of the second integral in eq.(18) does not lead to a pole of the integrand, and for $z\to 1$ a pole at the denominator $1-f_R^*(s)$ cancels out. Hence only $f_{SE}^*(-s)$ and $f_{R_1}^*(s)/s$ contribute to the set of poles. Now applying Jordan's Lemma, the path of integration is expanded in the clockwise direction and utilizing (17) and applying the residue theorem, the PGF for the number of handoffs is given by

$$P_{HO_Z}(z) = 1 + \sum_{poles\ s_P} \underset{s=s_p}{Res}\, \frac{f_{R_1}^*(s)}{s} \frac{1-z}{1-z \cdot f_R^*(s)} f_{SE}^*(-s), \tag{19}$$

where $Res_{s=s_p}$ denotes a residue of $f_{SE}^*(-s)$ at a pole $s = s_p$ in the right half of the complex plane.

Example1: Model 1 with hyper-Erlang distributed call holding time and general residence times
In this example the CHT follows a hyper-Erlang distribution, which has Laplace transform

$$f_{SE}^*(s) = f_S^*(s) = \sum_{i=1}^{I} \alpha_i \left(\frac{\mu_i}{s + \mu_i} \right)^{m_i}, \quad \alpha_i \geq 0, \quad \sum_{i=1}^{I} \alpha_i = 1. \tag{20}$$

Then, with (3) - (5) and (17) we have the complex contour integral

$$P_{HO_Z}(z) = 1 - \sum_{i=1}^{I} \alpha_i \frac{1}{2\pi j} \int_{\sigma - j\infty}^{\sigma + j\infty} \frac{1 - f_R^*(s)}{s^2 E\{R\}} \cdot \frac{1-z}{1 - z \cdot f_R^*(s)} \cdot \left(\frac{\mu_i}{\mu_i - s} \right)^{m_i} ds. \tag{21}$$

Poles of order m_i are given for $s \to \mu_i$, thus using (19) the integral can be evaluated as

$$P_{HO_Z}(z) = 1 + \sum_{i=1}^{I} \frac{\alpha_i}{E\{R\}} (-\mu_i)^{m_i} \frac{1-z}{(m_i - 1)!} \lim_{s \to \mu_i} \frac{d^{m_i - 1}}{ds^{m_i - 1}} \frac{1 - f_R^*(s)}{s^2 \left(1 - z \cdot f_R^*(s)\right)}. \tag{22}$$

Now utilizing $P_{HO}(n) = \frac{1}{n!} \frac{d^n}{dz^n} P_{HO}(z)\big|_{z=0}$ the distribution can be derived and follows to

$$P_{HO}(0) = 1 - \sum_{i=1}^{I} \alpha_i (-1)^{m_i - 1} \frac{(\mu_i)^{m_i}}{(m_i - 1)! E\{R\}} \lim_{s \to \mu_i} \frac{d^{m_i - 1}}{ds^{m_i - 1}} \frac{1 - f_R^*(s)}{s^2} \tag{23a}$$

$$P_{HO}(n) = \sum_{i=1}^{I} \alpha_i (-1)^{m_i - 1} \frac{(\mu_i)^{m_i}}{(m_i - 1)! E\{R\}} \lim_{s \to \mu_i} \frac{d^{m_i - 1}}{ds^{m_i - 1}} \left(\frac{1 - f_R^*(s)}{s} \right)^2 \left(f_R^*(s) \right)^{n-1}, n = 1, 2, \ldots \tag{23b}$$

As an example we will give closed form results for the cases $m_i = 2$ and $m_i = 1$:

$$P_{HO}(0) = 1 - \sum_{i=1}^{I} \alpha_i \frac{m_i / \mu_i}{E\{R\}} \left[1 - f_R^*(\mu_i) + \frac{m_i - 1}{2} \mu_i f_R^{*(1)}(\mu_i) \right], \tag{24a}$$

$$P_{HO}(n) = \sum_{i=1}^{I} \alpha_i \frac{m_i / \mu_i}{E\{R\}} \left[\left(1 - f_R^*(\mu_i)\right)^2 - \frac{m_i - 1}{2} \mu_i f_R^{*(1)}(\mu_i) \left((n+1) f_R^*(\mu_i) - 2n + \frac{n-1}{f_R^*(\mu_i)} \right) \right] \left(f_R^*(\mu_i) \right)^{n-1}, n \geq 1. \tag{24b}$$

For $m_i = 1$ (hyper-Exponential CHT) we have a geometric law, which involves one geometric distribution for each phase of the hyper-exponential distribution of the call holding time, whereas for $m_i = 2$ (hyper-Erlang CHT) this behaviour is overlaid by the additional term $n \cdot f^{n-1}$.

2.4 MEAN NUMBER OF HANDOFFS AND THEIR VARIANCE

From the PGF (18) all moments as well as the binomial moments can be derived, but only for lack of space we restrict to the mean and variance. First, the mean number of handoffs is given by

$$E\{HO\}=\frac{d}{dz}P_{HO_Z}(z)\big|_{z=1}=\frac{1}{2\pi j}\int_{\sigma-j\infty}^{\sigma+j\infty}\frac{f_{R_1}^*(s)}{\left(1-f_R^*(s)\right)}\frac{f_{SE}^*(-s)}{s}ds. \tag{25}$$

This result is valid for a large class of probability distribution functions. Under the previous assumptions $\left|f_R^*(s)\right|<1$ the residue theorem can be used to solve eq.(25). Since it is widely known that any distribution function with finite mean and variance can be approximated by a Coxian distribution, this method is our convenient as Coxian distributions have rational Laplace transforms. It is now also straightforward to derive higher moments, e.g. the variance can be derived as

$$Var\{HO\}=\frac{d^2}{dz^2}P_{HO_Z}(z)\big|_{z=1}+E\{HO\}(1-E\{HO\}), \tag{26}$$

where, from (18) we have

$$\frac{d^2}{dz^2}P_{HO_Z}(z)\big|_{z=1}=\frac{1}{2\pi j}\int_{\sigma-j\infty}^{\sigma+j\infty}\frac{2f_{R_1}^*(s)f_R^*(s)}{\left(1-f_R^*(s)\right)^2}\frac{f_{SE}^*(-s)}{s}ds.$$

Now we study eq.(25), for the three different session behaviour models we proposed earlier. Note that for session model 3 (i.e., the case of triggered sessions), eq.(25) is the simplest solution form, with $f_{R_1}^*(s)=f_R^*(s)$ and $f_{SE}^*(s)=f_S^*(s)$. Hence, one cannot avoid evaluating the residues.

Proposition 1: Session model 1:

For session model 1 the mean number of handoffs is given by

$$E\{HO\}=\frac{E\{S\}}{E\{R\}}=\rho. \tag{27}$$

Thus the mean number of handoffs is equal to the call to mobility ratio ρ, defined by the mean CHT divided by the mean CRT. While this result was derived in many other publications, it was always restricted by various assumptions, e.g. see [8] where the sequence of residence times must form an equilibrium renewal or a renewal Poisson process. In our case, we only require the existence of rational Laplace transforms. It should be noted that this simple result only holds for session model1 and not in general as assumed in [8, 10].

Proof: Let us fist calculate the mean CHT as follows:

$$E\{S\}=\int_{t=0}^{\infty}t\cdot f_S(t)dt=\int_{t=0}^{\infty}f_S(t)\left(\frac{1}{2\pi j}\int_{\sigma-j\infty}^{\sigma+j\infty}\frac{e^{st}}{s^2}ds\right)dt=\frac{1}{2\pi j}\int_{\sigma-j\infty}^{\sigma+j\infty}\frac{1}{s^2}\int_{t=0}^{\infty}f_S(t)e^{st}dtds=\frac{1}{2\pi j}\int_{\sigma-j\infty}^{\sigma+j\infty}\frac{f_S^*(-s)}{s^2}ds \tag{28}$$

Utilizing (3) and (25), the mean number of handoffs is given by

$$E\{HO\}=\frac{1}{E\{R\}}\frac{1}{2\pi j}\int_{\sigma-j\infty}^{\sigma+j\infty}\frac{\left(1-f_R^*(s)\right)}{\left(1-f_R^*(s)\right)}\frac{f_S^*(-s)}{s^2}ds=\frac{1}{E\{R\}}\frac{1}{2\pi j}\int_{\sigma-j\infty}^{\sigma+j\infty}\frac{f_S^*(-s)}{s^2}ds=\frac{E\{S\}}{E\{R\}}. \tag{29}$$

Notice that the factor $\left(1-f_R^*(s)\right)$ in the denominator of the integral cancels out because the residual of the residence time is used for the first cell and hence we do not require that $\left|f_R^*(s)\right|<1$.

Proposition2: Session model 2

Substituting the residual of the service time in (7) with (25) we get

$$E\{HO\}=\frac{-1}{2\pi j}\int_{\sigma-j\infty}^{\sigma+j\infty}\frac{f_R^*(s)}{1-f_R^*(s)}\frac{1-f_S^*(-s)}{s^2E\{S\}}ds=\frac{1}{2\pi j}\frac{1}{E\{S\}}\int_{\sigma-j\infty}^{\sigma+j\infty}\frac{f_R^*(s)}{1-f_R^*(s)}\frac{f_S^*(-s)}{s^2}ds, \tag{30}$$

where we have assumed that $|f_R^*(s)|<1$, hence $s^2(1-f_R^*(s))$ does not have a zero in the right half plane. If the contour of integration is closed in the right half plane, the first part of the left integral evaluates to zero and hence we only calculate the right integral. From the residual theorem we have

$$E\{HO\}=\frac{-1}{E\{S\}}\sum_p \underset{s=s_p}{Res}\frac{f_R^*(s)}{s^2(1-f_R^*(s))}\cdot f_S^*(-s) \tag{31}$$

Proposition3: Negative exponential distributed residence time

For this special case the mean number of handoffs for all session models is given by

$$E\{HO\}=\frac{E\{SE\}}{E\{R\}} \tag{32}$$

Proof: For the negative exponential distribution we have

$$f_{R_1}^*(s)=f_R^*(s)=\frac{\theta}{\theta+s} \text{ and } \frac{f_{R_1}^*(s)}{(1-f_R^*(s))}=\frac{\theta}{s}, \tag{33}$$

where the mean residence time is denoted by $E\{R\}=1/\theta$. For the exponential distribution the assumption $|f_R^*(s)|<1$ is always valid as long as $\sigma=Re\{s\}>0$. Note that this assumption is assured by the line of integration in (25). With eq.(28) and (25) we have

$$E\{HO\}=\rho=\frac{1}{2\pi j}\int_{\sigma-j\infty}^{\sigma+j\infty}\frac{f_{R_1}^*(s)}{(1-f_R^*(s))}\frac{f_{SE}^*(-s)}{s}ds=\frac{\theta}{2\pi j}\int_{\sigma-j\infty}^{\sigma+j\infty}\frac{f_{SE}^*(-s)}{s^2}ds=\frac{E\{SE\}}{E\{R\}}.$$

For model 1 and 3 we have $E\{SE\}=E\{S\}$ and (32) is equivalent to (27), as expected. In contrast for model 2 the ECHT is the residual of the call holding time S given by $E\{SE\}=0.5E\{S^2\}/E\{S\}$. Thus, unlike non-roaming users, the second moment of the call holding time must be considered for roaming users. Now, following the derivation given for (28), it is simple to show that, $E\{SE^2\}=\frac{1}{2\pi j}\int_{\sigma-j\infty}^{\sigma+j\infty}2f_S^*(-s)/s^2ds$. Utilizing (26) the variance is given by the simple closed formula

$$Var\{HO\}=\frac{E\{SE^2\}}{E\{R\}^2}+E\{HO\}(1-E\{HO\}). \tag{34}$$

Proposition 4: Negative exponential distributed call holding time

Let us assume that $|f_R^*(s)|<1$ and that $f_R^*(s)$ has only poles in the left half plane. Then, if the call holding time is negative exponentially distributed with mean $E\{S\}=1/\mu$, the mean number of handoffs and the variance for all session models is given by

$$E\{HO\}=\frac{-1}{2\pi j}\int_{\sigma-j\infty}^{\sigma+j\infty}\frac{f_{R_1}^*(s)}{1-f_R^*(s)}\frac{\mu}{s(s-\mu)}ds=\frac{f_{R_1}^*(\mu)}{1-f_R^*(\mu)}. \tag{35}$$

$$Var\{HO\} = 2\frac{f_{R_1}^*(\mu) f_R^*(\mu)}{\left(1 - f_R^*(\mu)\right)^2} + E\{HO\}(1 - E\{HO\}) \tag{36}$$

For the special case of model 1 we have $f_{R_1}^*(\mu) = (1 - f_R^*(\mu))E\{S\}/E(R)$ and the general result (27) for the mean number of handoffs is only a special case here. Note that similar results for the variance as in (34) and (36) are given in [8], but only for session model 1 where the mean number of handoffs is equal to the mobility ratio.

Example 2: Model 3 and deterministic residence times

Let us assume an application, where the session has started at a cell boundary and the residence times are given by fixed time intervals, e.g. a highway model. In this case we have $f_{R_1}(t) = f_R(t) = \delta(t - T)$ with Laplace transform $f_R^*(s) = e^{-sT}$. For $\sigma = Re\{s\} > 0$ we have $\left|f_R^*(s)\right| = \left|e^{-\sigma T}\right| < 0$, and $(1 - f_R^*(s))$ does not lead to a pole in the right half complex plane. Thus, if the residual theorem is applied to eq.(25), we will have

$$E\{HO\} = -\sum_p \operatorname*{Res}_{s=s_p} \frac{e^{-sT}}{\left(1 - e^{-sT}\right)} \cdot \frac{f_S^*(-s)}{s},$$

which is simple to evaluate for all examples given above including the hyper-Erlang CHT.

3. NUMERICAL RESULTS

In this section, we show numerical results for the probability mass function (pmf) and the variance of the number of handoffs for the three session scenarios. In this regard, we assume that the Erlang-2 for the CHT (i.e., $i = 1$, $m_i = 2$ in (24)), with a mean of $E\{S\} = 2/\mu$ and variance of $Var\{S\} = 2/\mu^2$. The CRT follows a gamma distribution with mean $E\{R\} = \gamma\theta$, variance $Var\{R\} = \gamma\theta^2$. In this case, the call-to-mobility ratio is calculated as $\rho = 2/(\mu\gamma\theta)$ and the corresponding probability distribution is obtained from (24). The variance of N is given by (26) where the second derivative of the PGF simplifies to $\frac{d^2}{dz^2} P_{HO_Z}(z)\big|_{z=1} = \frac{2 \cdot \rho \cdot f_R^*(\mu)}{1 - f_R^*(\mu)}\left(1 + \frac{1}{\rho + 2/\gamma} \frac{1}{1 - f_R^*(\mu)}\right)$. For our results, we use an average CHT of $E\{S\} = 1.76$ min and relevant values for γ out of $\{0.01, 0.1, 1.5, 10\}$ [6]. The pmf of N for several values of ρ is shown in Fig. 2. First, we observe that the pmf of handoffs differs considerably depending on the session behaviour. The effect is very clear for the value of probability of no handoffs. We also observe that the decay of the pmf in all models is slower when the mobility ratio is larger. This is due to the fact that higher ρ means larger average number of handoffs. We can also see that the pmfs of model 1 and 3 are more similar to each other, whereas the pmf of model 2 has a different behaviour. Especially its pmf has a slower decay. This comes from the fact that model 2 describes the roaming case, where the residual of the session times has to be considered. In case of the Erlang-2 CHT the coefficient of variation of the residual CHT is higher than for the CHT itself. This results in a slower decay of the distribution of the number of handoffs. In Fig.3, we study the variance of *N(t)* for model 1 as a function of the mobility ratio ρ. We observe that as the coefficient of variation of the residence time $c_R = 1/\sqrt{\gamma}$ increases, the variance of the number of handoffs increases dramatically.

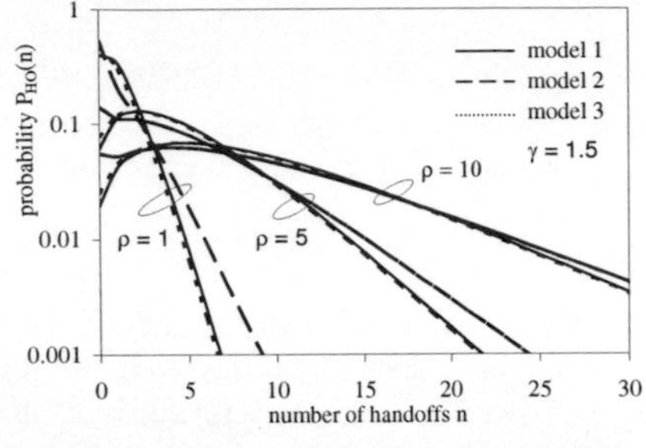

Fig. 2. $P_{HO}(n)$ of N(t) for different values of γ and ρ

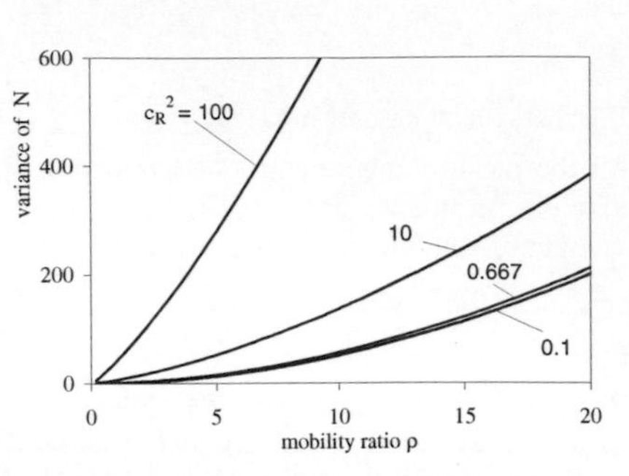

Fig.3. Variance of N for c_R = 10, 0.667 and 0.1..

4. CONCLUSIONS

In this paper, we have derived explicit forms for the pmf and the moments of the number of handoffs by means of the Laplace transform and the residue theorem for general distributed residence and call holding times and for different session models. We derived the mean number of handoffs for each session model and illustrated the applicability of our framework using several examples and numerical results.

REFERENCES

[1] Fang Y. and Chlamtac I., *Teletraffic analysis and mobility modelling for PCS networks,* IEEE Trans. Communi., vol. 47, 1999, pp. 1062-1072.

[2] Orlik P. and Rappaport S. S., *A Model for Teletraffic Performance and Channel Holding Time Characterization in Wireless Cellular Communication with General Session and Dwell Time Distributions*, IEEE J. Select. Areas Commun., vol. 16, no. 5, 1998, pp. 788-803.

[3] Lin Y. B., Mohan S. and Noerpel A., *Queueing priority channel assignment strategies for handoff and initial access for a PCS network,* IEEE Trans. Veh. Technol., vol. 43, 1994, pp. 704-712.

[4] Jordan J. and Barcelo F., *Statistical modelling of channel occupancy in trunked PAMR systems,* in Proc. 15th Int. Teletraffic Conf. (ITC 15), 1997, pp. 1169-1178.

[5] Y. Fang and I. Chlamtac: *A New Mobility Model and its Application in the Channel Holding Time Characterization in PCS Networks,* Infocom 99, vol. 1, 1999, pp. 20-27.

[6] Y. Fang and I. Chlamtac: "*Call performance of a PCS network*", IEEE J. Selec. Areas Commun, vol. 15, no. 8, pp. 1568-1581, 1997.

[7] Fang Y. and Chlamtac I., *Analytical Generalized Results for Handoff Probabilities in Wireless Networks*, IEEE Trans. Communications, Vol. 50, No. 3, 2002, pp. 396-399.

[8] Rodriguez-Dagnino R. M. and Takagi H., *Counting Handovers in a Cellular Mobile Communication Network: Equilibrium Renewal Process Approach*, Performance Evaluation, vol. 52, no. 2, 2003, pp. 153-174.

[9] S. Zaghloul, W. Bziuk, A. Jukan, "*Signaling and Handoff Rates at the Policy Control Function (PCF) in IP Multimedia Subsystem (IMS),*" IEEE Communications Letters Journal, to appear in June 2008.

[10] Nanda S., *Teletraffic models for urban and suburban microcells: Cell sizes and handoff rates*, IEEE Trans. nn Vehicular Technology, vol. 42, no. 4, 1993, pp. 76-84.

wireless ad hoc network, selfish behaviour, reputation system

Jerzy KONORSKI*, Rafał ORLIKOWSKI*

DISTRIBUTED REPUTATION SYSTEM FOR MULTIHOP MOBILE AD HOC NETWORKS

The paper discusses the need for a fully-distributed reputation system dedicated for multihop wireless ad hoc networks whose nodes may exhibit selfish forwarding behavior. We propose a reputation-based system with anonymous packet forwarding and a congestion avoidance mechanism.

1. INTRODUCTION

Mobile ad hoc networks (MANETs) as well as ad hoc wireless sensor networks (WSNs) are collections of mobile stations that exchange packets over a wireless transmission medium. There may be pairs of stations out of each other's reception range, for which the only way of exchanging data is via in-range nodes acting as packet forwarders. i.e., agreeing to relay packets on behalf of other nodes. However, packet forwarding costs extra energy, a scarce resource in wireless ad hoc devices, and consumes the bandwidth a node could use for its own traffic. If one accepts that nodes can be rational (meaning selfish) then it is not surprising that they may try to save energy and bandwidth, and the most obvious way of doing it is refusing to relay packets. Without a mechanism preventing such selfish misbehavior MANETs or ad hoc WSNs become unreliable.

We propose a reputation-based system suited for secure *Anonymous Packet Forwarding* (APF) whereby a packet's source address is invisible to intermediate nodes (e.g., is encrypted end-to-end along with the message it carries). The reputation system is able to enforce cooperation among nodes and render noncooperative behavior unprofitable. The controlling of packet forwarding is done through implementing interacting agents in a certain node subset whose task is to monitor and evaluate other nodes' behavior using *Exponentially Moving Weight Average* (EWMA) method. Nodes are then assigned ratings in the form of reputation metric values that reflect the perception of their packet forwarding functionality. As APF hides pakets' source addresses as well as because the network topology need not be known to a particular node, the proposed reputation metrics have to be used in connection with hop-by-hop routing such as *Greedy Perimeter Stateless Routing* (GPSR) [1], where the knowledge of immediate neighbourhood is sufficient to determine packet paths.

The rest of the paper is organized as follows. Section 2 discusses related work and outlines some of the well-known reputation systems. Section 3 describes a concept and implementation of

* Gdańsk University of Technology, ul. Narutowicza 11/12, 80-952 Gdańsk, Poland, jekon@eti.pg.gda.pl

our approach to reputation-based systems, with a detailed description of the constituent modules in a network node and their interoperability. Sample performance evaluation results are reported in Section 4.

2. RELATED WORK

Enforcement of cooperative behavior in MANETs has recently received considerable attention. Two types of solutions for coping with misbehaving nodes are being proposed. Schemes based on virtual currency e.g., *Nuglets* [2] or *Sprite* [3], use a form of micropayments to build incentives for cooperation. They are usually quite complex and hard to implement as they typically require tamper-proof hardware in every node or a trusted third party to control the flow of the currency and ensure transaction security.

Another, more promising type of solution are reputation-based schemes. The cooperation goals are achieved by way of determination and sharing reputation values among all the network nodes (global reputation) or within groups of nodes (local reputation). Schemes of this kind for ad hoc networks include *Cooperation of Nodes Fairness in Dynamic Ad-Hoc NeTworks* (CONFIDANT) [4], *Collaborative Reputation Mechanism* (CORE) [5], *Secure and Objective Reputation-based Incentive Scheme* (SORI) [6] and *Observation-based Cooperation Enforcement in Ad Hoc Networks* (OCEAN) [7]. Reputation-based systems are also implemented in other environments like *Peer-to-Peer* (P2P) networks e.g. *P-Grid* [8] or *PeerTrust* [9].

Our system is based on hop-by-hop routing. There are several proposals of this kind of routing in the literature. The most popular is GPSR [1]. It is a geographical routing protocol based on greedy forwarding using information only about a node's immediate neighbours. Each node receiving a packet to forward looks up local routing table for a suitable next-hop node. The only criterion for this is the geographical distance to the packet's destination address. The nearest-to-destination node is then selected as the next-hop. In some situations greedy forwarding can fail e.g., the next-hop node may have no connectivity to the destination at a given time. To cope with such problems, planar graph theory is used.

Currently proposed reputation-based schemes mostly decrease reputation values whenever they detect misbehavior, whether selfish or caused by node congestion. Moreover, they often rely on extracting source node identities from the contents of packets. Our approach differs from existing solutions in the following respects:

- it attempts to discern selfish node misbehavior from congestion by means of a *Congestion Control Mechanism* (CCM), introducing additional messages to be exchanged between neighbor nodes to announce congestion,
- it takes advantage of APF and its hiding of packet's source node identities to preclude selective packet forwarding and make network topology invisible to any particular node, only aware of the immediate neighbourhood; in this way a number of cheating techniques are rendered useless,
- it is dedicated for hop-by-hop routing, thereby more resistant to scalability problems, and can work equally efficiently in networks consisting of only a few nodes as well as in large-scale topologies.

3. PROPOSED REPUTATION SYSTEM

Before we describe our approach, we give a few details and secure APF and suggest how its limitations can play into the hands of a reputation system.

3.1 FORWARDING MODEL

Assuming that a public-key cryptographic infrastructure, e.g., RSA, is in place, the main steps of inter-node communication are as follows (Figure 1). A pair of adjacent nodes A and B establish a neighborhood relationship by exchanging their addresses in the form of public keys, key_A and key_B, as well as routing tables (each node can have multiple keys). For an GPSR-like routing protocol, a generic routing table entry may contain key_D, key_C, and path_cost (where D is a destination node, C is designated as the next-hop node, and path_cost is the cost of path towards node D via node C.

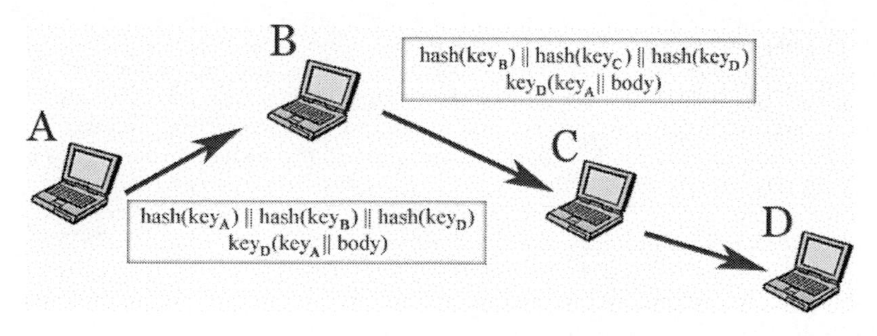

Fig. 1. Anonymous packet forwarding

When a source node A wishes to send a packet to node D, whose public key key_D it knows, it first determines an appropriate next-hop station i.e., looks up the (key_D, path_cost) entry in the routing table to find the next-hop node's key, key_B, then uses key_D to encrypt the packet body along with key_A, appends to the packet the values $h_A = hash(key_A)$, $h_B = hash(key_B)$, and $h_D = hash(key_D)$ using a public hash function, and finally sends the packet.

Node B compares h_A, h_B, and h_D with locally stored hashed values of public keys to check that it has a neighborhood relationship with node A and to recognize itself as the receiver and possibly destination of the packet. If $h_B = h_D$, the packet body is decrypted using node B's private key, key_B^{-1}. Otherwise h_B is replaced by $h_C = hash(key_C)$ with C determined by node B as the next-hop node, and the forwarding toward node D continues.

Note that packet anonymity, ensured by end-to-end encryption of a source node's public key along with the packet body, prevents traffic analysis attacks [10]. At the same time, packets overheard or received at network nodes can no longer be tied to their source nodes, therefore a reputation scheme can easily be confused as to whether a given node is overloaded by its local application traffic or misbehaves by forcing its own (source) packets instead of relaying other nodes'. As a result, nodes may easily overestimate a neighbor node's reputation. To cope with this form of "throughput cheating", our reputation system incorporates a CCM mechanism of provably legitimate refusal to forward packets.

From the viewpoint of a reputation scheme, APF has gentle sides, too. A notorious abuse of such schemes is selective packet forwarding. Sharing of nodes' reputation values like in CONFIDANT, SORI, or CORE allows a selfish node refusing to perform packet forwarding on behalf of certain node pairs to nevertheless keep high reputation, hence be recognized as a well-behaved node, if it continues to forward packets on behalf of other node pairs. It is hard to identify such selfishness in a standard way. APF hides source nodes – no node can determine a packet's source node other than the destination node and the source node itself – which makes selective packet forwarding impossible or at least risky. This is taken advantage of in the design of our reputation system.

3.2 REPUTATION SYSTEM ARCHITECTURE

By itself, APF and CCM do not enforce cooperation of selfish nodes; it needs to be complemented with incentives that render packet forwarding on behalf of others more profitable than refraining therefrom. Formation and dissemination of node reputation values provide one building block for such incentives. Essentially, reputation values reflect how much nodes trust one another to perform the desired functions, such as routing and forwarding, based on their past performance. The other incentive-building block is a routing mechanism that leverages the reputation values to form source-to-destination paths only passing through – and therefore keeping alive the neighborhood relationships of – trusted, well-behaved nodes. In the process, selfishly misbehaving nodes are gradually isolated and prevented from injecting their source packets into the network, unless they follow the incentive and "reform" i.e., resume packet forwarding.

In our solution, every network node implements a set of collaborating *Reputation System Components* (RSC). The four main RSCs are: *Reputation System Interface* (RSI), *Reputation Output Queue* (ROQ), WATCHDOG, and *Reputation Manager* (RM).

RSI acts as an interface for data exchange between RSC components and any standard node mechanism e.g., routing.

WATCHDOG is responsible for gathering direct information about neighboring nodes' behavior. If node A (Figure 1) sends a packet to node B, to be further forwarded to node C, it stores the encrypted packet body, next-hop station's public key, key_B, and current time in the *Watchdog Buffer* (WB). WATCHDOG at node A listens promiscuously to neighbor nodes. If any packet is overheard, its encrypted body is checked with packets stored in the local WB. If a match is found, a WdgEvt event is associated with key_B and communicated to RM with the parameter *state* set to 1. WATCHDOG periodically checks WB to remove outdated packets, with WdgEvt, associated with appropriate key and *state* = 0 communicated to RM.

Standard FIFO output queue in *Logical Link Control Layer* in our solution is replaced by a *Reputation Output Queue* (ROQ) based on *Priority Queuing* (PRIQ). ROQ consists of several FIFO buffers, each of them assigned a different priority value. Incoming packets are classified to appropriate buffer based on reputation values for their previous-hop node if only the value is higher than certain retputation threshold R_T, else the packet is dropped. It is meant as cooperation enforcement: the lower the reputation value, the longer packet forwarding delays become, which all rational nodes are interested in preventing. The only way to achieve it is cooperation with other nodes to increase its own reputation. On the other hand dropping packets causes additional packet retransmissions, which increases nodes' energy consumption, while each rational node in ad hoc network is interested in saving.

The main CCM mechanism concept is based on *Random Early Detection* (RED) mechanism determining a node's current congestion condition. RED monitors the average FIFO queue length. If it exceeds a maximum ROQ_{MAX}, the node notifies its neighborhood via a special FULL frame.

Neighbor nodes then refrain from sending packets via this node for a certain time (FULL validity time) and do not reduce its reputation value when forwarding misbehavior is detected.

RM consists of the *Reputation Calculator* (RC) and *Reputation Table* (RT). RC uses exponentially weighted moving average to calculate reputation values based on communicated WdgEvt events, to be next stored in RT:

$$R_{A,B}^{t} = rR_{A,B}^{t-1} + (1-r)state \quad , \tag{1}$$

where $R_{A,B}^{t}$ and $R_{A,B}^{t}$ are current and previous reputation values calculated by node A for node B, and $r \in [0, 1]$ represents the influence of node B's past behavior on the current reputation value; the higher r, the more emphasis is put on past behavior. Every node with a reputation value below R_T is considered misbehaving (selfish). Reputation values are used by the routing protocol as metrics for the selection of a packet's next-hop node. As was mentioned, during next-hop node selection process, standard GPSR protocol only looks at the geographical distance to the destination. Our implementation also cares about route security, so the next-hop node is selected based on both its geographical location and reputation value. Firstly, all neighbours of a node are filtered according to their reputation values. Each one with a reputation value less then R_T is disregarded. The remaining set of neighbour nodes is searched for the nearest-to-destination one subsequently selected as the next-hop.

4. SIMULATION

In this section we investigate via simulation the robustness and efficiency of the proposed reputation-based system for a mobile ad hoc network. We try to address the questions how long it takes to detect all misbehaving nodes as well as what profits arise from CCM and APF.

The proposed reputation system is implemented and evaluated using the j-sim tool [11], in a simulation environment of IEEE 802.11-based ad hoc networks. The simulated scenario features 100 nodes arranged on a grid with each node pair's reception range of one hop. Reputation values are calculated as shown above, with $r = 0.9$ and $R_T = 0.75$. If there are no selfish nodes in the network, there is nothing for our reputation-based system to do. The overall network efficiency (defined as the proportion of packets delivered to destination) then oscillates around 90%, as shown in Figure 2.

To demonstrate the robustness of our reputation system, we let 10% of the network nodes behave selfishly by refusing to forward with per packet probability $p = 0.25$. To emphasize the benefits of CCM it is also assumed that 5% of the network nodes are congested (not able to forward packets due to source traffic overload).

The presence of selfish nodes decreases network efficiency (Figure 2), since packets are now lost due to refused forwarding. When our reputation system detects such selfish nodes, the network efficiency increases, since more packets traverse nodes with a higher reputation, hence are correctly delivered to their destinations. Selfish nodes are gradually isolated from the network routing topology. The reputation system needed about 30 seconds and an average of 32 observations per node to find all selfish nodes in the analyzed scenario.

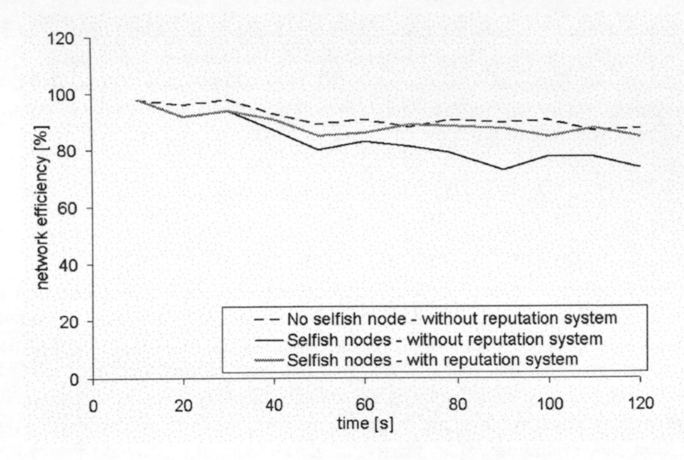

Fig. 2. Network efficiency

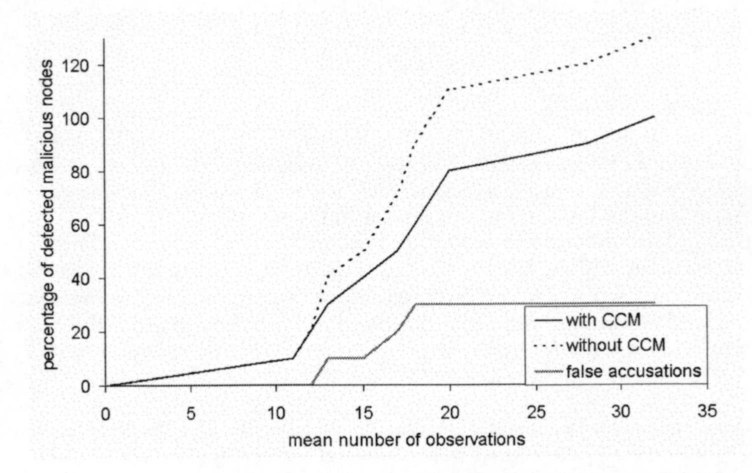

Figure 3. Role of CCM in selfish node detection

As mentioned above, existing reputation-based mechanism does not recognize if a refusal to forward packets results from selfish behavior or from congestion caused by behaving too well i.e., relaying too many packets for other nodes. All these nodes are treated by the system in the same way and all of them are considered selfish only if their reputation value falls below R_T. This brings about occasional *false accusations* (Figure 3) and it is why the system without CCM detects more selfish nodes then there are in the network. Using CCM, our reputation system is able to better distinguish between selfish and congested nodes.

APF does not have a direct influence on malicious nodes detection. However, by hiding packets' source nodes, it prevents selective packet forwarding.

5. CONCLUSION AND FUTURE WORK

This paper investigates some aspects of detecting and preventing selfish behavior in multihop mobile ad hoc networks. A novel solution coping with selfish refusal to forward packets, selective packet forwarding, and nodes' congestion has been proposed, retaining simple, well-known methods of calculating network node's forwarding reputation. Preliminary simulations have shown that the proposed system does contribute to the improvement of network efficiency, while diminishing the risk of false reputation rating.

Nevertheless more work has to be done especially on reputation data gathering mechanisms and their thorough evaluation in complex scenarios. Currently watchdog is the most popular mechanism and is used in all existing reputation-based systems, even though it requires omnidirectional transmission and is vulnerable to confusion due to packet collisions ane nodes' power control. We plan to replace this mechanism by a more transaction-oriented one.

6. ACKNOWLEDGMENT

Effort sponsored by the Air Force Office of Scientific Research, Air Force Material Command, USAF, under Grant FA8655-08-1-3018, also supported in part by the MNiSW Grant PBZ-MNiSW-02/II/2007.

REFERENCES

[1] KARP B., KUNG H. T., *GPSR: Greedy perimeter stateless routing for wireless networks*, Proc. ACM/IEEE MobiCom, August 2000.
[2] BUTTYAN L., HUBAUX J.-P., *Nuglets: a Virtual Currency to Stimulate Cooperation in Self Organized Mobile Ad Hoc Networks*, Tech. Rep. DSC/2001, EPFL, Lausanne 2001.
[3] ZHONG S., CHEN J., YANG R.: *Sprite: A Simple, Cheatproof Credit-Based System for Mobile Ad hoc Networks*, Proc. IEEE Infocom'03, San Francisco CA, April 2003.
[4] BUCHEGGER S., Le BOUNDEC J-Y.: *Performance Analysis of the CONFIDANT Protocol: Cooperation Of Nodes – Fairness In Distributed Ad-hoc NeTworks*, Proc. IEEE/ACM MobiHOC, 2002.
[5] MICHIARDI P., MOLVA R.: *CORE: A Collaborative Reputation Mechanism to Enforce Node Cooperation in Mobile Ad Hoc Networks*, Proc. IFIP Comm. and Multimedia Security Conf., 2002.
[6] HE Q., WU D., KHOSLA P.: *SORI: A Secure and Objective Reputation-Based Incentive Scheme for Ad Hoc Networks*, Proc. IEEE Wireless Comm. and Networking Conf., 2004.
[7] BANSAL S., BAKER M., *Observation-based cooperation enforcement in ad hoc*, Technical report Arxiv preprint cs. NI/0307012, 2003.
[8] ABERER K., DESPOTOVIC Z., *Managing trust in peer-2-peer information system*, In Proc. of the Nith International Conference on Information and Knowledge Management (CIKM), 2001.
[9] XIONG L., LIU L., *PeerTrust: Supporting Reputation-Based Trust in Peer-to-Peer Communities*. IEEE Transaction on Knowledge and Data Engineering (TKDE), 16(7):843-857, July, 2004.
[10] RAYMOND J., *Traffic analysis: Protocols, Attacks, Design Issues and Open Problems*, Proc. Int. Workshop on Design Issues in Anonymity and Unobservability, LNCS 2009, pp. 10–29, Springer-Verlag, 2001.
[11] http://www.j-sim.org

VII. ROUTING AND TRAFFIC ENGINEERING

Key words – wireless, non-interfering routes, Radio Disjoint Multipath, multipath performance

K. Kuladinithi, C. An, A. Timm-Giel and C. Görg
{koo|chunlei|atg|cg}@comnets.uni-bremen.de
ComNets, TZI, University of Bremen, Germany

PERFORMANCE EVALUATION OF RADIO DISJOINT MULTIPATH ROUTING

Multipath routes in wireless multihop ad hoc networks have proven to increase performance compared to single path routing when providing backup paths in case of path failures and also distributing flows (i.e. data traffic) among several paths. The drawback of using multipath simultaneously in wireless multihop ad hoc networks is that the other nodes in the network located in the other active paths may interfere with their own communications, degrading the effective throughput.

This work proposes an analytical model to find multiple non-interfering routes or routes with minimum interference between a source and a destination in a given network topology. These types of routing paths are called Radio Disjoint Multipath (RDM) routes. Compared to previous work, this work pioneers the use of models to determine interference free routes in a wireless multihop ad hoc network considering two criteria: the mutual interference of each path and the background traffic load of the path. Further, simulation results are taken by implementing RDM routes in 802.11 based multihop ad hoc networks. Both analytical and simulation results are obtained and analyzed for different network scenarios that use RDM routes by distributing packets of a single flow among multiple RDM paths. Analytical results show that the sustainable troughput can be doubled when using RDM paths for unidirectional flows. Simulation results also show that throughput can be improved with RDM routes by a factor of 1.6 and 4 in some cases, when using uni and bidirectional UDP flows respectively.

1. INTRODUCTION

Previous research in wireless multihop ad hoc networks mostly focuses on the utilization of multiple paths as backup paths in case of failures in the main routing path [1]. Multipath routing can also be used to improve communication efficiency and promote quality of service by utilising different paths simultaneously to improve the application performance, load balancing between paths, increased reliability, optimal utilization of otherwise non used paths. Multipath routes can be utilized simultaneously by distributing independent flows [2] or by splitting (distributing) packets of a single flow [3] or by replicating packets of a flow among all paths. The replicating increases the reliability in lossy environments such as a firefighting scenario, where reliability of the data communication is the most important aspect [4].

However, simultaneous use of several paths (routes) can result in reduced performance in multihop ad hoc networks due to the fact that all devices are using the same radio channel. This means, all devices within the interference range are competing for the channel using the well-known CSMA/CD schemes. Therefore, routes should be chosen in a manner giving a higher preference to non-interfering paths. Finding purely radio disjoint (no nodes in the interfering range of each other's path) paths is not realistic. However, selecting node disjoint paths with a minimum radio interference helps to avoid performance degradation (e.g. the flow in the middle problem in WLAN) [2]. In this work, non-interfering or with minimum interference multiple routes are called Radio Disjoint Multipath (RDM) routes.

This work proposes an analytical model to determine the RDM routes in a given wireless multihop network by analyzing the interference between paths and additionally considering the Background Traffic Load (BTL) of paths. RDM routes are implemented in the OPNET simulator for 802.11 multihop ad hoc networks. The next subsection discusses the related work highlighting what is newly proposed in our work. Section 2 details the analytical model giving a simple example. Section 3 elaborates the results obtained and gives a comparison of analytical and simulation results. Section 4 concludes the paper.

1.1. RELATED WORK

Research in the area of estimating network throughput in wireless multihop environments analytically are mainly focused in single path (SP) multihop ad hoc networks, considering the effect of hidden node and exposed node problems [5 , 6 , 7 , 8]. Simultaneous use of multipath, in contrast to SP, may result in further throughput degradation due to the mutual interference between paths. There are a few research papers published that assess network throughput behaviour analytically by taking interference into account when using multipath routes. Most of the research uses the max-flow problem [9] as a basis, which considers the flow allocation in wired networks. Reference [10] presents how to find sustainable throughput by using the max-flow problem and extending it to add interference related constraints. These constraints are found using the conflict graph theory. It gives upper and lower bounds for the computation of a sustainable throughput. Reference [11] also uses the max-flow problem with interference constraints. In contrast to [10], interference constraints are computed using Farkas' Lemmas, giving only a tighter upper bound of the throughput. Reference [12] presents a stochastic framework to model the impact of interference between nodes. Instead of the max-flow problem, it uses a non-linear programming model incorporating the interference to compute the network throughput. In contrast to [10] and [11], this paper investigates the variation of sustainable throughput in regular topologies with different types of multipath routes. It concludes that use of multipath with moderate interference performs better than SP, whereas SP performs better than multipath with heavily interfering nodes.

In summary, the above mentioned papers analyze the network throughput for a given set of (fixed) multipath routes. As of this writing, there is no research done to investigate how to model the determination of routing paths (i.e. to select multiple routes) which are completely free of mutual interference or paths that have minimum mutual interference. The work presented here proposes the modeling of such a routing mechanism described as RDM routing. The determination of RDM paths are based on mutual interference between paths and the BTL of a path. The mutual interference refers to the impact of the interference between links in different paths. BTL determines the number of packets that have already been transmitted in a path. In summary, this work proposes the following new features, compared to the previous work:

- Previous work [10-12] does not provide a solution to discover multiple paths, though it discusses how to determine sustainable throughput when using multipath. This paper introduces two different criteria to select multiple paths: interference between paths and BTL in a path.
- Previous work [10-12] computes the sustainable throughput only considering the effect of interference. This paper extends the computation to consider both the interference and BTL.

Furthermore, this model considers the effect of interference from other links which already carrying traffic, but are not part of the selected routes.

- This paper analyzes scenarios considering different types of multipath routes (interfering routes: Non RDM, non-interfering routes: Full RDM and partially interfering routes: PRDM routes). In previous work, analysis of behavior of different types of multipath routes is not done except in [12], but it analyzes scenarios without considering BTL.
- In any of the work, no simulation results are taken by implementing their respective proposed models to compare against analytical results. In [10], multiple paths are setup with static routes and SP is evaluated using the AODV protocol. This work compares analytical results with simulation results taken considering the effect of interference and BTL when selecting paths. Previous work analyzes network performance by considering only unidirectional traffic. This work provides the comparison for both uni and bidirectional traffic using the simulation results, showing a difference in the performance gain.

2. ANALYTICAL MODEL: DETERMINATION OF RDM ROUTES

This section explains how to determine RDM routes analytically based on two criteria: mutual interference of a path and the BTL of a path. These explanations are given based on a simple 3x3 grid network for the simplicity of understanding. It begins with an overview to the model description providing some background and terminology.

2.1. MODEL DESCRIPTION

Fig. 1(a) shows the connectivity graph of a 3x3 grid topology. The distance between 2 adjacent nodes is set as 1 unit of the transmission range and 1 unit of the interference range. Nodes n_S and n_D are chosen as the sender and the destination respectively. It is visually obvious that two possible combinations of node disjoint paths exist which can be used as RDM routes. They are either "P_1 {consisting of nodes n_3, n_6 & n_7} & P_2 {consisting of nodes n_1, n_4 & n_5}" or "P_1 & P_3 {consisting of nodes n_1, n_2 & n_5}" which can be used simultaneously to distribute the traffic among RDM routes.

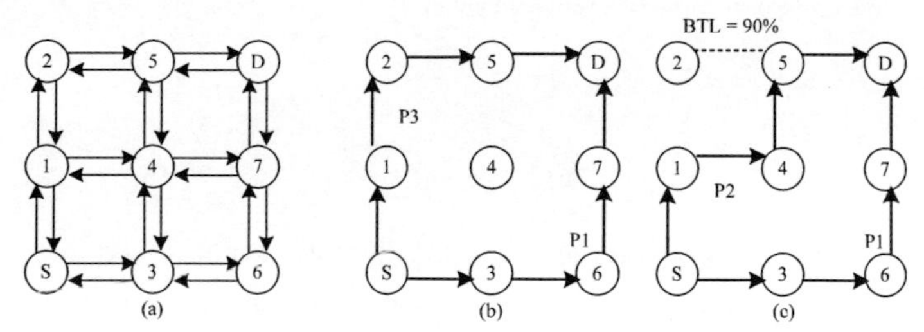

Fig. 1 3x3 grid (a) all possible links (b) no BTL (c) with 90% of l_{25} is utilized for BTL

As shown in fig. 1, case (b) there is no BTL on any link and in case (c) 90% of link 2-5 (l_{25}) has been used for background traffic in P_3. The selection of paths "P_1 & P_3" gives a higher throughput than selecting P_1 & P_2 for case (b), while the selection of P_1 & P_2 causes throughput degradation due to the mutual interference. For the case (c), the selection of P_1 & P_3 causes more throughput degradation due to the existing BTL on P_3.

The motivation of this work is to determine the RDM routes by analyzing the mutual interference of a path, I_r standing for *interference index* of the rth path and the BTL of a path, T_r standing for the *BTL index* of the rth path. Discovered routes are used to distribute the packets of a single flow among RDM routes and the performance is compared by computing the sustainable throughput of the network. The sustainable throughput is computed assuming that each node can schedule the packets for transmissions with the aid of a central entity. This model can be used to compute the throughput only for unidirectional transmissions. The total number of paths that can be used to distribute the traffic is restricted to 2 since use of more paths results in more mutual interference between paths. The total load of a path, called as (*PathLoad, PL*) is computed by combining I_r and T_r as shown in (1). The weight factor α can have values between 0 and 1.

$$PL_r = \alpha I_r + (1-\alpha)T_r \qquad 1 < r \leq m \tag{1}$$

Paths with the least PL are chosen as RDM paths. When several paths have equal PL, the paths with the lower number of hops are chosen. m is the total number of possible paths between the source and the destination. RDM routing first selects the least congested path (path with lowest interference and BTL), as the primary path, denoted by P_1. The procedure of determining RDM paths based on the values of I_r and T_r is explained later using the theory and terminology explained in section 2.1.1. The max-flow problem and graph theory used in [10] is extended to determine the RDM routes in this work. More details about these theories can be found in [10, 13].

2.1.1 TERMINOLOGY

The ***conflict graph*** shows the connectivity between links that are interfering at the same time (i.e. conflicting links) in a network. The conflicts between links are assessed based on the physical interference model. Suppose node n_i wants to transmit a packet to node n_j, while node n_p is sending packets to n_q simultaneously. (2) shows how to compute the weight factor ω_{ij}^{pq}, indicating how strong the interference between the link l_{pq} and link l_{ij} is.

$$\omega_{ij}^{pq} = \frac{SS_{pj}}{\dfrac{SS_{ij}}{SNR_{thresh}}} \tag{2}$$

SS_{pj} and SS_{ij} denote the signal strength at the node n_j due to the transmissions from node n_p and node n_i respectively. SNR_{thresh} is the SNR threshold. The signal strength measured at the receiver, node n_j (i.e. SS_{pj} and SS_{ij} due to both transmissions) is determined by the *transmitting power* P_{Tx} at the sender, the *path loss* P_{Loss} during the transmission and the *receiver sensitivity* Sen_{Rx}. P_{Loss} *is* computed by using a free space propagation model as shown in (3), in which f stands for wireless frequency in GHz and d stands for the distance between the transmitter and the receiver in *km*. If P_{Tx} at node n_i, P_{Loss} between node n_i and n_j, and Sen_{Rx} at node n_j are known, the signal strength, SS_{ij} can be computed as in (4).

$$P_{Loss} = 92.45 + 20\log_{10} f + 20\log_{10} d \tag{3}$$

$$SS_{ij} = (P_{Tx})_i - (P_{Loss})_{ij} - (Sen_{Rx})_j \tag{4}$$

A group of links, that can be active simultaneously without interfering with each other can be computed by discovering the ***Independent Sets*** (ISs) of a given conflict graph [13]. Finding all ISs of a given graph may be *NP-hard*. A difficulty is to find the *maximum independent sets*, which have the maximum size. The algorithm used in [10] is applied to find ISs in this work.

The standard ***Max-Flow problem*** [9] has been extended by 2 additional constraints to consider the wireless interference. It has been used to find the sustainable throughput of a given mutihop wireless network. The sustainable throughput is computed by maximising the outgoing data flows from the source to all its neighbouring nodes, assuming the given network has no packet losses. All links belonging to an IS are scheduled to be active simultaneously to guarantee the optimal scheduling.

2.2 INTERFERENCE COMPUTATION: 3x3 GRID TOPOLOGY

As the first step, all possible paths from the sender to the destination should be found together with the interference of each link in the network. Here, the interference between links is computed using a conflict graph. The weighted conflict graph is generated, assuming all the nodes are transmitting simultaneously. The weight factors are computed using a physical interference model as explained in section 2.1.1. As shown in fig. 1a, there exist 24 links due to the size of the corresponding conflict graph. The equivalent conflict matrix is given in Table 1. The computation of weight factors for this example should be considered for mainly 3 different types of links as shown below.

Simultaneously Not Active Links: If there are two links that cannot be active simultaneously, the weight factor is not computed. This is indicated as "-" in the conflict matrix. There are two possibilities for having such links.

Case 1: if 2 links have one common node, then they cannot be active at the same time. The link l_{S3} and link l_{S1} (node n_S cannot transmit simultaneously to both node n_1 and node n_3) or the link l_{1S} and link l_{3S} (node n_S cannot receive simultaneously from both node n_1 and node n_3). The latter case is similar to reality when avoiding the hidden node problem in WLAN with RTS/CTS messages.

Table 1 Part of conflict matrix of a 3x3 grid

link	S-1	S-3	1-S	1-2	1-4	2-1
S-1	-	-	-	-	-	-
S-3	-	-	-	-	-	2.51
1-S	-	-	-	-	-	-
1-2	-	-	-	-	-	-
1-4	-	-	-	-	-	-
2-1	-	0.63	-	-	-	-
2-5	2.51	0.63	-	-	-	-
3-S	-	-	-	0.63	2.51	0.63
3-4	-	-	2.51	0.63	-	0.63
3-6	-	-	2.51	0.63	2.51	0.63
4-1	-	2.51	-	-	-	-
4-3	2.51	-	-	-	-	2.51
4-5	2.51	2.51	-	-	-	2.51
4-7	2.51	2.51	-	-	-	2.51
5-2	0.63	0.63	0.63	-	2.51	-
5-4	0.63	0.63	0.63	2.51	-	-
5-D	0.63	0.63	0.63	2.51	2.51	-
6-3	0.63	-	0.63	0.63	0.63	0.63
6-7	0.63	2.51	0.63	0.63	0.63	0.63
7-4	0.63	0.63	0.63	0.63	-	0.63
7-6	0.63	0.63	0.63	0.63	2.51	0.63
7-D	0.63	0.63	0.63	0.63	2.51	0.63
D-5	0.63	0.63	0.63	0.63	0.63	0.63
D-7	0.63	0.63	0.63	0.63	0.63	0.63
num of Interfering links	4	4	2	2	6	4

Case 2: If senders of both links are within each other's carrier sensing range, they cannot be active simultaneously due to the CSMA/CD mechanism used in the MAC protocol. Both node n_s and node n_1 are in each others CS range, so link l_{S3} and link l_{12} cannot be active simultaneously.

Table 2 Possible Paths of a 3x3 grid topology in fig. 1(a)

Path id	Used Links	Interference	BTL	Hop Count
1	S-3, 3-6, 6-7, 7-D	4	0	4
2	S-1, 1-2, 2-5, 5-4, 4-7, 7-D	6	0	6
3	S-1, 1-2, 2-5, 5-4, 4-3, 3-6, 6-7, 7-D	6	0	8
4	S-1, 1-4, 4-5, 5-D	6	0	4
5	S-1, 1-4, 4-7, 7-D	6	0	4
6	S-1, 1-4, 4-3, 3-6, 6-7, 7-D	6	0	6
7	S-3, 3-4, 4-1, 1-2, 2-5, 5-D	6	0	6
8	S-3, 3-4, 4-5, 5-D	6	0	4
9	S-3, 3-4, 4-5, 7-D	6	0	4
10	S-1, 1-2, 2-5, 5-D	4	0	4
11	S-3, 3-6, 6-7, 7-4, 4-1, 1-2, 2-5, 5-D	6	0	8
12	S-3, 3-6, 6-7, 7-4, 4-5, 5-D	6	0	6

Simultaneously Active Interfering Links: There are links that can be active simultaneously, but they can be interfering with each other's transmission. For example, one link's destination is within another sender's interference range. For this case, the weight factor (e.g. ω_{S1}^{25}) is computed according to (2) with values used in simulated scenarios (physical layer data rate is set to 1M, Sen_{Rx} is set to *-76dBm*, P_{Tx} is set to *100mw, f* is set to *2.45 GHz* and SNR_{thresh} is set to *4dB)*. P_{Loss} at n_1 due to the communication between n_2 and n_5 can be computed using (3). The distance between n_2 and n_1 is set to 1 unit equal to 0.6 km in the simulator. All nodes in the network have identical properties. P_{Loss} is computed as 95.79dB. Using (4), SS_{21} and SS_{S1} due to the transmission from n_2 and n_s is computed as 1mW ($SS_{21} = 20dBm - 95.79dB - (-76dBm) = 0.21dBm = 1mW$). Since the distance between n_S and n_1 is also 1 unit, value of SS_{S1} is equal to 1mW. Therefore, ω_{s1}^{25} between link l_{S1} and link l_{25} using (2) is computed as 2.51.

Table 3-a Conflict matrix – to compute mutual interference between P2 and P1

link	L_{21}(S-1)	L_{22}(1-S)	L_{23}(1-4)	L_{24}(4-1)	L_{25}(4-5)	L_{26}(5-4)	L_{27}(5-D)	L_{28}(D-5)
L_{11}(S-3)	-	-	-	1	0	0	0	0
L_{12}(3-S)	-	-	1	-	-	1	0	0
L_{13}(3-6)	-	1	1	-	-	1	0	0
L_{14}(6-3)	0	0	0	0	0	0	0	0
L_{15}(6-7)	0	0	0	0	0	0	0	0
L_{16}(7-6)	0	0	1	-	-	1	1	-
L_{17}(7-D)	0	0	1	-	-	1	-	-
L_{18}(D-7)	0	0	0	0	1	-	-	-

Table 3-b Conflict matrix - to compute mutual interference between P3 and P1

link	L_{31}(S-1)	L_{32}(1-S)	L_{33}(1-2)	L_{34}(2-1)	L_{35}(2-5)	L_{36}(5-2)	L_{37}(5-D)	L_{38}(D-5)
L_{11}(S-3)	-	-	-	1	0	0	0	0
L_{12}(3-S)	-	-	1	-	0	0	0	0
L_{13}(3-6)	-	0	0	-	0	0	0	0
L_{14}(6-3)	0	0	0	0	0	0	0	0
L_{15}(6-7)	0	0	0	0	0	0	0	0
L_{16}(7-6)	0	0	0	-	0	0	1	-
L_{17}(7-D)	0	0	0	-	0	0	-	-
L_{18}(D-7)	0	0	0	0	1	0	-	-

Simultaneously Active Non-Interfering Links: If both the sender and the destination of the link under study are out of the interference range, these links do not interfere with each other. For example, node n_1 cannot interfere with n_7's transmission. P_{Loss} at n_1 due to the transmission from n_7 can be computed as 101.817dB for the distance of 2 units (i.e. 1.2 km). The received signal strength at n_1 due to the transmission from node n_7 is computed as 0.262mW ($SS_{71} = 20dBm - 101.817dB - (-76dBm) = -5.817dBm = 0.262mW$). As computed earlier, $SS_{S1} = 1mW$. Therefore the interference to the link, l_{S1} from link l_{7D} can be computed as 0.63 using (2).

Table 1 shows the part of the conflict matrix. Here the weight factor of 2.51 refers to the interfering links and the weight factor of 0.63 refers to the non interfering links. The conflict matrix is used to calculate how many links are interfering with each single link.

There are 12 possible combinations of paths for the connectivity between n_S and n_D (Table 2). The primary path, P_1 out of 12 paths is selected avoiding paths which have a higher number of interfering links, a higher BTL and a higher number of hop counts (the interference and the BTL of a path are equal to the maximum interference and BTL of the link that forms the path). Considering these 3 factors, path 1 or 10 can be selected as P_1. It can be observed that since the node n_4 is located in the middle of the grid network, the links connected to this node are interfering more. Once P_1 has been selected, all the other disjoint paths w.r.t P_1 have to be evaluated further to select more RDM paths.

Computation of Mutual Interference w.r.t. Primary Path: After knowing P_1, there are 2 more possible paths that exist between n_S and n_D, which do not share nodes with P_1. They are P_2 consisting of nodes n_1, n_4 and n_5 and P_3 consisting of nodes n_1, n_2 and n_5. But, both P_2 and P_3 have nodes n_1 and n_5 in common. Since RDM paths have to be node disjoint, either P_2 or P_3 has to be selected to be used together with the P_1. This is done by measuring the mutual interference of P_2 and P_3 w.r.t. P_1. The mutual interference index of rth path w.r.t. P_1 is denoted as I_{1r} and can be computed as given in (5). k_r refers to the total number of links in the rth path. L_{1j} denotes j^{th} link of the primary path, P_1. Similarly L_{ri} denotes i^{th} link of the rth path. $L_{1j}L_{ri}$ becomes one if both links are interfering and becomes zero if they are not interfering. That means, for the above example, if the weight factor between L_{1j} and L_{ri} is 2.5, $L_{1j}L_{ri}$ becomes one since they are interfering links in the conflict matrix.

$$I_{1r} = \sum_{j=1}^{k1} \sum_{i=1}^{k_r} L_{1j} L_{ri} \qquad (5)$$

For the 3x3 grid, an expansion of (5) can be shown in a matrix format, where both k_1, k_2 and k_3 are having 8 links in each path. The earlier generated conflict matrix is used to compute the mutual interference of P_2 and P_3 as

Table 4 Computation of PL_2 and PL_3

α	$1-\alpha$	Case (b) $I_{12} = 1$, $I_{13} = 0.33$, $T_2 = 0$, $T_3 = 0$		Case (c) $I_{12} = 1$, $I_{13} = 0.33$, $T_2 = 0$, $T_3 = 0.9$	
		PL_2	PL_3	PL_2	PL_3
0.7	0.3	0.70	0.23	0.70	0.50
0.3	0.7	0.30	0.09	0.30	0.72

shown in Table 3. These tables are sub sets of the complete conflict matrix, showing only the links in P_1, P_2 and P_3. Here the 1 represent the interfering links (weight factor = 2.5) and 0 represents the non interfering links (weight factor = 0.63). The mutual interference between "P_1 & P_2" and "P_1 & P_3" can be computed as, $I_{12} = 12$ and $I_{13} = 4$ respectively.

2.3 SELECTION OF RDM PATHS

BTL of a path can be computed as in (6). T_r denotes the background traffic of rth path given as a percentage w.r.t the available link capacity. T_{i_r} is equal to the traffic load of ith link in the rth path. k_r represents the number of links in the rth path. T_r is considered as the maximum BTL of an individual link.

$$T_r = \max(T_{1r}, T_{2r}, ..T_{ir}.T_{k_r}) \qquad (6)$$

By knowing the interference index and BTL index of each available paths, only node disjoint paths with less PL are chosen as RDM paths using (1). As shown in fig. 1, case (b) there is no BTL on any path and in case (c) 90% of l_{25} has been used for background traffic in P_3. Table 4 shows the computed PL_2 and PL_3 for different values of α. In this computation, the interference index has been normalized w.r.t. the maximum mutual interference of a path, i.e. $I_{12} = 12$. Different values of α gives different selections as shown in Table 4. Values of α has been considered based on the conclusions made using the simulation results; if the BTL of a path is lesser than 50% of the available link capacity, α has to be weighted more to represent the behaviour of mutual interference. Therefore, case (b) selects P_3 as the other node disjoint path to be used with the primary path P_1 and case (c) selects P_2 as the other path avoiding the highly congested path of P_3.

Table 5 ISs in 3x3 grid (for P_1 and P_3)

IS id	Links
IS_1	l_{S1}, l_{5D}, l_{67}
IS_2	l_{S1}, l_{7D}
IS_3	l_{12}, l_{36}
IS_4	l_{12}, l_{67}
IS_5	l_{12}, l_{7D}
IS_6	l_{S3}, l_{25}, l_{7D}
IS_7	l_{25}, l_{36}
IS_8	l_{25}, l_{67}
IS_9	l_{S3}, l_{5D}
IS_{10}	l_{36}, l_{5D}

2.4 COMPUTATION OF SUSTAINABLE THROUGHPUT

Once the RDM routes are found, the ISs of given RDM paths can also be computed using the same conflict matrix. After knowing the ISs, all constraints in extended max-flow problem as given in [10] can be completed to compute the sustainable throughput. The sustainable throughput is computed as 0.667, i.e. 66.7% of the link capacity. This is achieved by scheduling the links in 3 ISs (IS_1, IS_3 & IS_6) out of total 10 ISs discovered as given in Table 5. IS_1 and IS_6 are maximum ISs, containing the maximum number of links that can be active simultaneously. The traffic is evenly distributed between the RDM paths of P_1 and P_3. ISs in selected RDM paths are assigned an equal active time of 1/3 when scheduling. How packets are transmitted with scheduling is shown in fig. 2-(a). Similarly, ISs and sustainable throughput can be computed for the use of P_1 and P_2 (fig. 2-(b)). Different paths lead to different scheduling. Sustainable throughput in this case is computed as

0.429. It uses 5 ISs. When the paths are not fully RDM, less links are included in IS, which lead to a lower sustainable throughput.

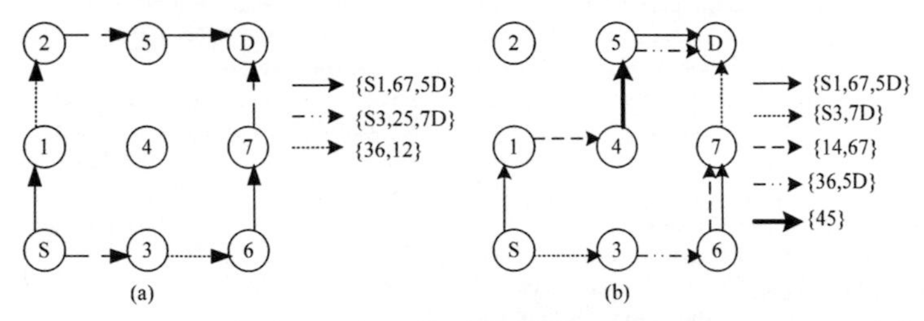

Fig. 2 Scheduling used (a) P_1 and P_3 (b) P_1 and P_2

3. EVALUATION OF ANALYTICAL RESULTS

The comparison of the results is done by computing the percentage of the sustainable throughput. This is measured in simulated scenarios by computing the maximum number of bits that an application can send until there is no packet loss at the link layer. For example, if an application can send a maximum of 20 packets of 1000 bytes each in a second, without loosing any packet at the link layer, the sustainable throughput is computed as "(1000 x 20)x8 bps". Since the PHY mode is set to 1M, the percentage of the sustainable throughput is computed as 16%. For analytical results, this is computed using the extended max-flow problem as explained in section 2.1.1. The OPNET simulator has been extended to implement the RDM protocol [14], by extending one of the reactive MANET protocols (DYMO [15]). The route discovery process of DYMO has been changed to compute the BTL at the WLAN MAC layer. The mutual interference is computed by counting the number of route request packets that are heard by each node. Implementation details are not discussed here due to the space limitation and can be found in [14]. In this simulation, the data of a UDP stream (generated as a video stream) is distributed among selected RDM routes. Single Path (SP) routes are found using the standard DYMO protocol by sending data without any distribution.

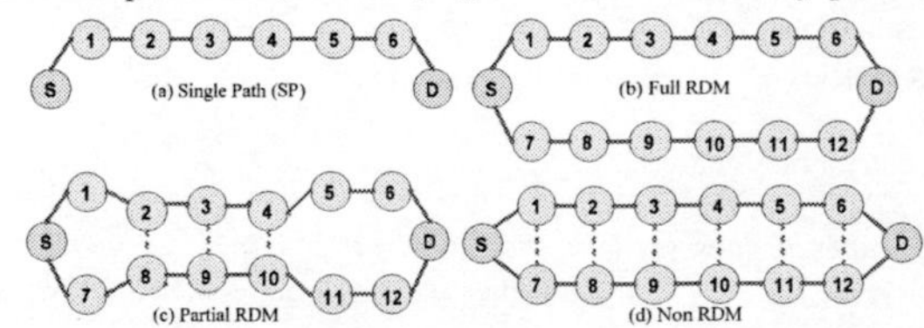

Fig. 3 Basic Topologies

Following terms are used when creating the scenarios to compare the results in different types of networks. ***SP:*** this uses only one path between source (S) and destination (D) to send UDP data. SP is selected based only on the hop counts. ***Full RDM (FRDM):*** This uses multiple paths to send UDP data between S and D. In this case, the mutual interference between all the links of each selected path are zero. ***Non RDM (NRDM):*** This also uses multiple paths to send UDP data. In this case, all or some of the links of each selected path are not free of interference between each other. If all are interfering, it is called as NRDM. If some are interfering, it is called as ***Partial RDM (PRDM).***

In general, the simulation results show a lower sustainable throughput than the analytically computed throughput. This is mainly due to following reasons:

- There is no scheduling performed in the simulation. This means the contention, defined in the 802.11b MAC layer (CSMA/CD) can reduce the network throughput further. Therefore, the simulated throughput is always below the analytical throughput.
- The analytical model assumes that there are no packet collisions, while this can happen in the simulation environment due to the hidden node problem. In order to reduce the hidden node problem, the simulations are run with RTS/CTS messages. Though the data packet are not lost due to hidden node, WLAN ACK and RTS/CTS packets can be lost, causing a degradation of the throughput due to retransmissions and finally dropping the data packet after few unsuccessful tries of RTS. Further, the simulation throughput can be reduced since some possible transmissions can be blocked with RTS/CTS messages.
- The analytical model computes the throughput without considering the overhead of the headers at the different layers. In the simulation, the overhead of UDP, IP and MAC headers occupy part of the bandwidth. Moreover, the analytical model does not consider the bandwidth taken to exchange WLAN layer ACK packets and RTS/CTS control packets.

Table 6 % of Sustainable Throughput

Scenarios	Analytical	Simulation	
	Uni UDP		Bi UDP
SP	33.3	23.1	5.76
FRDM	66.7	36.4	15.2
PRDM	50.0	28.8	5.44
NRDM	50.0	27.2	5.28

3.1 BASIC TOPOLOGIES

Fig. 3 shows the use of SP and different types of RDM routes in a simple network, where S denotes the sender, n_S and D denotes the destination node, n_D. Broken lines in the PRDM and NRDM scenarios show the interfering links.

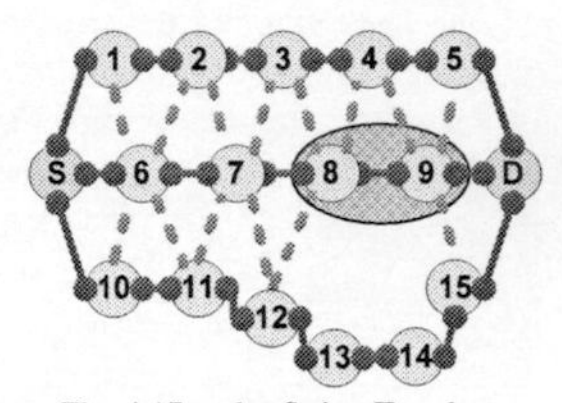

Fig. 4 17 nodes String Topology

Table 6 shows that the sustainable analytical throughput of the SP scenario is 33.3%, 1/3 of the given link capacity, whereas with FRDM routes it goes up to 66.7%. This shows that the simultaneous use of FRDM together with optimized scheduling can double the sustainable throughput. In the SP scenario, all links are divided into 3 groups (ISs), and each group is assigned the equal amounts of time, utilizing 1/3 of the link capacity. In the FRDM scenario, 2 paths are scheduled exactly like in SP, showing double the throughput due to the use of two paths. In the NRDM scenario, where each intermediate node in one path interferes with the one in the other path, there are 4 ISs that have been chosen for scheduling. Since the links are conflicting (i.e. interfering), each link belongs to only one IS. The active time for one IS is given as 25%. This computes the throughput of one path as 25% and since 2 paths are used, the overall throughput can go up to 50%. PRDM scenario also gives 4 ISs.

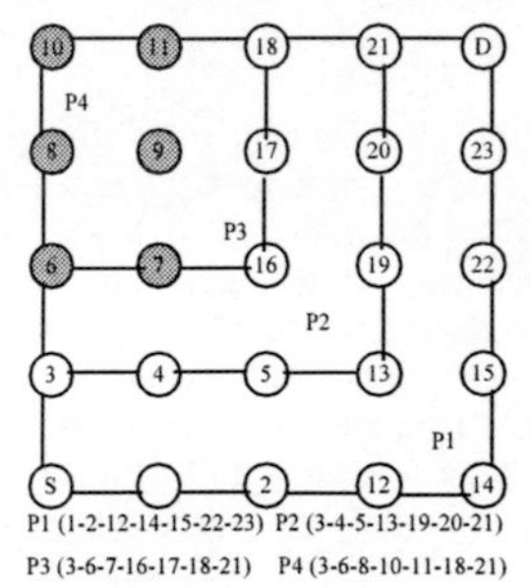

Fig. 5 5x5 Grid Topology

The simulation results also show the same behavior even without optimal scheduling as in analytical results. The simulated throughput is always lower compared to the analytical throughput, as mentioned earlier (no scheduling, contention delay, overhead of headers and RTC/CTS control messages). Simulations have been done to evaluate the sustainable throughput for bidirectional UDP data transmission too. Compared to FRDM, simultaneous use of NRDM limits the sustainable throughput due to interference between paths. But, NRDM still performs better than SP in unidirectional communications. With bidirectional traffic, there are more contention delays in the network. This is due to the fact that the traffic arriving from the opposite direction may increase the probability of having collisions between links. The consequence is the exponentially increased contention delay at the WLAN layer. Therefore, the performance gain of FRDM w.r.t. bidirectional communication is 2.62, whereas it is 1.6 for unidirectional communications. With the use of bidirectional communications for NRDM and PRDM, the results show that there is no performance gain w.r.t. SP. Since there are already many interfering links and additionally there is no scheduling in the simulations, the dramatically increased contention delay may degrade the sustainable throughput drastically, when using interfering routes simultaneously. Additional simulation results which are not detailed here show that the performance of NRDM scenarios gets worse with the use of bidirectional traffic and the increase of hop lengths.

3.2 STRING TOPOLOGY

The scenarios discussed above only consider RDM paths with the same hop count and no BTL. Fig. 4 shows a 17 node network, where RDM paths have different hop counts. SP is having a BTL of 8% of the available link capacity due to the already existing unidirectional data traffic between node n_8 and node n_9. Without RDM routing, SP is chosen as the path with the lowest hop counts, i.e. the middle path. RDM routing avoids the selection of this as the primary path due to the BTL. It selects only the upper and lower paths as FRDM paths. In order to create NRDM paths, all three paths (i.e. upper, lower and the middle) have been used to distribute packets in this scenario.

Table 7 String Topology - % of Sustainable Throughput

Scenarios	Analytical	Simulation			
		Uni	Per. Gain	Bi	Per. Gain
SP	33.3	20.8	-	7.68	-
FRDM	60.0	32.0	1.54	14.4	1.875
NRDM	56.0	15.2	0.73	5.92	0.77

Table 7 shows that the sustainable throughput of SP in fig. 4 degrades compared to SP used in fig. 3. This is mainly due to the BTL used. These results show that it is better to avoid the path with BTL even though it is the shortest path. Use of all three routes results in creating NRDM routes for this topology and shows a degrading performance for the simulation results due to the mutual interference between paths. On the other hand, analytical results show relatively higher throughput for the NRDM scenario. This is due to the distribution of traffic among three paths together with optimum scheduling by avoiding the simultaneous use of interfering links. Table 7 computes the performance gain of RDM routes w.r.t SP throughput for simulation results. It shows that FRDM always has a throughput gain when using real 802.11 based multihop ad hoc networks.

3.2.1 5x5 GRID TOPOLOGY WITH BTL

This scenario consists of 25 nodes. The discovered paths between the sender (n_S) and destination (n_D) are shown in fig. 5. A unidirectional UDP stream is added to each link connected to the 6 nodes: (n_6 & n_7), (n_8 & n_9) and (n_{10} & n_{11}) as BTL.

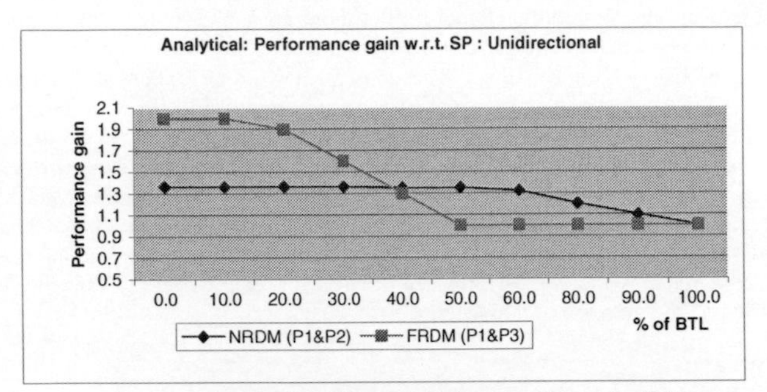

Fig. 6 Analytical Results - Performance Gain vs % of BTL

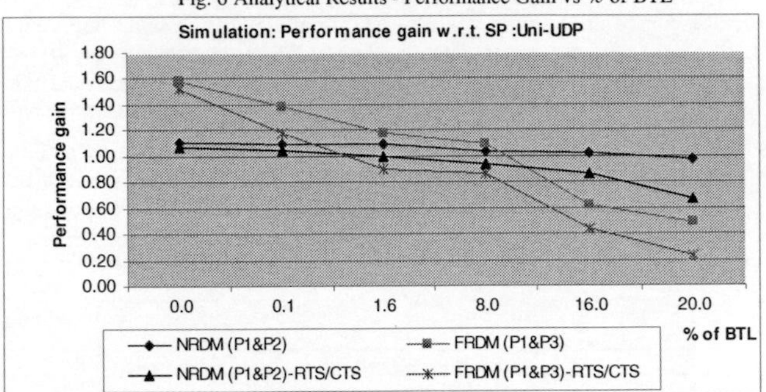

Fig. 7 Simulation Results (Uni UDP) - Performance Gain vs % of BTL

This scenario is used to compare the sustainable throughput with the increase of the BTL. When using RDM routing, P_1 is selected as the primary path (with zero BTL and having least interference). Computation of mutual interference w.r.t. P_1 gives the highest mutual interference for path P_2 (NRDM routes) and no mutual interference for paths P_3 and P_4 (FRDM routes). This scenario represents 3 different types of routing; SP with only P_1, NRDM with P_1 and P_2, and FRDM with either (P_1 and P_3) or (P_1 and P_4). Analytical results (fig. 6) show that shows that the sustainable throughput of FRDM is doubled compared to SP when the link l_{67} is carrying lesser BTL (< 40%). In this scenario, NRDM paths do not carry any BTL, but the performance is degrading with the increase of the BTL on the link l_{67}. When using FRDM, the primary path (P_1)

is always used to send the data at its maximum rate (i.e. 33.33% without any BTL in the P_1). For example, when the BTL is 40% of the link capacity, sustainable throughput of FRDM is computed as 43.33%. This is achieved by sending 33.33% on path P_1 and 10% on path P_3. The link l_{67} on P_3 has been used by BTL and therefore available link capacity is only 60%. With optimum scheduling, 10% of the available capacity has been used in P_3 for the current communication.

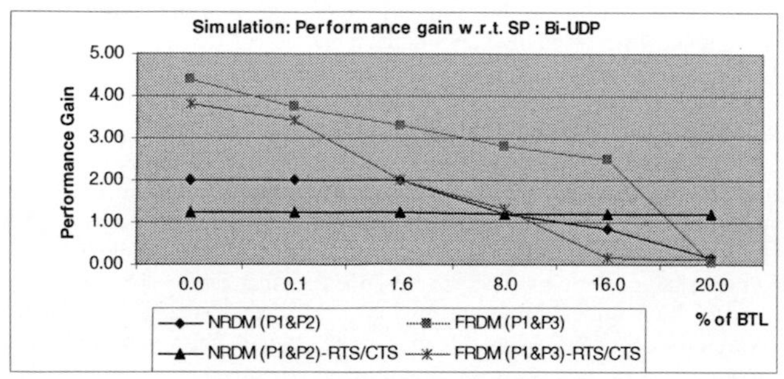

Fig. 8 Simulation Results (Bi UDP) - Performance Gain vs % of BTL

Simulation results for unidirectional UDP as shown in fig. 7, compare the performance gain with and without enabling RTS/CTS messages. This shows that both, for NRDM and FRDM the performance degrades with the use of RTS/CTS messages. Use of RTS/CTS increases blocking of nodes due to exposed node problem. The throughput degrades as more nodes are blocked, when there exists more interfering links as in NRDM. There are no interfering links in FRDM, use of RTS/CTS reduces the sustainable throughput due to BTL in P_3, which uses RTS/CTS for the BTL as well. These results show that when the links are congested with a BTL beyond 8% of the link capacity, there is no performance gain that can be achieved when using FRDM for unidirectional communications.

Fig. 8 shows the performance gain of FRDM and NRDM routes when using bidirectional UDP data transmissions. These results also show that RTS/CTS degrade the performance since the blocking probability caused by the exposed node problem is much higher when using bidirectional transmissions. In summary, this shows that FRDM can gain much higher performance (a factor between 2 & 4) with very low BTL values. Simultaneous use of FRDM routes show performance improvements with a BTL of 8% or less (when using RTS/CTS) and 16% or less (without RTS/CTS).

4. CONCLUSION

This paper has discussed an analytical model to assess the interference aware multipath routing scheme called RDM routes. Analytical results are compared together with simulation results by computing the sustainable throughput of different network topologies. Compared to previous work of analyzing network throughput considering the impact of interference, this work pioneers determination of multiple routes with minimum interference by considering two factors: assessing the mutual interference between paths and background traffic of the paths. None of the previous work has compared analytical results with simulation results taken from a real implementation of interference aware routing. This work gives a detailed analysis of the behavior of interference aware multipath routing considering both unidirectional and bidirectional communications. Furthermore, the computation of sustainable throughput when using multipath routes has been extended to consider the impact of BTL.

Several groups of scenarios have been examined, to evaluate the sustainable throughput for the use of RDM routes simultaneously. Analytical results give the optimum throughput with scheduling of interfering links, whereas simulation results computes the throughput, considering the contention in 802.11 networks. Analytical results show that simultaneous use of RDM routes perform better than SP. It gives the maximum sustainable throughput when using FRDM routes.

Simulation results also show that FRDM routes always perform better than SP, NRDM and PRDM. If the SP is congested or RDM paths have more mutual interference (as in fig. 4), sustainable network throughput degrades. The degradation of the sustainable throughput of interfering routes and SP is more when using bidirectional data transmissions since the data traffic is originated by nodes from both directions introducing more contention between links. The simulation results prove that FRDM routes perform better than SP even without any optimal scheduling as used in the analytical model. The performance gain that can be achieved with FRDM routes is much higher with bidirectional data transmissions (factor between 2 & 4 with lower BTL), because FRDM alleviates the contention by distributing packets among non-interfering links.

REFERENCES

1. S.J. LEE and M. GERLA, *AODV-BR: Backup Routing in Ad hoc Networks*, in *Proceedings of IEEE WCNC*. 2000: Chicago.
2. K. KULADINITHI, M. BECKER, C. GÖRG, and S. DAS. *Radio Disjoint Multi-Path Routing in MANET*. in *CEWIT (Center of Excellence in Wireless and Information Technology) 2005 Conference*. Dec 2005. Stony Brook.
3. S.J. LEE and M. GERLA, *Split Multipath Routing with Maximally Disjoint Paths in Ad hoc Networks*, in *Proceedings of IEEE ICC 2001*. 2001: Helsinki, Finland. p. 3201-3205.
4. A. TIMM-GIEL, K. KULADINITHI, P. HOFMANN, and C. GOERG, *Wireless and Ad Hoc Communications Supporting the Firefighter*, in *16th IST Mobile Summit*. June, 2006: Myconos, Greece.
5. P. GUPTA and P.R. KUMAR, *The Capacity of Wireless Networks*. IEEE Transactions on Information Theory, 1998.
6. J. LI, C. BLAKE, D. S. J. DE COUTO, H. I. LEE, and R. MORRIS, *Capacity of Ad Hoc Wireless Networks*, in *ACM SIGMOBILE*. July 2001: Rome, Italy.
7. P. NG and S. LIEW, *Throughput analysis of IEEE802.11 multi-hop ad hoc networks*. IEEE/ACM Trans. Networking, June 2007.
8. KAIXIN XU, MARIO GERLA, and S. BAE. *How Effective is the IEEE 802.11 RTS/CTS Handshake in Ad Hoc Networks?* in *IEEE GlOBECOM, November 2002.*
9. H. P. WILLIAMS, *Model Building in Mathematic Programming*. 4th ed. Nov 2006: John Wiley & Sons Ltd. pp 59-92.
10. K. JAIN, J. PADHYE, V. PADMANABHAN, and L. QIU, *Impact of Interference on Multi-hop Wireless Network Performance*, in *Mobicom 2003*. Sep 2003: San Diego, California, USA.
11. T. COENEN, M. DE GRAAF, and R.J. BOUCHERIE, *An Upper Bound on Multi-hop Wireless Network Performance*, in *20th International Teletraffic Congress, ITC 20*. June 2007: Ottawa, Canada.
12. R. DE HAAN, R. J. BOUCHERIE, and J.V. OMMEREN, *The Impact of Interference on Optimal Multi-path Routing in Ad Hoc Networks*, in *20th International Teletraffic Congress, ITC 20*. June 2007: Ottawa, Canada.
13. GIBBONS, A.M., *Algorithmic Graph Theory*. 1985: Cambridge University Press. pp 1-119.
14. K. KULADINITHI, A. TIMM-GIEL, and C. GÖRG. *Implementation and Analysis of Radio Disjoint Multi-path Routing over DYMO in OPNET*. in *OPNETWork 2007*. Aug 2007.
15. I. CHAKERES and C. PERKINS, *Dynamic MANET On-demand (DYMO) Routing*. 2007, draft-ietf-manet-dymo-10, IETF Draft.

Key words – effective bandwidth, network traffic modeling, routing

Adam KOZAKIEWICZ[*†‡], Krzysztof MALINOWSKI[*†]

NETWORK TRAFFIC ROUTING USING EFFECTIVE BANDWITH THEORY

Optimal routing in a packet network is a difficult task – the amount of connections makes per-connection routing impractical and even using aggregated connections global optimization in the network is infeasible. The currently used algorithms are therefore suboptimal heuristics, using shortest path computation instead of e.g. throughput optimization. The quality of obtained results depends on the selected link weights, but optimization of those is again a difficult mixed integer programming problem. In this paper we discuss some proposed heuristics using the results of large deviation analysis (effective bandwidths) for throughput optimization. The more advanced approach to traffic modeling uses more of the available information and potentially allows for finding better solutions. The presented algorithms are tested through simulation and discussion of the results is included.

1. INTRODUCTION

Routing in a packet network is a simple task only as long as the only objective is to avoid unnecessary detours. Fast shortest path algorithms exist. Unfortunately, the shortest path is not always the best. For example, a direct route through a congested, low-bandwidth link is usually worse than a path with three hops, but using fast and almost empty links. This is just a simple example, showing a more complicated truth – for most quality metrics, optimal routing depends on a lot of information – from the basic structure of the network as a graph, through the capacities of the links, to the requirements of each connection (note that some of this information can be used in the shortest path algorithms by introducing weights for each link). If the path for a connection is selected without judging its influence on all other connections, optimal routing cannot be achieved. The routing for the whole network is a difficult optimization problem, far beyond the capabilities of existing exact solvers, so in most cases we have to rely on heuristics.

Most of the heuristics proposed so far use rather simple traffic models. While simplicity can certainly be an advantage, we felt that using a more complete traffic description could result in better routing, using the multiplexing potential of the traffic streams more effectively. For this reason we adapted the effective bandwidth theory, developed mostly for connection admission and pricing, to the routing problem. In this paper we propose three heuristic algorithms using this theory, show the results of tests in a simulated network and comment on their advantages and disadvantages.

* Warsaw University of Technology, Institute of Control and Computation Engineering
ul. Nowowiejska 15/19, 00-665 Warszawa, Poland
† NASK – Research and Academic Computer Network
ul. Wąwozowa 18, 02-796 Warszawa, Poland
‡ adam.kozakiewicz@nask.pl

In Section 2 we describe in more detail the routing problem and present the simple heuristic used as a base for our new algorithms. Section 3 introduces the effective bandwidth mathematical framework, used in algorithms presented in Section 4. Section 5 presents the setting of the simulation experiment, with results of the experiment, along with some discussion, presented in Section 6. Finally, in Section 7 we give a summary, noting the most important conclusions and outlining further research plans.

2. HEURISTIC ROUTING IN IP NETWORKS

The approach used currently in the Internet is shortest path routing – instead of solving a global optimization problem for all demands, the shortest path in the network according to some metric is chosen for each connection. Protocols using this technique include the (mostly historical) RIP family of protocols, using Bellman-Ford shortest path algorithm and hop count as a metric, and OSPF and IS-IS protocols, using Dijkstra's shortest path algorithm with a user-defined metric, integer in both protocols, but with a different range of allowed values. With a metric as simple as hop count these protocols are definitely suboptimal, with no focus on quality of the resulting path and performance of the whole network.

In the case of OSPF and IS-IS this behavior can be changed by modifying link weights. Doing this well is, however, quite difficult. Simple heuristics can offer some improvement by promoting faster links, but taking into account actual demands is more difficult. The problem can be formulated mathematically as a two-level optimization problem (see [7]): a mixed integer programming problem optimizes the weights, but the paths (and therefore actual bandwidth allocation) for given weights need to be found using the shortest path algorithm – basically a special case of a linear integer programming problem. In theory, this formulation defines optimal weights, but the problem is very difficult. Solving it exactly even for medium-sized networks is not possible using off-the-shelf equipment and solvers; for large networks it is not solvable at all. Heuristic optimization algorithms must again be used. Luckily the results obtained from them are often very good, but even these algorithms are slow.

Quality of OSPF routing can also sometimes be improved by using ECMP (Equal Cost Multiple Paths): if a router cannot identify the next hop for a given destination, because several paths have equal cost and start with different outgoing links, then the router splits traffic equally between those links. Note that this task is not trivial, as allowing packets from one connection to travel using different paths introduces unwanted delay jitter and requires packet reordering.

One of the alternatives is the algorithm proposed by Jaskóła and Malinowski [3] [4]. This algorithm also assigns paths to connections one at a time, but each decision takes into account the previous ones. This mechanism uses link prices resulting from the Optimization Flow Control algorithm proposed by Low [6]. This algorithm is an application of price-based decomposition of the fixed-routing bandwidth allocation problem:

$$\begin{aligned} &\max_{x_s \in [m_s, M_s]} \sum_{s \in S} U_s(x_s) \\ &\forall l \in L \quad \sum_{s \in S_l} x_s \leq c_l \end{aligned} \tag{1}$$

where the source rates x_s with given bounds m_s and M_s are the decision variables (source rates), S is the set of sources, L – the set of links, c_l – the capacity of link l, S_l – the set of sources using link l, and U_s is the utility function for source s. The algorithm after decomposition is an iterative procedure, in which each link has an associated nonnegative price p_l, modified depending on the load on this link and each source collects the prices on its path and uses the total to find a rate by using the inverse utility function.

Note that the authors of this paper have developed a similar algorithm using effective bandwidth theory. In this case loss probabilities are calculated for each link and the difference between the computed and expected loss probability is used to modify the price. Details can be found in [5] and will soon be published elsewhere. This version of the algorithm allows more precise control over packet loss probability and uses more information about traffic, but is slower.

Jaskóła's algorithm has a simple structure:

Heuristic Routing Algorithm [HRA]

1. Build a list of all sources and sort it in descending order with respect to the upper bound on the offered traffic.
2. Remove the first source from the list and find a shortest route for it using the Dijkstra algorithm with a dynamic metric.
3. Find new metrics by applying Optimization Flow Control to find a new steady state of the network.
4. If the list of sources is not empty, return to step 2 (note that the metrics have changed – the next source can have a different path even if it enters and exits the network through the same routers).

The most important part of the algorithm is a proper selection of a dynamic metric. Jaskóła and Malinowski propose the following metric d^{price} for link l in the network:

$$d_l^{price} = p_l + \kappa M e^{-c/c_{max}} \tag{2}$$

where $\kappa \in [0,1]$ is a weighting factor, p_l is the the price for this link determined by the Optimization Flow Control algorithm (basically Lagrange multipliers of problem (1), initially all 0), c_{max} is the capacity of the largest link and M is the estimated highest possible steady-state link price.

The prices p_l in the metric provide a penalty for assigning already overloaded links to new routes, this penalty grows with the load on the link and is 0 on links with any free capacity. The second part of the metric ensures that larger links are preferred. The weight κ should be small, as in typical networks links are not heavily overloaded and maximal prices are never reached.

The algorithm is faster than exact optimization and gives good results. However, it has a number of weak points. An iterative traffic control algorithm has to be run after each step. This makes the algorithm rather slow for problems with many sources. Furthermore, the algorithm is based on rather limited information about the sources – only their maximal transfer rate is used. The utility function could contain more information, but it is notoriously difficult to identify and therefore likely arbitrarily chosen as a logarithmic function or similar. Even the average rates of the sources – easy to measure and important – are not used. These issues can be addressed with an application of effective bandwidth theory.

3. EFFECTIVE BANDWIDTH THEORY

The theory of effective bandwidth is an application of large deviation analysis to the case of computer network traffic. The goal of such methods is to estimate the asymptotic probabilities of rare events. In this respect it is a counterpart of central limit theorems, which analyze the small deviation asymptotic behavior, while the law of large numbers provides the expected value. Large deviations is a relatively new part of probability theory, developed in the XX century. The *large deviation principle* (LDP) is a family of results showing an exponential asymptotic behavior of distribution tails for different families of distributions and (in the *sample path* version) non-i.i.d. stochastic processes. The seminal reference for this theory is the book by Dembo and Zeitouni [2]. Good introduction to the theory is also presented in the second part of Cheng-Shang Chang's book [1]. Damon Wischik's dissertation [8] also provides a terse and difficult but deeper introduction.

Definition 1 For a given $t \geq 1$ and $\theta \geq 0$ the *effective bandwidth* α_X of process X (model of a traffic stream) at (θ,t) is defined as

$$\alpha_X(\theta,t) = \frac{1}{\theta t} \log \mathrm{E}\, e^{\theta X(0,t)} \tag{3}$$

where $X(0,t)$ is the total amount of traffic arriving in the interval $[0,t]$. E is used to denote the expected value of a random variable.

The effective bandwidth answers either of two question: what would the rate of a constant rate source have to be to have the same influence on other traffic as the analysed stream, or how much bandwidth of a multiplexed channel must be available for this connection. Since multiplexing with other streams is dependent on the characteristics of those streams, which may depend on the time scale, effective bandwidth is not a constant, but a function of two parameters, θ and t, called the space parameter (spacescale) and the time parameter (timescale) respectively. Even with this complication the effective bandwidths are quite useful due to the following theorem:

Theorem 1 (The inf-sup formula) The probability of overflow in a queue with a buffer of size B servicing a link with capacity C, fed by input flows $X_1, \ldots, X_n$ can be approximated as

$$-\log \mathrm{P}(overflow) \approx \inf_{t \geq 0} \sup_{\theta \geq 0} \left[\theta(Ct+B) - \theta t \sum_{i=1}^{n} \alpha_{X_i}(\theta,t) \right] \tag{4}$$

The approximation is good if n and C are large.

An important byproduct of the above approximation is the *critical point* of the queue – a pair of parameters $(\hat{\theta}, \hat{t})$ which describe the most probable scenario of overflow in the queue.

The critical timescale is easy to explain. The large deviation analysis provides the most probable sample path leading to an overflow, and $\hat{t}$ is the length of this path. In other words, $\hat{t}$ is the most probable time between the last time the queue was empty and the overflow. As such it is a very important information about the queue's state. If the critical timescale is very small, overflows are mostly caused by momentary culminations of traffic. If the critical timescale is large, reaching seconds, then long trends in traffic are the main cause of overflows.

The spacescale parameter can be interpreted as the sensitivity of traffic to buffer size changes. In other words, small values mean that the buffer size has little effect on loss probability, while large values suggest that the overflows are caused by relatively short peaks, that could be well controlled using a larger buffer. Critical spacescale is a good measure of the quality of multiplexing. If small, flows aggregate well and individual peaks are absorbed by stochastic averaging. Large values indicate that peaks cumulate often and are visible regardless of averaging.

In this work we use the effective bandwidths of two processes modeling the network traffic. One is fractional Brownian motion (FBM), described by three parameters: mean traffic rate μ, variance parameter σ^2 and the Hurst parameter H describing the self-similarity and long range dependence. The formulas for effective bandwidth and critical point (when all streams have the same parameters) are easy to obtain directly from equation (3) and simple (meaning of C and B is the same as in equation (4)):

$$\alpha_{\mathrm{fbm}}(\theta,t) = \mu + \sigma^2 \frac{\theta}{2} t^{2H-1} \tag{5}$$

$$\hat{t} = \frac{B}{C-\mu} \cdot \frac{H}{1-H} \tag{6}$$

$$\hat{\theta}=\frac{B+(C-\mu)\hat{t}}{\sigma^2\hat{t}^{2H}}=\frac{(C-\mu)^{2H}(1-H)^{2H-1}}{\sigma^2 B^{2H-1}H^{2H}} \tag{7}$$

The other model used in this paper is fractional Brownian motion with token bucket constraints. Unlike other results in this section this model is not found in reference sources and was developed by the authors [5]. Additional variables in this model are the parameters of the token bucket – c, the maximum sustainable rate and q – allowable burst size. Effective bandwidth of this process is much more complicated, but still available analytically. Critical point, however, must be found numerically. The effective bandwidth function is:

$$\alpha_{\text{fbm-tb}}(\theta,t)=\frac{1}{\theta t}\log\Big\{0.5\,\text{erfc}(\xi_0)+ \\ +0.5\exp\left[\theta t(\mu+0.5\sigma^2\theta t^{2H-1})\right]\cdot\left[\text{erf}(\xi_\eta-\xi_\theta)+\text{erf}(\xi_0+\xi_\theta)\right]+0.5e^{\theta(ct+q)}\text{erfc}(\xi_\eta)\Big\} \tag{8}$$

where $\xi_0=\mu t/\sqrt{2}\sigma t^H$, $\xi_\eta=[(c-\mu)t+q]/\sqrt{2}\sigma t^H$ and $\xi_\theta=\theta\sigma^2 t^{2H}/\sqrt{2}\sigma t^H$. The symbols erf and erfc denote the error function and complementary error function.

4. PROPOSED ALGORITHMS

In this work we considered three ways of introducing effective bandwidths in the heuristic routing algorithm proposed by Jaskóła and Malinowski. The first approach was trivial – using effective bandwidth instead of maximal rate in the sorting phase. This results in the following algorithm:

Heuristic Routing Algorithm with Effective Bandwidth Sorting [HRA-EBS]

1. Find an initial routing using a simple algorithm. If working routing tables are already available, use those.
2. Compute upper bounds for all sources using Optimization Flow Control algorithm.
3. Find a representative working point in the network by averaging the results from a number of links (all in general, but if some of the links are almost empty they may be ignored).
4. Compute effective bandwidths of all sources modeled as fractional Brownian motion using the representative working point.
5. Build a list of the sources sorted in decreasing order w.r.t. effective bandwidth.
6. Remove the first source from the list and find a shortest route for it using the Dijkstra algorithm with a dynamic metric.
7. Find new metrics by applying Optimization Flow Control to find a new steady state of the network.
8. If the list of sources is not empty, return to step 6.

A more advanced approach is to use the effective bandwidths in the traffic control algorithm. As was already mentioned in Section 2, such an algorithm was developed by the authors. This approach gives the following algorithm:

Effective Bandwidth Routing Algorithm [EBRA]

1. Find an initial routing using a simple algorithm. If working routing tables are already available, use those.
2. Compute upper bounds for all sources using the effective bandwidth based traffic control algorithm.
3. Find a representative working point in the network by averaging the results from a number of links (all in general, but if some of the links are almost empty they may be ignored).

4. Compute effective bandwidths of all sources modeled as FBM using the representative working point.
5. Build a list of the sources sorted in decreasing order w.r.t. effective bandwidth.
6. Remove the first source from the list and select a route for it using Dijkstra's algorithm using some metric defined by the current link load.
7. Calculate the total price as perceived by the source and set the source token bucket accordingly, as in the traffic control algorithm.
8. Recalculate loss probabilities after introducing the new source. If any of them exceed a given level P_{max}, run the traffic control algorithm (only for the sources with already selected routing).
9. If the list is not empty, return to step 6.

The traffic control algorithm using effective bandwidth gives more additional information about the state of the network than the original. Both algorithms provide prices, a good measure of the load level of a link, although the prices will obviously differ due to different traffic models. The effective bandwidth analysis also provides packet loss probability for each link. This gives us a choice of a link metric – we can either use the metric from the original algorithm, equation (2), or build a new metric based on the probabilities, as follows:

$$d_l^{\text{prob}} = \mathrm{P}(\textit{overflow on link } l) + \kappa \exp(-c_l / c_{\max}) \tag{9}$$

where κ is a very small positive weighting factor, ensuring that if links on several paths are almost empty and the probability of overflow is much lower than the desired value, then the wider path is preferred. With either metric, the additional information also allows us to avoid running the traffic control algorithm in every iteration.

The last algorithm proposed for comparison is based on the assumption that transport networks should not normally be overloaded and most links usually have significant headroom over simple sum of average loads from each flow. In this model the token bucket constraints are not very important, so they were eliminated from the model. This approach allows the use of simple mathematical formulas for the critical point and at the same time eliminates the need to keep track of the token bucket parameters, so the iterative traffic control algorithm is no longer necessary. Since some links can always become overloaded, we need to have a special case for them, but apart from that the algorithm can be kept very simple.

The special case is handled in the link metric. Normally, the metric defined in equation (8) is used, but overloaded links use a different metric:

$$d_l^{\text{cong}} = 2|L| + \exp\left(\sum_{s \in S_l} \mu_s - c_l\right) + \kappa \exp(-c_l / c_{\max}) \tag{10}$$

where $|L|$ is the number of links (L is the set of links) and S_l is the set of sources currently assigned to link l. This metric makes a path including that link worse than any path not including an overloaded link. At the same time it promotes paths with fewer congested links and depends on the amount of excess traffic, so that heavily congested links are avoided.

Fast Effective Bandwidth Routing Algorithm [FEBRA]

1. Find a representative working point in the network from formulas (6) and (7), assuming perfect distribution of the load and averaging the capacities of the links.
2. Compute the effective bandwidths of all sources modeled as FBM using the representative working point.
3. Build a list of the sources sorted in decreasing order w.r.t. effective bandwidth.
4. Remove the first source s from the list.
5. For each link compute the packet loss probability using and set the link's metric

according to (9). If the capacity of the link is exceeded, set the metric according to (10) instead.

6. Select a route for source s using Dijkstra's algorithm the the metric from step 5.
7. If the list is not empty, return to step 4.

5. EXPERIMENT SETTING

All algorithms were implemented in MATLAB and tested in a MATLAB-based simulator. This allowed us to identify candidates for actual implementation. The algorithms were tested extensively on a simple network with two bottlenecks. Tests on larger networks were also performed, with similar results, but the time required for simulation was rather prohibitive in this case, resulting in a small number of simulation runs. Since fractional Brownian motion is a long memory process, large number of simulations is necessary for acquiring meaningful results, especially if the number of sources is significant.

The algorithms proposed in this paper were compared with three others – the original HRA, OSPF with unit weights for all links and without ECMP (worst-case scenario, denoted as OSPF-1) and OSPF with optimized weights (denoted as OSPF-opt, simulations were run with ECMP enabled or disabled – the better result for each case is shown).

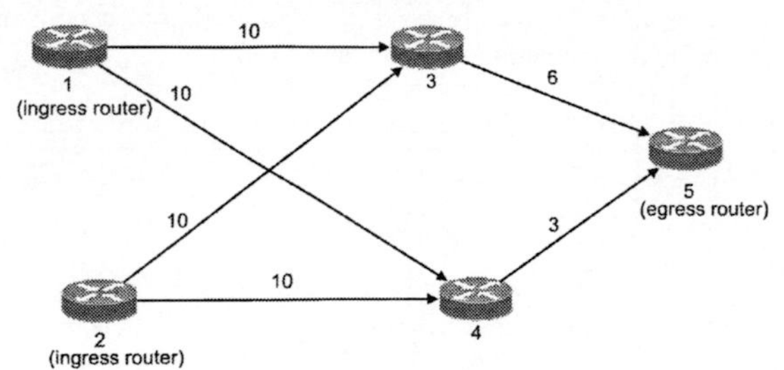

Figure 1: Layout of the network used for testing

The network used for testing is the same as in [3]. The layout of the network is presented in Fig. 1, together with link capacities. The buffer size on each link was set to 0.01. Since effective bandwidth is more useful in networks with many connections, the tests were done in two configurations: with 6 sources, as in [3], and with 36 sources with the same total sum of mean rates. Table 1 presents the parameters of the sources for the case with 6 sources. In the 36 sources case the each source was split into 6 sources with the same ingress node, equal mean rates (1/6th of the original source) and different standard deviations (two sources with each of three values: 0.08333, 0.01666 and 0.03333). The Hurst parameter of all sources was 0.8, the last mile link capacity was 10 for all sources and all sources used node 5 as the egress node.

Source number	1	2	3	4	5	6
ingress node	1	1	1	2	2	2
mean	2	1.5	1	1	5	1
standard deviation	0.5	0.1	0.2	0.1	0.5	0.2

Table 1: Source parameters for simulation

Each test was performed with three sets of source parameters. One set, attempted 115% utilization, that is the sum of mean rates of offered traffic was equal to 1.15 of the sum of bottleneck capacities, as in the experiment in [3] – losses were therefore unavoidable. The second set had the mean rates reduced proportionally to ensure 70% utilization, but without changing the variances (the high variance case) and in the third one both means and variances were reduced in proportion (the low variance case). All test cases were simulated with 120 different sets of fractional Brownian motion samples as input, the results in the tables are averages.

Algorithms using effective bandwidths use full information about each source – its mean rate, variance and Hurst parameter. HRA normally uses the maximum transmission rate as the only parameter of a source. This parameter is not available when traffic is modeled as a fractional Brownian motion. For that reason we compare four routing configurations generated by this algorithm with four different values of this parameter – equal to the mean rate, or mean rate plus one, two or three standard deviations – the best result is shown in the tables. The same approach was used during link weight optimization for OSPF.

6. SIMULATION RESULTS

All tests were performed with sources limited by token bucket constraints according to the effective bandwidth-based traffic control algorithm, so the packet loss probability was under 1% in all cases. Tables 2 and 3 present the total end-to-end throughput after 5 seconds of simulated time – the results are averages from 120 simulation runs for each case.

Load:	115%	70%, high variance	70%, low variance
OSPF-1	25.76	23.23	24.21
OSPF-opt	35.31	29.47	29.54
HRA	34.86	28.97	29.54
HRA-EBS	34.96	30.64	28.47
EBRA-prob	36.26	30.09	28.47
EBRA-price	37.41	28.97	33.05
FEBRA	35.73	31.06	30.99

Table 2: Results of simulation with 6 sources

Load:	115%	70%, high variance	70%, low variance
OSPF-1	24.81	25.95	26.2
OSPF-opt	34.59	36.75	36.7
HRA	35.02	35.27	34.58
HRA-EBS	35.11	34.93	32.95
EBRA-prob	37.2	31.0	37.46
EBRA-price	35.34	37.23	37.9
FEBRA	35.11	36.02	34.78

Table 3: Results of simulation with 36 sources

First of all, we see that all new traffic based algorithms perform much better than OSPF-1. This is obvious, as the OSPF-1 algorithm does not use one of the available connections at all.

The most interesting results concern the effective bandwidth routing algorithms. As the results show, they are all reasonably good, reaching solutions comparable to HRA and OSPF-opt. However, their quality differs.

HRA-EBS was shown to be a poor heuristic. The routing resulting from this algorithm is even worse than that generated by normal HRA, the algorithm we were trying to improve. This is unexpected, but explainable. Typical working point of the network is estimated in the first steps of this algorithm using a completely different routing. For this reason it is not as close to the real one as expected. Since the estimation is not a complete guesswork, the computed effective bandwidths do contain some real information about the sources, so the resulting routing is not completely worthless, but using the available source parameters directly, even in an arbitrary way, as in our tests of HRA, provides better information. HRA-EBS is also more complicated than HRA and slightly slower (although the difference in execution time is not very significant), so further development of this algorithm was abandoned.

We can make up for the errors in sorting caused by the wrong estimation of effective bandwidth in the first steps by using the entire model of the source later in the algorithm. This approach is used in EBRA. The tests confirm that this algorithm is indeed good, resulting in a high utilization of the network. However, the choice of the metric used in the Dijkstra algorithm is quite important in this case.

Using the estimated loss probability directly leads to many problems. This variant of EBRA is rather erratic – in some cases the routing generated by EBRA-prob is very good indeed, but in other cases it falls behind, sometimes very much so. The calculation of the metric is impaired by the traffic control algorithm used in the 8th step of the algorithm. The price-based metric is a much better choice for EBRA. This variant, EBRA-price is arguably the best of the tested routing algorithms in terms of achieved utilization. While sometimes (for example in case of 6 sources attempting 115% utilization) it falls behind, the results are always at least good, and in many cases the best of all tested algorithms.

It seems clear, that EBRA-price is indeed a good proposal. However, this algorithm is the slowest of all, and significantly so. Even though the traffic control algorithm is not run in each iteration, it is still run too often. As an effect, routing usually takes more than twice as much time as, for example, HRA does. Since HRA is not commendably fast either, EBRA may not be acceptable in many cases. Finding a routing for a simple network with less than a hundred connections should not take seconds on a reasonably fast machine, even with a very suboptimal implementation in an interpreted language – and this happened in some tests. This time might as well be used optimizing OSPF weights, with similar results.

The biggest surprise in the results is the good performance of FEBRA. While the utilization is usually less than optimal and other algorithms often manage to find a better configuration, FEBRA is never far behind. The interesting point is that this holds even in the case of 115% utilization, while the algorithm was designed for low utilization. In those cases the extended metric works as designed, being applied 10 times in the case of 6 sources (36 possible – number of sources times number of links) and 27 times in the case of 36 sources (216 possible).

These results, while good, would not be enough to make FEBRA a commendable choice, were it not for the fact, that it is by far the fastest of the tested heuristics. Since it does not run any traffic control algorithms internally, it is not even comparable in speed with the alternatives, being hundreds of times faster. For a network in which EBRA with the price-based metric, the best alternative, computes the routing in 10 seconds, FEBRA returns a comparable result in milliseconds. Another advantage is that since the selection of a route with FEBRA has a goal of minimizing the loss probability, a uniform distribution of traffic over available links is preferred, allowing the traffic control algorithm to reach a high aggregate utility function.

7. SUMMARY

In this paper we presented three new routing heuristics and compared them with the heuristic proposed in [3] and OSPF. Of those new algorithms one, HRA-EBS, proved to be interesting only in theory, with no real advantage over the existing solution. Algorithm EBRA with a price-based metric is noticeably better than the other solutions, offering better quality of routing, but at a cost of a slow execution. Algorithm FEBRA is the opposite – it is a good heuristic, giving good routing configurations, and while it is sometimes less effective than the alternatives, it is orders of magnitude faster and easy to implement under the resource constraints. Both of the new algorithms share the weakness of requiring fast floating point operations and centralized routing infrastructure, however their (especially FEBRA's) simplicity makes up for these shortcomings.

Research plans of the authors in this area of interest are currently focused on FEBRA. Description in this paper, due to space constraints, ignores some difficult implementation problems, mostly due to numerical precision. The currently implemented workarounds are not fully satisfactory from the authors' point of view and require some work. Another interesting plan is to use effective bandwidths as a tool for selecting OSPF weights, possibly faster than optimization and at the same time better than simple heuristics.

REFERENCES

[1] CHANG C. S., *Performance Guarantees in Communication Networks,* Telecommunication Networks and Computer Systems, Springer Verlag, London, 2000.

[2] DEMBO A., ZEITOUNI O., *Large Deviations Techniques and Applications*, 2nd edition, Applications of Mathematics – Stochastic Modelling and Applied Probability 38, Springer Verlag, New York, 1998 (1st edition 1993 by Jones and Bartlett).

[3] JASKÓŁA P., MALINOWSKI K., *Two Methods of Optimal Bandwidth Allocation in TCP/IP Networks with QoS Differentiation*, Proceedings of the Summer Simulation Multiconference SPECTS 2004, San Jose, California, 2004, pp. 373–378.

[4] JASKÓŁA P., *System rezerwacji pasma oraz optymalizacji routingu w sieci TCP/IP*, in Polish, IV Sympozjum Modelowanie i Symulacja Komputerowa w Technice, Łódź, Poland 2008.

[5] KOZAKIEWICZ A. *Effective Bandwidth Theory for Pricing and QoS Control of Computer Networks*, PhD Thesis, Warsaw University of Technology, 2008.

[6] LOW S. H., LAPSLEY D. E., *Optimization Flow Control, I: Basic Algorithm and Convergence*, IEEE/ACM Transactions on Networking, vol. 7 (1999), no. 6, pp. 861-874.

[7] PIÓRO M., MEDHI D., *Routing, Flow, and Capacity Design in Communication and Computer Networks*, Morgan Kaufmann, 2004.

[8] WISCHIK D., *Large Deviations and Internet Congestion*, PhD dissertation, University of Cambridge, 1999.

Preemption, Traffic engineering, MPLS, Network topology

Sylwester KACZMAREK*, Krzysztof NOWAK†

PERFORMANCE OF LSP PREEMPTION METHODS IN DIFFERENT MPLS NETWORKS

Preemption in Multiprotocol Label Switching (MPLS) is an optional traffic engineering technique used to create a new path of high priority when there is not enough bandwidth available. In such case the path is admitted by removing one or more previously allocated paths of lower priority. As there are usually many possible sets of low priority paths which can be selected, a preemption algorithm is being started to select the best possible combination to be removed. In this paper we investigate if there is any influence of network topology on the performance of two different preemption algorithms. We perform series of simulations to check the performance of two preemption methods in several models of real backbone telecommunication networks.

1. INTRODUCTION

1.1. PREEMPTION

The Multiprotocol Label Switching (MPLS) enables using traffic engineering mechanisms in IP networks, what makes it possible to better manage the network resources and provide network users with Quality of Service (QoS). That in turn allows network operators to offer services with strict quality requirements.

In MPLS networks data are sent in tunnels called Label Switched Paths (LSP). The LSPs have to be created before any data can be sent. Although the MPLS architecture standard [1] does not require the LSPs to have any bandwidth reserved for them, for traffic engineered paths a specified bandwidth guaranteed on every link is a prerequisite to keep the Service Level Agreement (SLA).

LSP paths have two priorities associated with them. The setup priority is used when the path is set up. Since the path is established, its holding priority is used instead. It works as follows. When a new path is declared to be created in the network, it has the setup priority and the holding priority assigned to it. If the setup priority is greater than the holding priority of an existing path, the new path will be treated as a more important one. The both priorities are represented as integer numbers in range from 0 to 7, where 0 is the highest priority, i.e. greater priority corresponds to numerically smaller value.

* Gdańsk Univeristy of Technology, Gdańsk, Poland, e-mail: sylwester.kaczmarek@eti.pg.gda.pl
† Nokia Siemens Networks, Warsaw, Poland, e-mail: krzysztof.nowak@nsn.com

As long as there is enough bandwidth to admit the new path, the setup priority has no practical effect. But as soon as the bandwidth is exhausted, the path can be admitted at cost of preempting one or more existing paths of lower priority.

Preemption in MPLS networks is a process of removing some of the previously allocated paths to free resources for a new path. Basically, it is performed in five steps.
1. A route with insufficient bandwidth is selected for a new path.
2. One or more lower priority paths on the route are selected as candidates for preemption.
3. The candidates are removed.
4. The new path is allocated.
5. The removed candidate paths are allocated on new routes, if possible.

If the nodes are capable to perform so called soft preemption, then the make-before-break principle can be used. In that case the candidate paths are not just removed, but only the occupied bandwidth is released. Only after their recreation on the new routes is completed, the original candidate paths are deleted.

Regardless of the preemption method, the traffic on the preempted paths is disrupted, what can have negative influence on QoS level perceived by the users. Hence, in practice, unnecessary preemptions should be avoided. For that reason, it is important to know how different methods perform.

1.2. MOTIVATION

In the first widely available paper [2] on preemption in ATM networks the authors define the problem and provide the proof that both the problems of minimizing the number of paths and minimizing the preempted bandwidth are NP-complete. They present simple heuristic centralized algorithms for equal and unequal bandwidths of the existing paths. A universal, decentralized method has been presented in [3] and later published as informational RFC [4]. A similar method, but with centralized approach, has been presented in [5]. There are extensions of the decentralized version of the method available. In [6] it has been extended to enable shrinking of the existing paths instead of simply removing them and in [7] it has been simplified at the cost of flexibility.

In previous works we introduced our preemption method [8] and checked its performance in three sample networks [9]. In current paper we focus on performance difference of two selected algorithms when run in different networks. Additional condition is that the networks are similar to those used in real MPLS networks.

2. RESULTS

2.1. NETWORK TOPOLOGIES

As the base for our topology analysis we chose the SND network library gathered at Zusse Institut Berlin [10], which currently includes more than 20 topologies. Most of them are real topologies published by network operators, what makes the library suit well to our needs. We picked up 2 sets of topologies. The first one consists of networks of similar connectivity index, but different size in means of number of nodes. This will be used to check the influence of network size. The second set consists of networks of different connectivity index, to check if that factor has the impact as well. In table 1 we show the characteristics of every network. For the sake of readers convenience, we also show the topologies which we are discussing in this paper in Fig.1. The criteria for selecting the networks were diversity in the values of investigated parameters and closeness to the real network provider's topologies.

Id.	Network	Nodes #	Links # (unidir.)	Connectivity index	Avg. route length	Name in ZIB library
1	Poland	12	36	3.00	2.23	polska
2	Atlanta	15	44	2.93	2.15	atlanta
3	France	25	90	3.60	2.30	france
4	USA	26	84	3.23	2.89	janos-US
5	Europe I	37	114	3.08	2.84	cost266
6	Germany I	50	176	3.52	3.01	germany50
7	Europe II	28	82	2.93	2.63	nobel-eu
8	Telecom Austria	24	102	4.25	2.16	ta1
9	New York	16	98	6.13	1.83	newyork
10	Di-Yuan	11	84	7.64	1.25	di-yuan
11	Germany II	10	90	9.00	1.00	dfn-bwin

Table 1. Characteristics of topologies chosen with regard to different network size (1-6) and connectivity index (7-11)

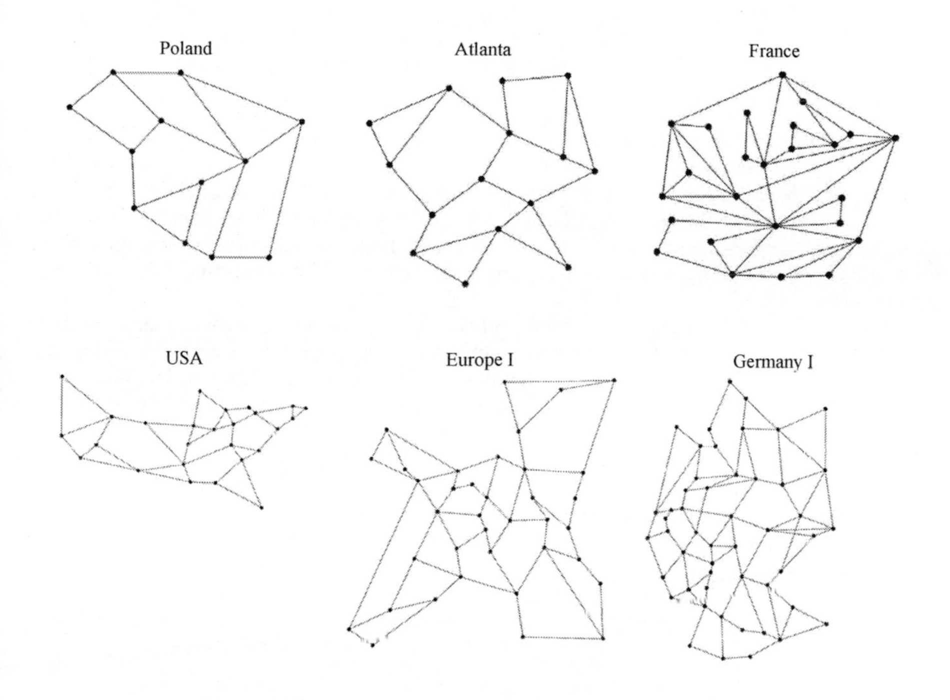

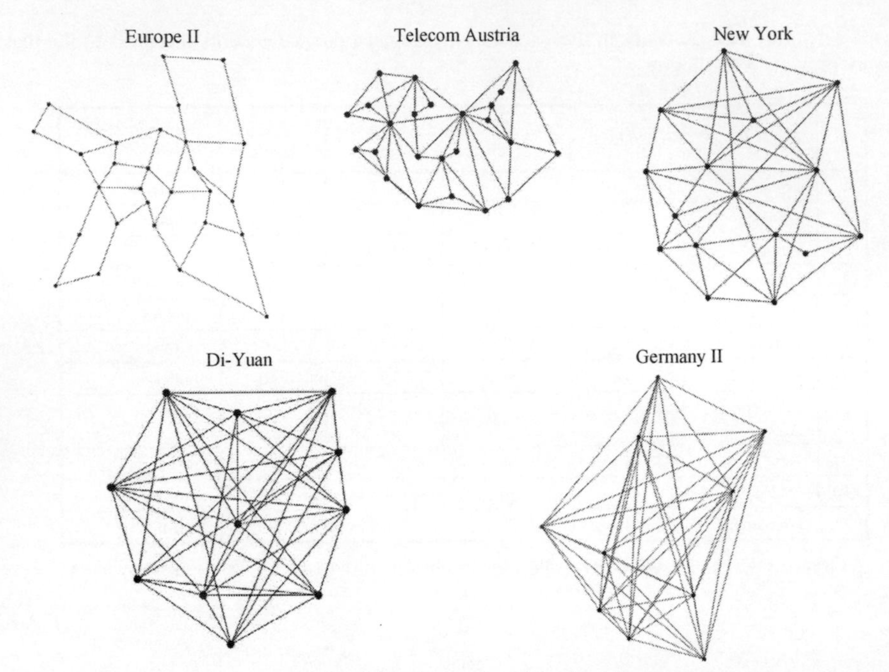

Fig. 1. Topologies selected for simulations

2.2. SIMULATION CONDITIONS

We used our own simulation tool MPLSsim written in C++. Topology sections of the input files used by the tool have been automatically created by a dedicated filter program from the XML files published at the ZIB webpage [10]. Every link in simulated networks is bidirectional, i.e. there are two links in opposite directions for every arc in the network graph. The simulations have been performed on call level.

As call generators, we introduced number of sources in every node, assuming that every node can work as edge router, what is the scenario for most of the real topologies we simulated. The statistical amount of bandwidth requested and terminated at every node is the same. The sources are created and deleted during the simulation to enforce creation and deletion of the LSP paths. The requested bandwidth values are generated by the random generator of exponential distribution. The average bandwidth c_{src} has been calculated based on estimated network traffic volume A using the formula (1), where k is the number of traffic classes, λ is the call intensity, and h is the average call duration. The simple formula is based on assumption, that the traffic amount is equally divided between the sources of different classes and the classes have the same characteristics in terms of call duration.

$$c_{src} = \frac{A}{k\lambda h} \tag{1}$$

The traffic volume A is calculated as the sum C of the bandwidths available on all the links in both directions divided by the route length $\bar{l}$ calculated as the number of hops in the shortest path and averaged over every pair of unequal source and destination node. For equal bandwidths of every link, what is the case for the simulated networks, we get the simple equation (2), in which e stands for the number of links.

$$A = \frac{eC}{\bar{l}} \tag{2}$$

Such procedure of traffic source dimensioning leads to high bandwidth reservation and high preemption ratio, what enables us to examine the preemption methods. The side effect of such calculations is that for the same λ the average number of paths per link differs between networks. In examined topologies it varies between 3.3 and 18.6 with the general tendency of being greater in smaller networks, but there is no simple dependency. There is, though, a dependency that in denser networks number of paths per link is smaller.

2.3. MEASURES

There is no standardized set of measurements to evaluate preemption methods. Different authors use different metrics, but two most common are the relocation count and the relocated bandwidth or the bandwidth wastage. The relocation count r is the number of paths which are removed to admit a new path. Usually we want to keep the number as low as possible, because even if the removed paths are allocated immediately on alternative routes, the connections which follow the paths will most likely be broken. The second metric – relocated bandwidth is the sum of bandwidth of the paths which is removed as the result of preemption. The value is sometimes multiplied by the hop count of the path, so as the longer paths will generate higher value.

A more practical modification of the relocated bandwidth metric is the bandwidth wastage w. This is the amount of bandwidth, which is unnecessarily removed because of preemption. We say of the wastage if the removed bandwidth is greater than the bandwidth really needed to allocate the higher priority path. The main reason from that is the finite set of paths to select from. If the bandwidths of the paths are randomly distributed, then it is unlikely to have the ideal choice, where the selected paths sum up to exactly the bandwidth which is needed on every link. Still, the average value of wasted bandwidth is a good performance measure of a method.

Usually preemption methods don't perform best in both measures. For that reason and to provide more flexibility, some methods allow the user to choose the preferred criterion. Having said that, it would be useful to have a metric, which combines both measures to be able to say how universal the method is. We developed for that purpose the combined metric q using the formula (3), which is the inverse of the relocation count r multiplied by the bandwidth wastage w. For the selected route R the bandwidth wastage is defined as the needed bandwidth divided by the preempted bandwidth. The needed bandwidth is the sum of differences between the bandwidth c_N of the new path and the free bandwidth c_f. The preempted bandwidth is calculated as the sum of bandwidths b_p occupied by a path p multiplied by the hop count l_p of the path.

$$q = \frac{\sum_{v \in R} \max(0, c_N - c_f)}{r \sum_{p \in P} l_p b_p} \tag{3}$$

The values of combined metric span in period of (0, 1]. The bigger the value is, the better the method performs, but the measured values rarely exceed 0.25. This is due to the restrictive metric definition, for which the value of 1 is reached only if exactly one path is preempted and its bandwidth fits exactly into the missing bandwidth value on every link along the selected path. This is very unlikely to happen and there are even situation when the value of 1 simply cannot be reached. One example is the case when the needed bandwidth differs on at least two links on the route. Another case happens when there is not enough bandwidth on at least two links on the route which are not contiguous.

2.4. METHODS

We examined the centralized preemption method (KN), and compare it with the decentralized method (RFC) [4]. The KN method is taking advantage of the knowledge of the network topology. It uses a heuristic algorithm to select a paths which fit best into the missing bandwidth amount. It works in "add-and-optimize" manner, in which every potential candidate is added to a list, and after every addition an optimization routine checks the list and removes from it some of the candidates, if necessary. There method can be adjusted by changing the way the list is sorted, what in turn has influence on the results of the optimization. For detailed description of the method the reader should refer to [8].

The RFC method is a local approach which only uses knowledge of the paths traversing the current node. It is using a simple sorting of all the potential candidate paths and selecting the paths from the head of the list, until the missing bandwidth is collected. As this is a local method, the same procedure must be repeated on every node on the route, as long as there is still not enough bandwidth. The method can be adjusted by changing the sorting criteria by setting several parameters used in calculating the universal metric $H(l)$ (4). In the equation r is the missing bandwidth, l denotes the current path, $y(l)=8-p(l)$ is the priority reversing function, $p(l)$ is the priority, and $b(l)$ is the bandwidth reserved for the path. The coefficients α, β, χ, and θ are used to adjust the metric.

$$H(l) = \alpha\, y(l) + \beta \frac{1}{b(l)} + \chi \left(b(l) - r\right)^2 + \theta\, b(l) \qquad (4)$$

Both methods are adjustable and we performed separate series of methods, first time using the relocation count priority (RC) and second time with preempted bandwidth priority (BW) set. For the RFC method we set the priorities by using method-specific variables. The RC variant corresponds to $\alpha=\chi=\theta=0$, $\beta=1$, whereas the BW variant corresponds to $\alpha=\beta=\chi=0$, $\theta=1$. By combining the two methods with two priorities we get four series of results: RFC-RC, RFC-BW, KN-RC and KN-BW.

2.5. SIMULATION RESULTS

On the first set of graphs (Figures 2, 3 and 4) we compared the networks 1 to 6, selected with regard to different size, or more precisely, by the number of nodes and the results are ordered in the same way, with smallest network on the left hand side. If we look at the combined metric results, we can clearly identify the trend which shows that the value is higher in small networks. That means that better performance can be reached in smaller networks, as can be seen in Fig. 2. Interesting fact is that we observe the same trend for both methods, even if the algorithms have not much in common. That suggests that the behaviour is a property of the network and the paths allocated on them, rather than the result of the difference in performance of the methods. The reason behind the phenomenon is the fact that in smaller networks paths more often follow common directions. In result the candidates selected on the first link more often provide the required bandwidth also on the other links, where the bandwidth is not enough. That exactly means more effective preemption.

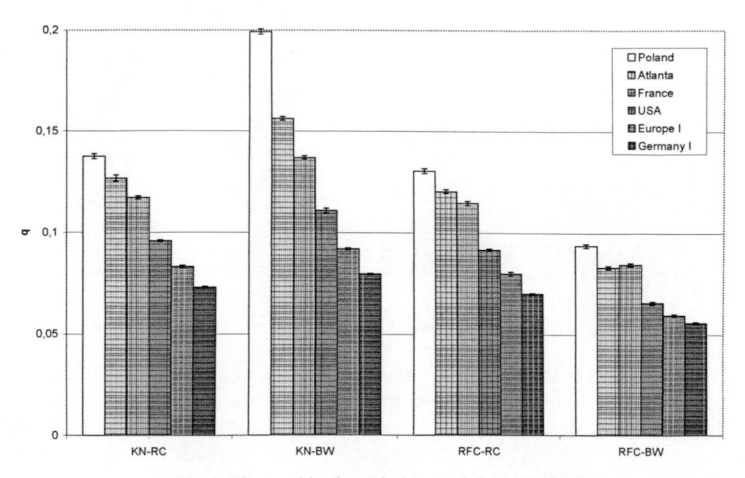

Fig. 2. The combined metric (greater is better) – the first set

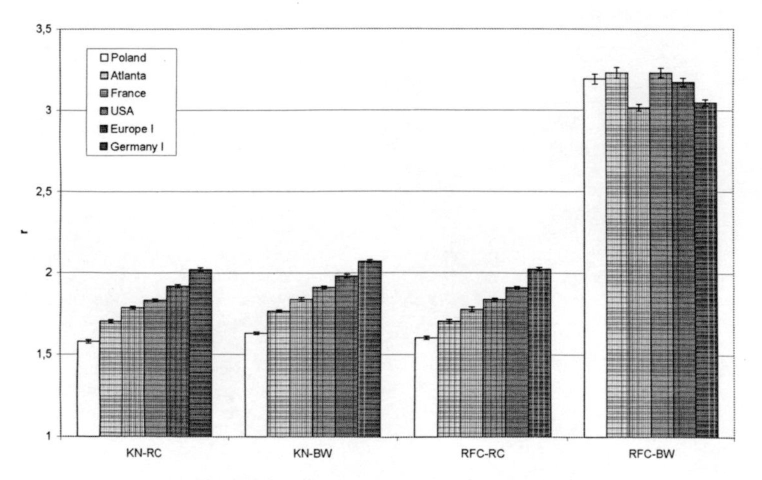

Fig. 3. Relocation count (smaller is better) – the first set

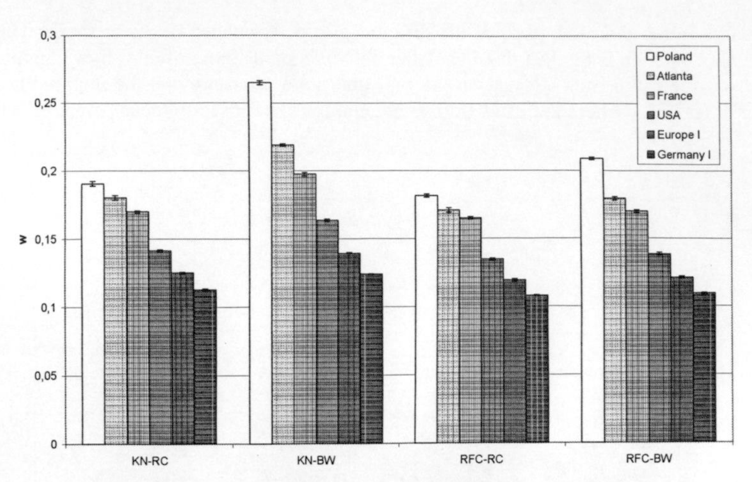

Fig. 4. The bandwidth wastage index (greater is better) – the first set

The described results base on the value of combined metric q, which in fact is a combination of the relocation count r and the bandwidth wastage w. However, we cannot ignore the results of the factors alone. The corresponding results, are shown in Fig. 3 and 4, follow the rule observed by the combined metric, which shows that in the smaller networks the methods can perform better. The difference in not that big for the relocation count, whereas we observe strong topology dependency for the bandwidth wastage.

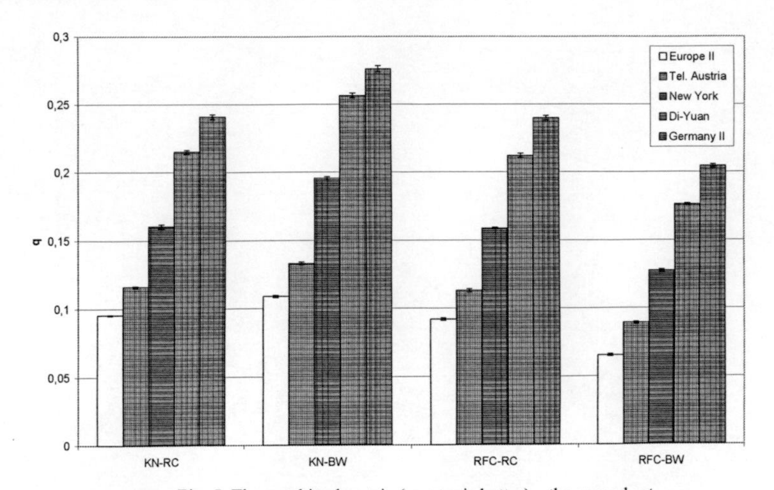

Fig. 5. The combined metric (greater is better) – the second set

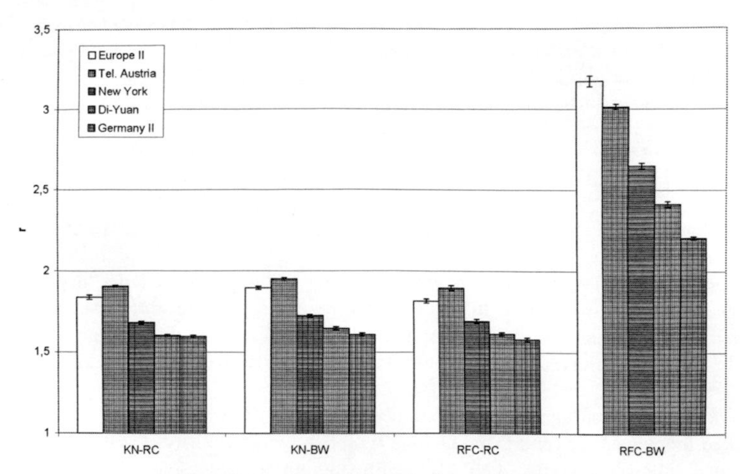

Fig. 6. Relocation count (smaller is better) – the second set

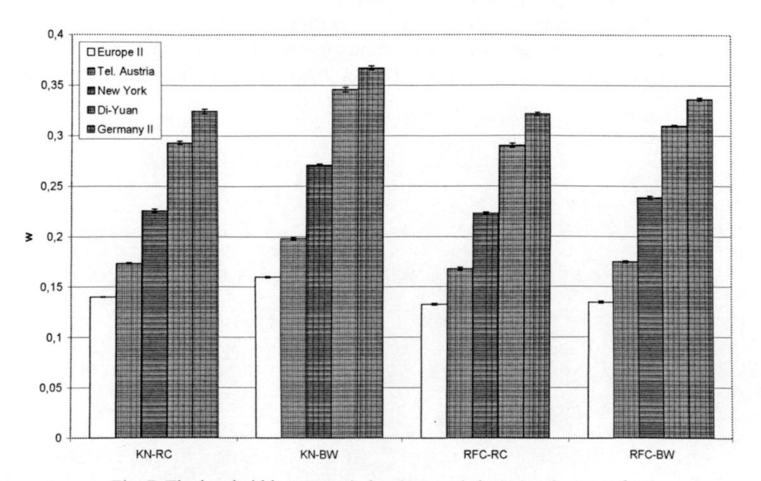

Fig. 7. The bandwidth wastage index (greater is better) – the second set

The Figures 5, 6, and 7 present the results for networks of different connectivity index. The networks are ordered respectively, with sparser networks on the left hand side. Also for this criterion the comparison demonstrate clear trends. The combined index q is greater in dense networks, with relocation count r and bandwidth wastage w alone following the rule as well. What is also similar to the results from the first set, the relocation count doesn't change that much. The main difference is in the relocated bandwidth. In general, we can say that in the dense networks the performance of the methods is better.

The results of that part can be quite easily interpreted. In dense networks the nodes are better interconnected and the routes are shorter. In such conditions it is easier to select the set of paths which fits better to the gap between available and needed bandwidth. That makes the selection more accurate and the influence on the links outside the selected route is minimized.

3. CONCLUSIONS

We performed simulations of two preemption methods in several networks similar to real telco backbone networks.

The results confirmed that the performance of the methods does change depending on the network size and connectivity index. In smaller networks the routes are shorter and the paths more often overlap, which makes effective preemption easier to achieve. In result, both methods follow the principle. We also observed that the methods perform better in dense networks than in sparse ones.

Considering the performance of the simulated methods, the differences remain similar for every checked topology. The centralized method with bandwidth priority (KN-BW) scores the best results, keeping relocation count small while allowing for the lowest bandwidth wastage. Both methods with relocation count priority (RFC-RC and KN-RC) score similar results having little better relocation count, at cost of higher bandwidth wastage.

In future works we want to implement other preemption methods and focus on the ways to improve their performance. There are also some interesting aspects of preemption which we want to examine deeper like chain effect of preemption in the networks where more different priorities are implemented.

REFERENCES

[1] ROSEN E., VISWANATHAN A., CALLON R., *Multiprotocol Label Switching Architecture*, RFC 3031, January 2001.

[2] GARAY J. A., GOPAL I. S., *Call Preemption in Communication Networks*, Proc. INFOCOM '92, pp. 1043-1050, Florence, Italy.

[3] OLIVEIRA J. C. de., SCOGLIO C., AKYILDIZ I. F., UHL G., *A new preemption policy for DiffServ-aware traffic engineering to minimize rerouting*, Proc. INFOCOM 2002, New York, USA, June 2002.

[4] OLIVEIRA J. de, Ed., *Label Switched Path (LSP) Preemption Policies for MPLS Traffic Engineering*, RFC 4829, April 2007.

[5] OLIVEIRA J. C. de, SCOGLIO C., AKYILDIZ I. F., UHL G., SMITH J., *A New Topology -Aware LSP Preemption Policy for DiffServ-MPLS Networks,* Proc. Networks 2002, May 2002.

[6] OLIVEIRA J. C. de, SCOGLIO C., AKYILDIZ I. F., UHL G., New preemption policies for DiffServ-aware traffic engineering to minimize rerouting in MPLS networks, IEEE/ACM Trans. Networing, vol 12, issue 4, pp. 733-745, August 2004.

[7] BLANCHY F., MELON L., LEDUC G., Routing in a MPLS network featuring preemption mechanisms, Proc. 10th ICT, 2003.

[8] KACZMAREK S., NOWAK K., *A New Heuristic Algorithm for Effective Preemption in MPLS Networks*, Workshop on High Performance Switching and Routing, Poznań 2006, pp. 337-342.

[9] KACZMAREK S., NOWAK K., *Comparison of centralized and decentralized preemption in MPLS networks*, Polish Teletraffic Symposium, Zakopane 2007, pp. 259-268.

[10] SND network library, Zusse Institut Berlin, http://www.sndlib.zib.de/.

This work was supported in part by the Polish National Central for Research and Development under project PBZ MNiSW – 02/II/2007.

VIII. NETWORK PLANNING

Key words – measurement techniques, VoIP, QoS, QoE, E-model, PESQ

Tadeus UHL*

E-MODEL AND PESQ IN THE VOIP ENVIRONMENT: A COMPERASION STUDY

The aim of this paper is to sketch out the concepts, standards and properties associated with the measurement methods by Voice over IP (VoIP). First the methods for measuring the QoS are classified. They are followed by a precise description of the E-model and the PESQ algorithm. After this, the two techniques for measuring the QoS are introduced to a real VoIP environment. The core of the analysis carried out in this paper is the determination of the QoS values under changing network parameters (i.e. packet loss and jitter delay) and under using different voice codecs. The received QoS measures (i.e. factor R and PESQ values) are compared to each other and then discussed. Studies show that the E-model has room for improvement regarding the VoIP service. This paper contains first suggestions towards this problem. The paper concludes with a summary and a future prospects.

1. INTRODUCTION

The networks of the past and even some present-day networks belong to the group of networks known as dedicated networks (each service requiring its own network). This changed towards the end of the nineties, however, with the introduction of ISDN (Integrated Services Digital Networks) and a new brand of networks arose known as integrated networks, the word *integrated* referring to the multiple services that these networks support. This process of integration has continued to this day: modern GSM, UMTS and TV networks, and the World Wide Web are all integrated service networks.

The services themselves that the networks transmit have also undergone an equally rapid development, especially in the last twenty years. The variety of services offered today is very wide indeed. The most recent developments include high-speed data transfer, VoIP, SMS, MMS, home banking, WWW, audio/video on demand, teleconferencing, messaging and, most recently, an increasing trend towards IPTV. Many of these applications have become very popular indeed. Parallel to the development of these new services, we have witnessed the recent appearance of innovative systems designed to exploit the impressive multimedia capabilities of the IP platform. A good example of such systems are the so-called e-learning systems, that are widely used in education. In the near future, we can definitely expect a variety of other innovative services and high-capacity communications systems to flood the market.

* Flensburg University of Applied Sciences, Kanzleistr. 91-93, D-24943 Flensburg, tadeus.uhl@fh-flensburg.de

One new technology that has evolved rapidly in the last ten years is Voice over IP (VoIP). VoIP requires an IP transport platform. Several standards (e.g. H.323 [1], SIP [2]) have been published in the last few years, allowing this service to become a reality. There are several VoIP systems on the market, (Siemens' HiPath 5000 and OpenScape, Innovaphone PBX from TLK Computer GmbH, SIPSTAR IP-PBX from Nero AG, OmniPCX Enterprise from Alcatel AG, Cisco CallManager from Cisco Systems GmbH, ASTERISK from DIGIUM, etc.) that work in accordance with the standards named above. VoIP is becoming incredibly popular.

A fundamental aspect in practice is the Quality of Service in a network. The QoS must be determined permanently in a real environment. For this, special measurement systems are needed. These systems use different methods. The two most common techniques in telephony are the E-model and the PESQ algorithm. They differ greatly in the degree of complexity and use different QoS metrics: the E-model uses the transmission rating factor R while PESQ uses the PESQ value. With the use of different measurement techniques, varying results of QoS in the network may be obtained. This paper focuses on the verification of measurement results in the VoIP environment under the use of the two methods mentioned above.

2. QOS-METHODS FOR VOIP

In order to determine the Quality of Service in a network, generally two models are used: a) dual ended model and b) single ended model; cf. fig. 1. In case of the dual ended model two signals are used: a) original signal and b) degraded signal. These two signals are available uncompressed. For this reason, measurements can be carried out for both the Quality of Experience (subjective evaluation) and the Quality of Service (objective evaluation). In case of the single ended model only the impaired signal (compressed) is available. This allows only an objective evaluation (QoS). The QoS measurement is referred to as „intrusive measurement" (offline) in case of the dual ended model and as „non-intrusive measurement" (online) in case of the single ended model [3].

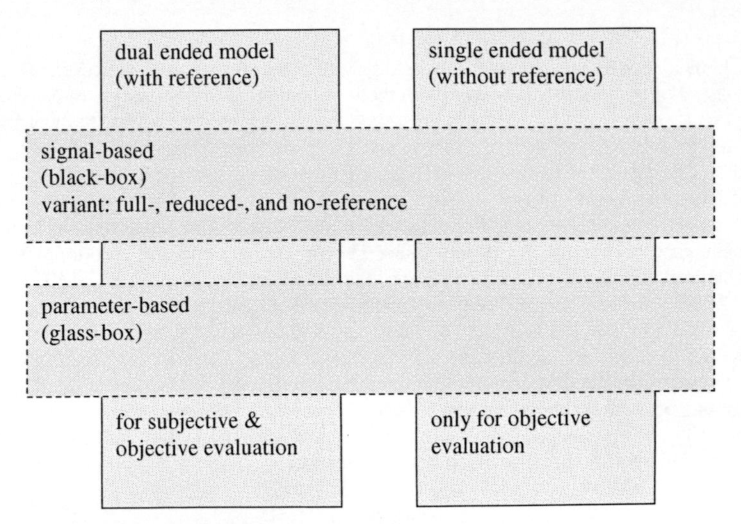

Fig. 1. Overview of QoS and QoE Measurement Techniques in General

Two measurement techniques can be used in the two models: a) signal-based and b) parameter-based measurement (see horizontal layers in fig. 1 and 2). The dual sided model uses specialised algorithms to compare the input and output signals of signal-based measurements. In case of the single sided model, the reference signal is assessed before making a comparison. In both cases the system to be assessed is treated as a "black-box". When carrying out parameter-based measurements, the "glass-box" principle is applied. In this case, both the structure of the system to be assessed as well as the reaction of the individual system components to the reference signal is known. This knowledge is then considered in a suited model. Additionally, the measured network parameters can be included in the calculation of the Quality of Service (cf.[3]).

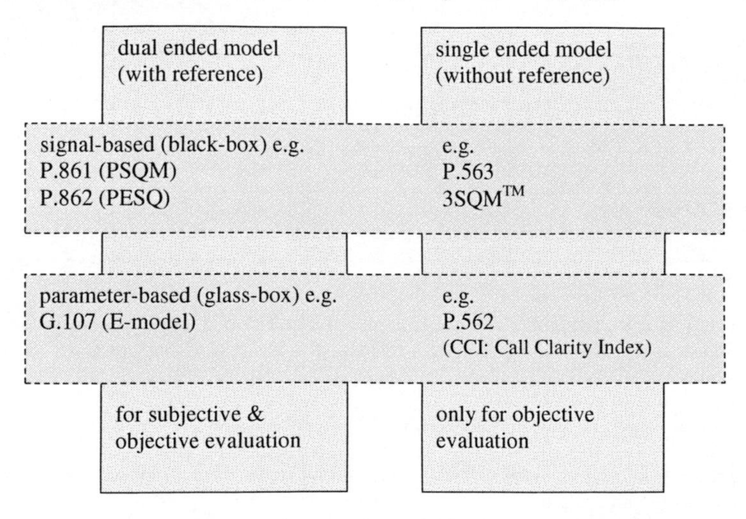

Fig. 2. Overview of QoS and QoE Measurement Techniques by VoIP

Fig. 2 shows the most common measurement techniques for determining the Quality of Service in a VoIP environment. The most important techniques in practice are: the E-model and the PESQ algorithm. These two techniques constitute the core of the analysis carried out in this paper and are therefore described in detail in the following sections.

2.1. THE E-MODEL

The E-model is in accordance with the ITU-T Recommendation G.107 [4]. This model is widely used to estimate quality during the planning of networks, and also during their operation (monitoring). The E-model is based on a parametric description of telephone networks. Psychological factors on the psychological scale are additive (cf. Table 1).

Table 1: G.107 – Default Values and Permitted Ranges for the Parameters [3, 4]

Parameter	Abbr.	Unit	Default value	Permitted range
Send Loudness Rating	SLR	dB	+8	0 … +18
Receive Loudness Rating	RLR	dB	+2	-5 … +14
Sidetone Masking Rating	STMR	dB	15	10 … 20
Listener Sidetone Rating	LSTR	dB	18	13 … 23
D-Value of Telephone, Send Side	Ds	-	3	-3 … +3
D-Value of Telephone, Receive Side	Dr	-	3	-3 … +3
Talker Echo Loudness Rating	TELR	dB	65	5 … 65
Weighted Echo Path Loss	WEPL	dB	110	5 … 110
Mean one-way Delay of the Echo Path	T	ms	0	0 … 500
Round-Trip Delay in a 4-wire Loop	Tr	ms	0	0 … 1000
Absolute Delay in echo-free Connections	Ta	ms	0	0 … 500
Number of Quantization Distortion Units	qdu	-	1	1 … 14
Equipment Impairment Factor	Ie	-	0	0 … 40
Packet-loss Robustness Factor	Bpl	-	1	1 … 40
Random Packet-loss Probability	Ppl	%	0	0 … 20
Burst Ratio	BurstR	-	1	1 … 2
Circuit Noise referred to 0 dBr-point	Nc	dBm0p	-70	-80 … -40
Noise Floor at the Receive Side	Nfor	dBmp	-64	--
Room Noise at the Send Side	Ps	dB(A)	35	35 … 85
Room Noise at the Receive Side	A	dB(A)	35	35 … 85
Advantage Factor	A	--	0	0 … 20

The result of any calculation with the E-model in a first step is a transmission rating factor R, which combines all transmission parameters relevant for the considered connection. This rating factor R is composed of:

$$R = Ro - Is - Id - Ie,eff + A \tag{1}$$

Ro: Represents in principle the basic signal-to-noise ratio, including noise sources such as circuit noise and room noise.

Is: This factor is a combination of all impairments which occur more or less simultaneously with the voice signal.

Id: This factor represents the impairments caused by delay and the effective equipment impairment.

Ie,eff: This value represents impairments caused by low bit-rate codecs. It also includes impairment due to packet-losses of random distribution.

A: This advantage factor allows for compensation of impairment factors when there are other advantages of access to the user.

The term *Ro* and the *Is* and *Id* values are subdivided into future specific impairment values. The explicit formulae used in the E-model can be found in [3, 4].

The transmission rating factor R predicated by the E-model can be converted to a *MOS*-rating using equation:

$$MOS = \begin{cases} 1 & for\ R < 0 \\ 1 + 0{,}035R + R(R-60)(100-R)\cdot 7\cdot 10^{-6} & for\ 0 \le R \le 100 \\ 4{,}5 & for\ R > 100 \end{cases} \tag{2}$$

2.2. THE PESQ ALGORITHM

Perceptual Evaluation of Speech Quality (PESQ) is a method of objective speech quality evaluation in telephony in the frequency range 300 – 3400 Hz. PESQ is described in ITU-T recommendation Q.862 [5] and is based on real conditions for end-to-end speech communication. Amongst those factors taken into account are packet loss, noise and the audio codec used.

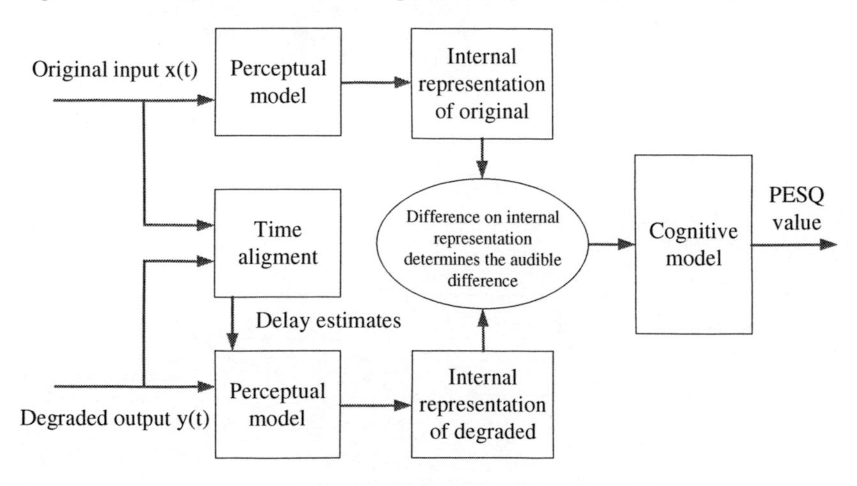

Fig. 3. Schematic Diagram of the PESQ Algorithm [5]

Fig. 3 shows a schematic diagram of the PESQ algorithm. PESQ compares an original signal x(t) with a degraded signal y(t) that is the result of passing x(t) through a communications system. In the first step of PESQ a series of delays between original input and degraded output is computed, one for each time interval for which the delay is significantly different from the previous time interval. For each of these intervals a corresponding start and stop point is calculated. Based on the set of delays that are found PESQ compares the original signal with the aligned degraded output of the device under test using a perceptual model. The key to this process is transformation of both the original and degraded signals to an internal representation that is analogous to the psychophysical representation of audio signals in the human auditory system. This is achieved in several stages: time alignment, level alignment to a calibrated listening level, time-frequency mapping, frequency warping, and compressive loudness scaling.

The internal representation is processed to take account of effects such as local gain variations and linear filtering that may have little perceptual significance. This is achieved by limiting the amount of compensation and making the compensation lag behind the effect. Thus minor, steady-state differences between original and degraded are compensated. More severe effects, or rapid variations, are only partially compensated so that a residual effect remains and contributes to the overall perceptual disturbance. This allows a small number of quality indicators to be used to model all subjective effects. In PESQ, two error parameters are computed in the cognitive model; these are combined to give an objective listening quality MOS (cf. Table 2).

Table 2: Evaluation of Speech Quality according PESQ [5]

PESQ Value	MOS Value	Speech Quality
4,5	5	excellent
4	4	good
3	3	fair
2	2	poor
1	1	bad

An exact determination of PESQ values is rather complex and therefore cannot be represented in a closed formula expression. Details are found in specification [5].

3. COMPARISION STUDY

Fig. 4 shows the analysis environment used in the analyses. It consists of the real-time communications system TraceSim_VoIP [6] with implemented measurement system TraceView_VoIP, the TraceSim_VoIPClient (to the reflection of RTP-packets) and the Wanulator (to the network impairments) [7].

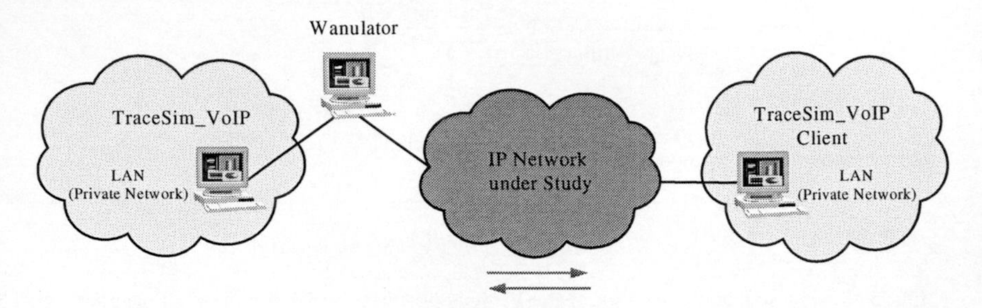

Fig. 4. Analysis Environment

TraceSim_VoIP establishes an RTP-connection to the TraceSim_VoIPClient. The client is able to mirror the incoming RTP packets back to TraceSim_VoIP. Before the QoS measurement is started, TraceSim_VoIP and the TraceSim_VoIPClient are synchronised using DFTM tones. After the synchronisation of the systems is completed, the reference signal (the file `Or105.wav from` ITU-T [5]) is sent. The RTP packets received by the TraceSim_VoIPClient are mirrored and sent back to TraceSim_VoIP, which includes the integrated TraceView_VoIP. This measurement system then compares the two signals by means of the E-model and the PESQ algorithm. By using the Wanulator impairments in the network, such as packet loss, network delay and jitter delay can be realised.

Further analyses have been carried out using the test environment shown in fig. 4. The results are presented and assessed in detail in further chapters.

3.1. QOS AS A FUNCTION OF PACKET LOSS

In the course of this analysis the following codecs have been taken into consideration: codec G.711 with A characteristic curve, G.723.1 and iLBC. Packet losses varied between 0% and 10% in steps of 1%. Fig. 5 graphically shows the dependency calculated according to E-model between QoS values and packet losses for codec type G.711 with different robustness factors, i.e. 4.3

(according to G.113 [8]), 8.8 (corrected for VoIP [3]) and 13 (assumed value for approximation used in this paper). Other parameters were taken from table 1 (default values). It becomes apparent that the QoS is increased the more sophisticated methods of error concealment are used. The latter is significantly influenced by the jitter buffer size and the scheduling within the jitter buffer. In this analysis a buffer size of 60 ms was applied. For scheduling the method „Silence Insertion" (according to recommendation by ITU-T G.113) was used. The speech samples had a length of 10 ms. Fig. 5 also contains a curve for measured PESQ values as a function of packet losses. The corresponding confidence intervals (probability of uncertainty is 5%) represent less than 10% of the calculated mean values. It becomes apparent that the QoS retains a relatively good value in the real environment. The curve only declines in case of great packet losses. Furthermore it becomes apparent that the PESQ values measured in practice are significantly superior to the QoS values calculated according to the E-model. Conclusion: the codec G.711 has proven to perform surprisingly well in the VoIP environment and can therefore be recommended as standard codec.

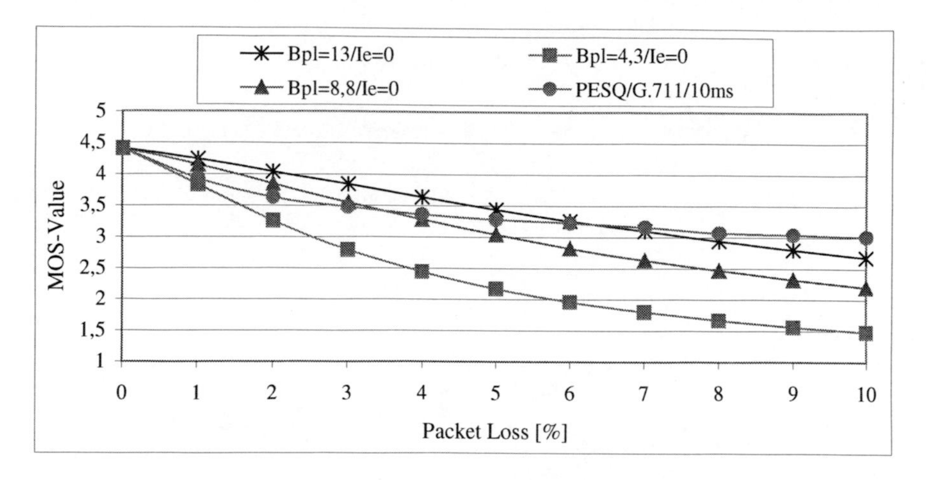

Fig. 5. QoS Value as a Function of Packet Loss for the Codec G.711

Fig. 6 shows the development of the theoretically derived QoS values (according to E-model) and the QoS values measured in the real environment as a function of packet loss for two further codecs relevant in practice (type G.723.1 and iLBC). The latter two codecs work on the principle of the LPC process and are therefore safer from packet losses (cf. Bpl values in [3]). The speech samples used were of a length of 10 ms (according to the specification for these codecs). Here, it becomes apparent that the theoretically derived QoS values are significantly superior to the QoS values measured in the real environment. This contradicts the results in connection with codec G.711. In this regard, the codec iLBC performs particularly poor. Conclusion: the theoretically derived QoS values (according to E-model) are rated too optimistic in comparison to the PESQ values measured in the real environment. Here, the parameters Bpl (packet loss robustness factor) und Ie (impairment factor) have to be corrected in correspondence with the E-model. This paper furthermore covers the applicable adaption of both parameters in the E-model. The results of the approximations performed are shown in fig. 7 and 8.

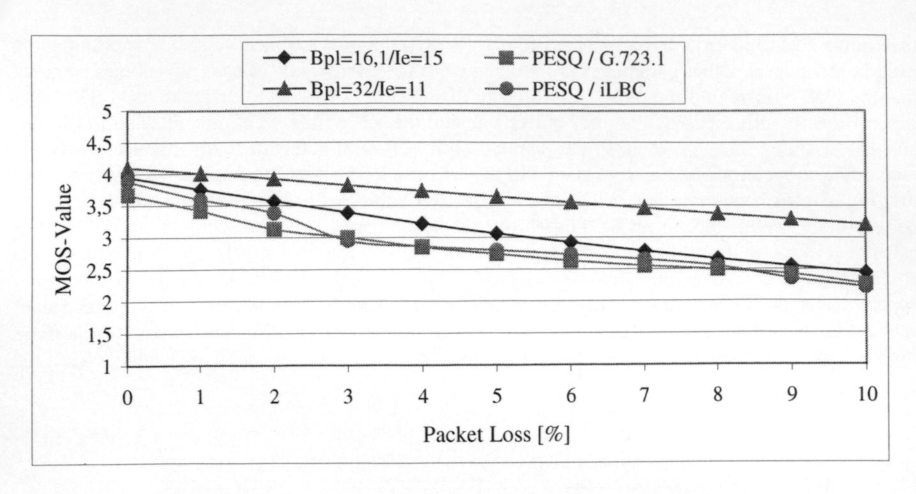

Fig. 6. QoS Value as a Function of Packet Loss for the Codec G.723.1 and iLBC

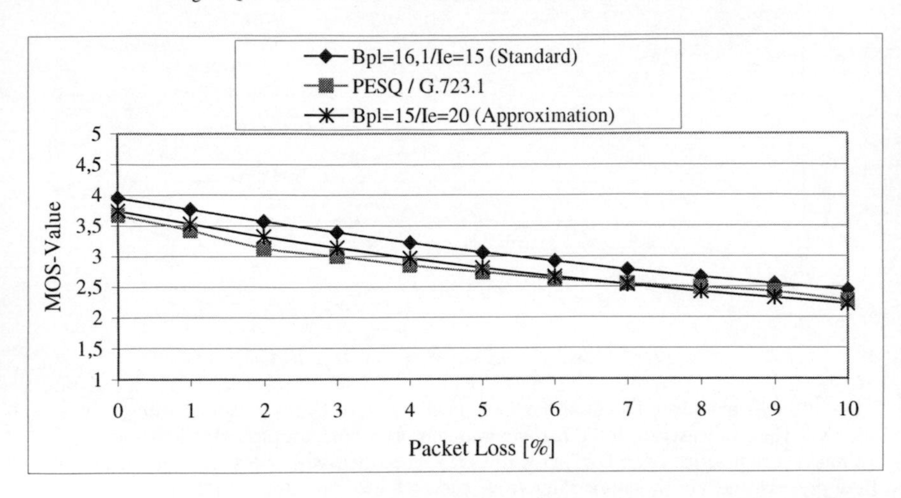

Fig. 7. QoS Value as a Function of Packet Loss for the Codec G.723.1 with and without Approximation

The output illustrated in fig. 7 shows that the impairment factor Ie for codec G.723.1 had to be severely increased to achieve an acceptable approximation of the PESQ curve. The packet loss robustness factor Bpl maintained its original value. The impairment factor Ie had also to be highly increased for codec iLBC. In addition the packet loss robustness factor Bpl had to be decreased by about the factor two. The applied changes allowed for a vary good approximation of the PESQ curve. As you can see the iLBC codec is still highly overvalued regarding packet loss robustness.

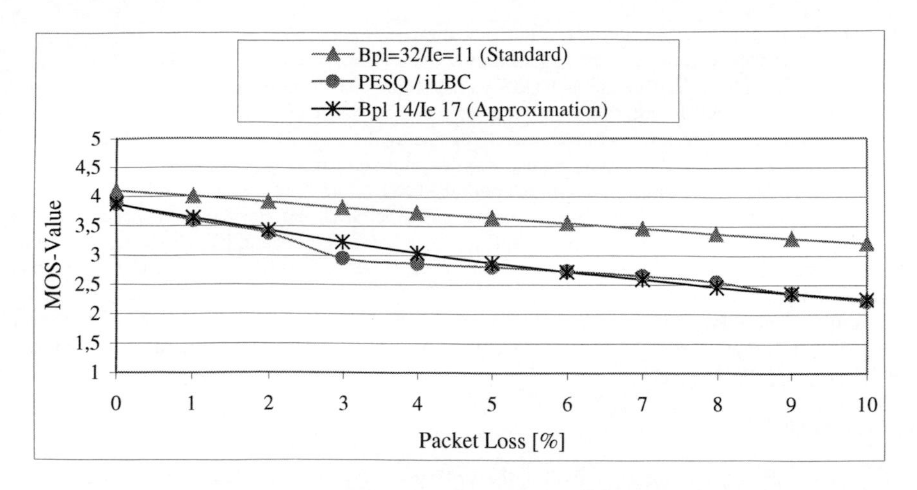

Fig. 8. QoS Value as a Function of Packet Loss for the Codec iBLC with and without Approximation

3.2. QOS AS A FUNCTION OF JITTER DELAY

In this analysis the same codecs as mentioned above (section 3.1) were considered: codec G.711 with A characteristic curve, G.723.1 and iLBC. Jitter delay varied between 0 and 200 ms in steps of 20 ms. Again a buffer size of 60 ms was applied. Also for scheduling the method „Silence Insertion" was used. The speech samples used were generally of a length of 30 ms, except for the codec G.711 where a length of 20 ms was used in addition.

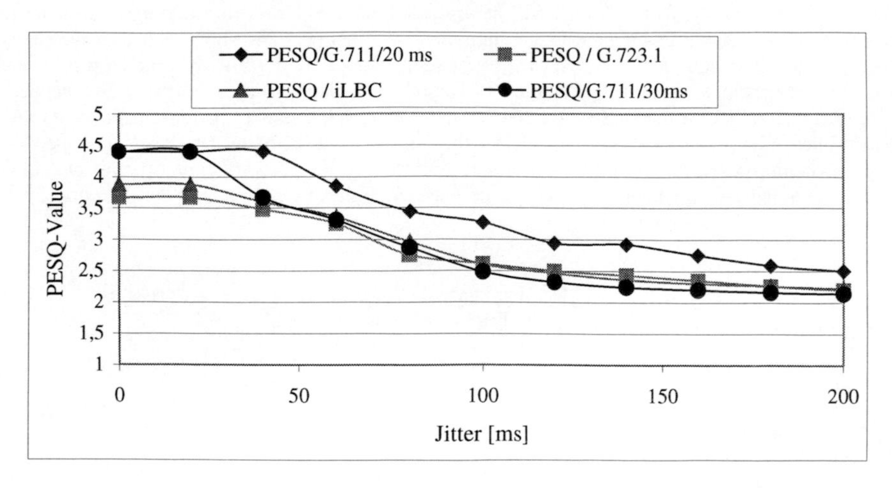

Fig. 9. PESQ Value as a Function of Jitter Delay for Different Codecs

Fig. 9 shows the PESQ values measured in the real environment as a function of jitter delay. The confidence intervals (probability of uncertainty is 5%) also represent less than 10% of the calculated mean values. The theoretical QoS values could not be derived in this case, as the E-model does not contain the jitter delay parameter (cf. table 1). It becomes apparent that the QoS retains the optimum basic level, as long as the jitter delay does not exceed the applied buffer size. However, with increasing jitter delay values a quick and significant deterioration of QoS can be noticed. Particularly remarkable are the much poorer PESQ values for the codecs G.723.1 and iLBC in comparison to the codec G.711. This is even more distinctive the shorter the used speech samples are. Conclusion: the codec G.711 again prevails against its competitors and can be recommended as standard codec for VoIP. However, one disadvantage must be mentioned: the codec G.711 requires a larger bandwidth (64 kbit/s) in comparison to the codec G.723.1 (6,3 kbps) and to the codec iBLC (13,8 kbps). The codec G.711 can, however, re-compensate this handicap by means of a shorter coding time in comparison to the codecs G.723.1 and iBLC. This also entails a reduction of end-to-end delay.

4. SUMMARY

In the course of this study, the QoS of the VoIP was considered in detail. First a classification of measuring methods for evaluating the QoS was conducted. Two essential measuring methods, i.e. E-model and PESQ algorithm were described in detail. The core of this paper focuses on the comparison between the two mentioned methods for determining the QoS. This was conducted in a real IP environment specially set up for this purpose.

The comparison study shows that the old, popular codec G.711 can well be recommended for a VoIP environment. The codec is robust against both packet losses and jitter delay. This applies particularly to speech samples sent as short sequences. It furthermore becomes apparent that a deeper evaluation of LPC codecs is necessary in order to adjust them more closely to the E-model. This work contains first suggestions towards this problem as well as promising solutions which are presented in several graphs and which will be discussed later. The resulting findings contribute to the perfectioning of the E-model which is important in practice. This shows that the absolutely objective but more complex and cost-intensive PESQ method is not always necessary in a real measuring environment. The well parameterised E-model can also be successfully used in practice.

In connection with the QoS within a VoIP environment it seems to be reasonable to develop and evaluate new parameter-based models for this service. The E-model has been developed for circuit switching networks. In order to successfully introduce the E-model into a packet switching network a number of adjustments are to be made. Further studies should point in this direction.

REFERENCES

[1] http://www.openh323.org/standards.html - H.323-Protocol suite; page last viewed August 2008

[2] http://www.sipcenter.com/files/sip2.pdf - SIP-Protocol suite; page last viewed August 2008

[3] Raake A., *Speech Quality of VoIP,* John Wiley & Sons, Chippenham, 2006

[4] http://www.itu.int/rec/T-REC-G.107/en - ITU-T Recommendation G.107 (E-model); page last viewed August 2008

[5] http://www.itu.int/rec/T-REC-P.862/en - ITU-T Recommendation P.862 (PESQ); page last viewed August 2008

[6] http://trafficlyser.de - Protocol analyser TraceSim_VoIP by ITD; page last viewed August 2008

[7] http://www.wanulator.de - Wanulator-Tool; page last viewed August 2008

[8] http://www.itu.int/rec/T-REC-G.113/en - ITU-T Recommendation G.113 (Bpl value); page last viewed August 2008

Key words – P2P, BitTorrent, modeling, optimization

Krzysztof WALKOWIAK

ON TRANSFER COSTS IN PEER-TO-PEER SYSTEMS: MODELING AND OPTIMIZATION

Peer-to-Peer (P2P) systems have become extremely popular in the Internet. One of the most important aspects of P2P systems is that the traffic produced by various P2P systems (e.g. BitTorrent) has been overloading the Internet. According to many works it would be decisive to reduce this load simultaneously maintaining the performance of P2P systems at acceptable level. Therefore, in this paper we take a detailed look at the problem of flows modeling in P2P systems with a special focus on aspects related to transfer costs. We formulate an optimization model of a P2P network. To solve this problem we apply an exact algorithm and heuristic approaches developed akin to real P2P systems. Optimal results produced by the exact algorithm can be applied as an effective benchmark to evaluate quality of heuristic approaches. Extensive numerical study is provided analyzing P2P networks in terms of transfer costs. To our best survey, this is the first study that provides optimal results of flow optimization in P2P systems.

1. INTRODUCTION

The concept of Peer-to-Peer (P2P) has gained much attention recently in the research community. Many applications including file-sharing, distributed computing, Internet based telephony (e.g. Skype), Internet television (e.g. Joost) were successfully implemented using P2P techniques. BitTorrent and other P2P systems generate more than 50% of consumer Internet IP traffic [1]. Consequently, a great number of research challenges in the area of P2P is still open [13], [18], [20]. The problem that this paper investigates, namely, modeling and optimizing of network flows in P2P systems, is one that arises naturally from the need to optimize performance of P2P systems from the network perspective. We propose an offline model that enables optimization of network flows in P2P systems. We are aware of the fact that P2P systems are very dynamic and stochastic, so offline modeling of such systems can seem incorrect. However, it is a common research approach that difficult and dynamic systems are reduced to simpler cases to enable creation of models that can be solved optimally in reasonable time. Having Linear Program of Integer Program formulations, network design problems can be solved by standard methods, and optimal or near optimal solutions can be found. Thus, we can obtain benchmarks that allow us to estimate performance of real systems. Finally, our work is a theoretical one and we hope that methodical analysis of the model we propose and obtained results will produce some inspirations in evaluation of various issues regarding P2P systems.

Our focus on transfer costs in P2P networks follows from the fact that currently used P2P systems do not take into account locality information. For instance, current implementations of BitTorrent ignore the underlying Internet topology or ISP (Internet Service Providers) link costs, and set up

data transfers among randomly chosen sets of peers distributed around the Internet. Consequently, since popularity of P2P systems is very large, the lack of locality-awareness in P2P systems leads to many unnecessary transfers crossing different ISPs, countries and continents what increases the operating cost of an ISP significantly [1], [5], [18], and [24].
The main contributions of the paper are: the formulation of an offline optimization model of the P2P network, obtaining optimal results regarding transfer costs in P2P network, comparing the effectiveness of simulation-based approaches developed according to real P2P systems against optimal results. The rest of the paper is organized as follows. In Section 2 the related work on optimization and simulation of P2P systems is discussed. Section 3 includes the optimization model of a P2P system. In Section 4 we briefly present the algorithm developed to simulate a BitTorrent-like P2P system. Section 5 contains results. Finally, last section concludes this work.

2. RELATED WORK

The authors of [25] study the performance characteristics of 2nd generation P2P applications, e.g., BitTorrent (BT). First, the deterministic model is examined and the average delay is calculated in an analytical way. Next, a branching process model for a P2P system in the transient regime is proposed. Lastly, a steady state analysis for a P2P service capacity based on Markov chain model is presented. Authors also examine several traces obtained on the real BitTorrent network. The authors of [23] examine several protocols developed for P2P based file distribution. Next a centrally scheduled file distribution (CSFD) protocol, to minimize the total elapsed time of a one-sender-multiple-receiver file distribution task is proposed. A discrete-event simulator is applied to study the performance of CSFD and other approaches (e.g. BitTorrent). Paper [8] focuses on the problem how to disseminate a large volume of data to a set of clients in the shortest possible time. A cooperative scenario under a simple bandwidth model is solved in an optimal solution involving communication on a hypercube-like overlay network. Furthermore, various randomized algorithms are examined. Also noncooperative scenarios based on the principle of barter are discussed. Arthur and Panigrahy present several routing algorithms to distribute data blocks on a network with limited diameter and maximum degree [3]. The time scale of the system is divided into steps. A special attention is put on upload policy – a randomized approach is proposed and examined. In [9] a performance study of BitTorrent-like P2P systems applying modeling, based on extensive measurements and trace analysis is presented. The study of representative BitTorrent traffic is analyzed. Motivated by the analysis and modeling results, authors propose a graph based model to study interactions among multiple torrents. Qiu and Srikant develop in [17] simple fluid models to study the performance of BitTorrent protocol. Also incentive mechanisms included in BitTorrent are analyzed. Experimental results comparing the fluid model against the discrete-event simulation of BitTorrent are provided. In [14] a probabilistic model of coupon replication systems is developed in order to study a P2P file swarming system based on BitTorrent. Markov processes are applied to obtain necessary and sufficient stability conditions.
Killian *et al.* examine the overlay network content distribution problem [12]. All content is in the form of unit-sized tokens – files can be represented as sets of tokens. The distributed schedule of tokens proceeds as a sequence of timesteps. There is a capacity constraint set on each overlay arc, i.e. only a limited number of tokens can be assigned to an arc for each timestep. Two optimization problems are formulated: Fast Overlay Content Distribution (FOCD) and Efficient Overlay Content Distribution (EOCD). The goal of the former problem is to provide a satisfying distribution schedule of minimum number of timesteps. The latter problem aims at minimizing the number of tokens' moves. Both problems are proved to be NP-complete. An Integer Program formulation of

EOCD is presented. Various online approximation algorithms for distributed version of overlay content distribution problem are proposed and tested.
In [22] a new selection strategy for BitTorrent-like P2P systems is proposed. The goal is to reduce the download time of BitTorrent. The proposed approach is based on the greedy strategy that a peer assigns each missing piece a weight according to total number of neighbor's downloaded pieces. Next, the peer selects the missing piece with the highest priority for next download. The simulation run on a discrete-event simulator show that the new strategy can improve more than 15% average download time and reduce in average 60% total elapsed time comparing to the BitTorrent system.
There are numerous papers that consider the problem of P2P-based streaming content distribution in overlay networks e.g. [2], [6], [21], and [27]. The common assumption of these works that significantly simplifies the analysis is that the content is distributed via a multicast tree, i.e. all subsequent blocks (pieces) of the same content stream are transported on the same paths. Thus, there is no need to model the time scale of P2P system as subsequent time steps and to incorporate to the model the constraint on block possession. In contrast, the parts of non-streaming content can be distributed in P2P systems autonomously, i.e. different blocks can follow different paths in the overlay network, what yields additional constraints in the model.
To survey other aspects of P2P systems refer to [18], [20].

3. MODELING

In this section we model network flows of P2P systems. Behavior of P2P systems currently used in the Internet (e.g. BitTorrent, eMule) is very stochastic and thus it is very difficult to model it in a deterministic way. However, our major goal is to formulate the optimization problem of network flows in P2P network as an Integer Program (IP) to provide optimal results of the transfer costs. The obtained solution can be thus used to benchmark other algorithms including centralized and distributed approaches.

3.1 ASSUMPTIONS

We apply several simplifying assumptions about the network model, as do most of the recent works on P2P modeling [2], [3], [8], [12], [15], [19], [21], [23], [25]. The time scale of P2P system that we consider is divided into time slots that can be interpreted also as subsequent iterations of the systems. We assume that each time slot has the same length. Moreover, all actions of the P2P systems completed in the previous time slot are updated in such a way that in the beginning of the next time slot this information is available to all elements of the system. For instance, if block b was transferred to vertex (peer) v in time slot t, then all other peers can try to get this block from v in time slot $(t+1)$. This follows from the fact that there is an index storing detailed information on current availability of blocks in the system. Our model is not limited to one exact implementation of the index, which can be organized in a centralized (e.g. similar to BitTorrent [7]) or decentralized (e.g. DHT [18]) manner. Such a P2P system can be called *synchronous* [21]. Obviously, in real P2P systems peers are mostly *asynchronous*, with different processing times and messaging latencies. However, the assumption on synchronous mode of P2P system is a common approach in many research works on P2P modeling [8], [12], [15], [21]. This follows from the fact that modeling and next optimization of asynchronous P2P system is very difficult in a deterministic way. Therefore, the most widespread method of asynchronous P2P systems evaluation is simulation [4], [5], [8], [12], [15], [17], [19], [23], [24], [25].

Data to be sent is divided into blocks (pieces). Each block has the same length, for instance 256 KB. Transfer of one block is completed within one time slot. Furthermore, we assume that each block must be delivered to each node (peer). However, the model presented below can be easily modified to include also more heterogeneous scenarios, e.g. transfer of one block takes more than one slot or only some selected blocks are required by peers. We assume that each peer has a limited upload and download capacity; we do not consider capacity constraint on overlay links. According to analysis presented in [27], node capacity constraints are typically sufficient in overlay networks. Furthermore, in the concept of overlay networks usually the underlay core network is considered as overprovisioned and the only bottlenecks are access links [2], [15], [23].
An important characteristic of P2P systems is dynamics – peers can frequently join or leave the network. To model this phenomenon in our approach we use constants a_{vt}, which equals 1 if peer v in time slot t is connected to the network (is available) and 0 otherwise. Consequently, although our model is deterministic, the stochastic nature of P2P system can be incorporated into our considerations.

3.2 COSTS

Currently used P2P systems ignore the underlying Internet topology and ISP link costs, since they are designed to randomly choose logical neighbors. Because the P2P system has no impact on the IP layer, the transfers are not optimized in terms of download paths length. Consequently, there are many cross-continental downloads that can congest backbone networks. Moreover, P2P systems affect strongly ISPs by generating a large volume of cross-ISP traffic what generates additional operating costs for ISPs. Therefore, many new approaches to enhance traffic locality in existing P2P protocols (mainly BitTorrent) are proposed, e.g. [5], [18], [24], and [26]. However, it was observed that a large part of file sharing traffic is local within regions and countries mostly due to such issues like language, culture differences, locality of media markets, etc. Nevertheless, P2P systems still provide some content that is world-wide popular [11], [18].
The key issue is to provide an effective mechanism for localization of peers. First, IP location databases can be applied. According to [11] less than 1.5% of the IP addresses could not be mapped to ISP, only 3-4% of the IP addresses could not be mapped to continents and/or countries, and as many as 21% of the IP addresses could not be mapped to a city. Main drawbacks of IP location databases are: time required to query such large databases and the fact that coverage of such databases is limited. Another relatively simple method is to match IP addresses with the same prefix. However, this approach could lead to selection of peers with the same prefix but in different geographical location or conversely, peers in the same ISP but with different IP prefix could be interpreted as remote peers. Authors of [26] propose a peer selection algorithm based on traceroute records – peers that share a common part of a path to a destination are picked. Both hop number and RTT (round-trip time) can be applied as a metric. Yamzaki et al. develop a method in which the peer selection mechanism includes additional information on costs of traffic distribution between ISPs. The goal is to minimize the cross-ISP traffic and thus reduce the operational costs of ISPs [24]. In [5] two ways to implement biased neighbor selection in the context of BitTorrent are proposed. First, the tracker and client are modified in order to include the ISP locality. Second, P2P traffic shaping mechanisms can be applied – ISP routers discover P2P responses and modify them to substitute outside peers with internal ones.
In our model we use a broad-spectrum constant ζ_{wv} that is defined as cost of one block transfer between peers w and v. It can be interpreted arbitrarily according to our needs, e.g. number of hops between w and v, number of ISPs between w and v, RTT between w and v, distance in kilometers between w and v, cost of cross-ISP transfers, etc.

3.3 MODEL

The model of the P2P transfer is formulated as an Integer Program (IP) with binary variables. The goal is to transfer to each node (peer) all blocks in a given number of iterations (time slots) minimizing the transfer cost. Prior to the process of blocks' transfer, blocks are located in some nodes, which next upload blocks to other peers. We use the notation proposed in [16].

indices

$b = 1,\ldots,B$ blocks to be transferred
$t = 1,\ldots,T$ time slots (iterations)
$v,w,s = 1,\ldots,V$ vertices (network nodes, peers)

constants

a_{vt} = 1 if vertex v is available in time slot t; 0 otherwise
ζ_{wv} cost of block transfer from node w to node v
d_v maximum download rate of node v
g_{bv} = 1 if block b is located in node v before the P2P transfer starts; 0 otherwise
u_v maximum upload rate of node v
M large number

variables

y_{bwvt} = 1 if block b is transferred to node v from node w in iteration t; 0 otherwise (binary)

objective

$$\text{minimize } F = \sum_b \sum_v \sum_w \sum_t y_{bwvt} \zeta_{wv} \quad (1)$$

constraints

$$g_{bv} + \sum_w \sum_t y_{bwvt} = 1 \quad b = 1,\ldots,B \quad v = 1,\ldots,V \quad (2)$$

$$\sum_b \sum_v y_{bwvt} \leq a_{wt} u_w \quad w = 1,\ldots,V \quad t = 1,\ldots,T \quad (3)$$

$$\sum_b \sum_w y_{bwvt} \leq a_{vt} d_v \quad v = 1,\ldots,V \quad t = 1,\ldots,T \quad (4)$$

$$\sum_v y_{bwvt} \leq M(g_{bw} + \sum_{i<t} \sum_s y_{bswi}) \quad b = 1,\ldots,B \quad w = 1,\ldots,V \quad t = 1,\ldots,T \quad (5)$$

The objective function (1) is the cost of block transfer using the P2P approach. The detailed discussion on the interpretation of the objective function and costs can be found in section 3.2. Constraints of the problem follow from the assumptions of our system presented in sections 3.1. To meet the requirement that each block must be transported to each network node we introduce the condition (2). Notice that either v is the seed of block b (g_{bv}=1) or block b is transferred to node v in one of iterations (variable $y_{bwvt} = 1$). Constraint (3) assures that the number of blocks uploaded by node w can not exceed a given threshold. Analogously, (4) bounds the download rate of node v. The constant a_{vt} used on the right-hand side of (3) and (4) enables to incorporate to the model stochastic nature of P2P clients that can frequently join and leave the network. Construction of (3) and (4) guarantee that if a peer v is not connected to the P2P network in iteration t ($a_{vt} = 0$) then v cannot upload and download any blocks in this time slot. Constraint (5) is in the model to meet the requirement that block b can be sent from node w to node v only if node w keeps block b in time slot t. We refer to (5) as to *possession* constraint [12]. Note that the right-hand side of (5) is a sum of constant g_{bw} (=1 if block b is located in node w) and $\sum_{i<t} \sum_s y_{bswi}$ (=1 if block b is transferred to node w from any node s in any iteration preceding the current time slot t). Consequently, the right-hand side of (5) equals 1 only if node w holds block b in time slot t. Constraint (5) enables the peer-to-peer transfer of blocks. Note that M must be larger than V. The proposed model is more generic than previously proposed in the literature e.g. [12], [15], [21], [27]. The major contribution is that

we incorporate to the model constants a_{vt} that indicates if peer v is available in time slot t. Consequently, dynamic and stochastic nature of P2P systems can be modeled.

4. HEURISTIC ALGORITHMS

In this section we describe briefly an heuristic algorithm that we developed in order to simulate a real BitTorrent-like P2P system. Our approach follows mainly from the BitTorrent protocol [7], [18] and ideas included in [4], [8], [22], and [24]. However, some simplifications had to be made in order to adjust the heuristic to the IP optimization model presented in the previous section. Since the goal of our research is to examine transfer cost aspects of P2P systems, we do not simulate all details – it is sufficient to mirror only the major characteristics of the BitTorrent-like P2P system. Fig. 1 presents the outline of the algorithm in pseudocode.

```
 1  for t=0 to T
 2   begin
 3    while (IsPossibleTranfer(t))
 4     begin
 5      v=RandomDownloadPeer(t)
 6      b=SelectBlock(v,t)
 7      w=SelectUploadPeer(b,v,t)
 8      TransferBlock(b,w,v,t)
 9     end (if)
10   end (for)
```

Fig. 1. Pseudocode of the P2P transfer algorithm

Since our model is synchronous, i.e. the system works in iterations, the main loop of the algorithm (lines 1-10) is repeated for each time slot t. Function `IsPossibleTranfer(t)` (line 3) returns 1 if there is a possible transfer in the P2P system, i.e. at least one block b can be transferred from a node w possessing block b to node v requesting block b not violating the constraints of the system (i.e. limits on upload and download capacity, possession of the block, availability of peers, etc.). Otherwise function `IsPossibleTranfer(t)` returns 0. The inner loop (lines 3-9) is repeated until there is at least one possible block transfer.

To model the stochastic nature of BitTorrent-like P2P system we randomly select the download peer among all feasible peers (line 5). A download peer v is feasible if it can upload at least one block from other peer satisfying all constraints of the system. Next, a block to be transferred is chosen among all feasible blocks (line 6). A block b is feasible if at least one node can upload this block to v satisfying all constraints of the system. Finally, the uploading peer is selected among all feasible upload peers (line 7). As previously, upload node w is feasible if it can upload block b to v satisfying all constraints of the system. Function `TransferBlock(b,w,v,t)` (line 8) transfers block b from w to v and updates state of the P2P system (upload and download limits, possession of the block, etc.).

We consider 4 versions of the algorithm. Thus, functions `SelectBlock(b,v)` and `SelectUploadPeer(b,v,t)` have different versions according to a particular strategy. The first algorithm – Random Strategy (RS) – selects the block (line 6) and upload peer (7) at random among all feasible blocks and upload nodes. To model the second strategy named Rarest First Strategy (RFS) function `SelectBlock` (line 6) returns the rarest feasible block in the network. Next, the upload peer is chosen at random. The third approach – Cost Selection Strategy (CSS) – takes into account transfer costs. The block to be transferred is selected at random (line 6), but the closest (in terms of the cost), feasible peer is chosen for upload. Finally, we model also the Weighty Piece Selection Strategy (WPSS) as in [21].

5. RESULTS

To solve the model (1-5) in optimal way we apply CPLEX 11.0 solver [10]. Heuristic algorithms simulating behavior of a real BitTorrent-like P2P system were coded in C++. Since our goal is to compare results of BitTorrent-like systems simulations against optimal results, we had to limit sizes of the problem instances in order to obtain optimal results approximately in one hour. After several experiments we decide to test networks consisting of 10 vertices (peers), 3 blocks to be transferred and 4 time slots. According to [11] we assume that peers are located in large cities worldwide. As mentioned above, unnecessary P2P transfers crossing different ISPs, countries and continents increase the operating cost of an ISP significantly [1], [5], [18], and [24]. To analyze this fact, we set the cost of one block transfer between two peers located in two cities just as the distance in kilometers between these two cites. We consider three networks: E5A5 – 5 peers in Europe and 5 peers in North America, E7A3 – 7 peers in Europe and 3 peers in North America, and E10 – 10 peers in Europe.
In the first experiment we assume that the seed (peer hosting blocks to be transferred) is located successively in each vertex. Three scenarios of access link capacity are considered: asymmetric (upload=4 blocks, download=2 blocks), symmetric (upload=2 blocks, download=2 blocks) and random (upload and download is selected randomly between 2 blocks and 4 blocks). Consequently, there are 30 different cases for each network. The CPLEX optimizer is applied to get optimal results. To simulate behavior of a real BitTorrent-like network, heuristic algorithms are run 1000 times and we for each case we record the minimum, maximum and average value of the objective function. In Table 1 we report for each network the average (over 30 cases) distance to optimal results of the following algorithms: RS, RFS, CSS and WPSS showing the minimum, average and maximum value for each algorithm.

Table 1. Average distance to optimal results for various networks

Net	AVG	RS			RFS			CSS			WPSS		
		min	avg	max	Min	avg	max	Min	avg	max	min	avg	max
E5A5	186%	90%	186%	283%	90%	185%	284%	21%	85%	175%	97%	186%	279%
E7A3	181%	78%	179%	297%	73%	180%	299%	13%	82%	194%	84%	179%	295%
E10	86%	44%	85%	130%	45%	85%	130%	18%	50%	90%	47%	85%	126%

In addition the second column of Table 1 shows the average value of the objective function calculated in the following way. For each downloading peer the average distance to other peers is found and multiplied by 3 (3 blocks be transferred). Next the sum over all downloaders is computed. We can easily notice that the CSS algorithm offers the best performance. On average, results yielded by CSS are greater than corresponding optimal results by 85%, 82% and 50% for networks E5A5, E7A3 and E10, respectively. The minimum results over 1000 tests reported for the CSS algorithm are in the same cases only 21%, 13% and 18% worse than optimal ones. Results of algorithms RS, RFS and WPSS are comparable to each other and on average very close to average values of the objective function calculated analytically (second column of Table 1). The experimental data is reasonably well explained by the fact that the CSS algorithm is the only one that incorporates transfer costs. Comparing results for different networks, we can notice that the smallest gap between heuristics and CPLEX is achieved in the case of network E10. This follows from the fact that network E10 includes peers located in Europe. In two other networks peers are divided to two continents: Europe and North America. To examine this, in Table 2 we present the number of cross-continents transfers. Analogously to Table 1, we report average values obtained for each heuristic. Additionally, in the second column we show the number of cross-continent transfers included in the optimal solution. We can easily notice a correlation between results reported in Table 1 and Table 2. Since the cost of cross-continent transfer is much larger than the cost of other transfers, algorithms agnostic of transfer costs (RS, RFS and WPSS) yields relatively high values of the objective function.

Table 2. Average number of cross-continent transfers for various networks

Net	OPT	RS			RFS			CSS			WPSS		
		min	avg	max	min	avg	max	min	avg	max	Min	avg	max
E5A5	3.0	8.0	15.0	22.1	7.8	15.0	22.3	3.3	7.8	14.6	8.3	15.0	22.0
E7A3	3.0	6.3	12.6	19.9	6.1	12.6	20.0	3.0	7.0	13.8	6.7	12.6	19.7

Fig. 2 plots the cost function yielded by CPLEX, RFS and CSS for the E5A5 network and random capacities of access links. We show only performance of RFS, since – as shown above – performance of two other methods: RS and WPSS is very similar to RFS. We can observe that values obtained for different seed location are comparable. In the case of optimal results, the gap between the "cheapest" city (Chicago) and "most expensive" city (Los Angeles) is 3588.

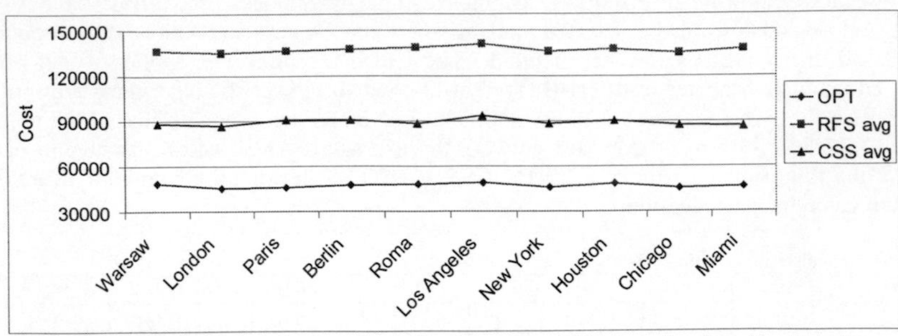

Fig. 2. Heuristics against optimal results obtained for network E5A5 as a function of seed location

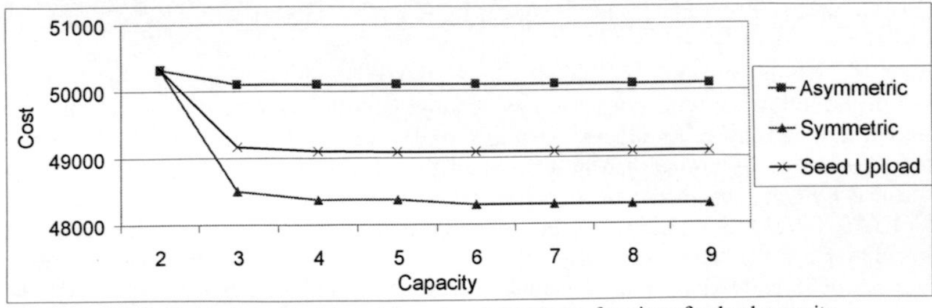

Fig. 3. Optimal results obtained for network E5A5 as a function of upload capacity

In Fig. 3 we present optimal results showing the impact of upload capacity on the cost function. Three scenarios are considered. In the first one – asymmetric – for each peer the download capacity is set to 2 blocks and the upload capacity varies from 2 to 9. The symmetric case assumes that for each peer both download and upload capacity is from 2 to 9. Finally, in the seed upload scenario only the upload capacity of the seed is increasing, all other capacities are set to 2. We can easily notice that in all cases increasing the capacity limits do not affect significantly the objective function.

Fig. 4 shows the optimal values of the transfer cost as a function of peers availability. The x-axis is the percentage availability of all peers in the network. Since there are 4 time slots and 10 peers, the availability is calculated according to the following formula: $(\sum_v \sum_t a_{vt}) / 40$. The general trend is that decreasing the peer availability increases the overall cost, since less resources are offered for P2P transfer.

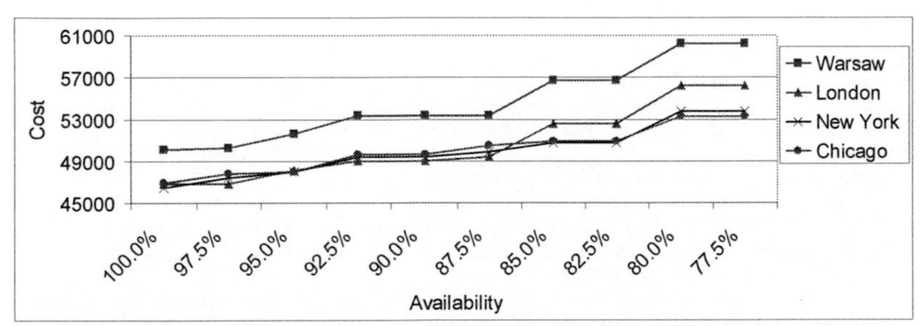

Fig. 4. Optimal results obtained for network E5A5 as a function of peer availability and seed location

In the last part of our experiments we run algorithms RS, RFS, CSS and WPSS for larger networks consisting of 100 peers, with 50 blocks to be transferred and the number of time slots set to 40. Three networks were considered: E50A50 (50 peers in Europe and 50 peers in North America), E70A30 (70 peers in Europe and 30 peers in North America) and E50A30A20 (50 peers in Europe, 30 peers in North America and 20 peers in Asia). For each network 30 different scenarios in terms of seed location and access link capacity were considered. Strategies RS, FRS and WPSS yield values of the cost function greater than corresponding values of CSS by 217%, 195% and 185% for networks E50A50, E70A30 and E50A30A20, respectively. Analysis of the number of cross-continent transfers shows similar correlation as above.

6. CONCLUSIONS AND FUTURE WORK

We formulated an optimization model of a P2P system as an Integer Program. Our model is generic and various additional constraints following from real P2P systems can be easily incorporated to the model. We applied the CPLEX solver to obtain optimal results. Several heuristic approaches were developed in order to simulate performance of a real BitTorrent-like P2P system. It was shown in simulations that applying information on location of peers the network transfer could be significantly reduced. However, there is still a large gap to optimal results, which encourages to work on new peer and block selection strategies. Therefore, in future we want to develop effective offline heuristics that produce results close to optimal and can be applied to larger networks than those tested in the case of CPLEX. Moreover, we are going to formulate new models of P2P systems that are aware of the underlying network and take into account several constraints related to the underlying network, e.g. capacity constraints of physical links, routes, etc.

REFERENCES

[1] AGGARWAL V., FELDMANN A., SCHEIDELE C., *Can ISPs and P2P Users Cooperate for Improved Performance?*, ACM SIGCOMM Computer Communication Review, Volume 37 , Issue 3 , 2007, pp. 31-40.

[2] AKBARI B., RABIEE H., GHANBARI M., *An optimal discrete rate allocation for overlay video multicasting*, Computer Communications, 31, 2008, pp. 551-562.

[3] ARTHUR D., PANIGRAHY R., *Analyzing BitTorrent and Related Peer-to-Peer Networks*, In Proc. of the seventeenth annual ACM-SIAM symposium on Discrete algorithm, 2006, pp. 961-969.

[4] BHARAMBE A., HERLEY C., PADMANABHAN V., *Analyzing and Improving BitTorrent Performance*, in Proc. of IEEE INFOCOM, Barcelona, Spain, Apr. 2006.

[5] BINDAL R., et al., *Improving Traffic Locality in BitTorrent via Biased Neighbor Selection*, In Proc. of . 26th IEEE International Conference on Distributed Computing Systems, ICDCS 2006, pp. 66- 66.

[6] BYUN S., YOO C., *Minimum DVS gateway deployment in DVS-based overlay streaming*, Computer Communications, 31, 2008, pp. 537-550.

[7] COHEN B., *Incentives Build Robustness in BitTorrent*, http://www.bittorrent.org/ bittorrentecon.pdf, 2003.

[8] GANESAN P., SESHADRI M., *On Cooperative Content Distribution and the Price of Barter*, In Proc. of the 25th IEEE International Conference on Distributed Computing Systems (ICDCS'05), 2005, pp. 81-90.

[9] GUO L., CHEN S., XIAO Z., TAN E., DING X., ZHANG X., *A Performance Study of BitTorrent-like Peer-to-Peer Systems*, IEEE Journal on Selected Areas in Communications, Vol. 25, No. 1, 2007, pp. 155-169.

[10] *ILOG CPLEX 11.0 User's Manual*, France, 2007.

[11] IOSUP A., GARBACKI P., PUWELSE J., ENEMA D., *Correlating Topology and Path Characteristics of Overlay Networks and the Internet*, in Proc. of Sixth IEEE International Symposium on Cluster Computing and the Grid Workshops, 2006. Vol. 2, pp. 10- 10.

[12] KILLIAN C., VRABLE M., SNOEREN A., VAHDAT A., PASQUALE J., *The Overlay Network Content Distribution Problem*, UCSD/CSE Tech. Report CS2005-0824, 2005.

[13] LEUF B., *Peer to Peer: Collaboration and Sharing over the Internet*, Addison Wesley, 2002.

[14] MASSOULIE L., VOJNOVIC M., *Coupon Replication Systems*, In Proc. of SIGMETRICS05, Vol. 33, No. 1. pp. 2-13.

[15] MUNIDGER J., WEBER R., *Efficient File Dissemination using Peer-to-Peer Technology*, Technical Report 2004--01, Statistical Laboratory Research Reports, 2004.

[16] PIÓRO M., MEDHI D., *Routing, Flow, and Capacity Design in Communication and Computer Networks*, Morgan Kaufman Publishers 2004.

[17] QIU D., SRIKANT R., *Modeling and performance analysis of bittorrent-like peer-to-peer networks*, In Proc. of ACM SIGCOMM'04, 2004.

[18] STEINMETZ R., WEHRLE K. (eds.), *Peer-to-Peer Systems and Applications*, Lecture Notes in Computer Science, Vol. 3485, 2005.

[19] STUTZBACH D., ZAPPALA D., REJAIE R., *Swarming: Scalable Content Delivery for the Masses*, University of Oregon, Computer and Information Science Technical Report, CIS-TR-2004-1, 2004.

[20] WU J. (ed.), *Theoretical and Algorithmic Aspects of Sensor, Ad Hoc Wireless and Peer-to-Peer Networks*, Auerbach Publications 2006.

[21] WU C., LI B., *On Meeting P2P Streaming Bandwidth Demand with Limited Supplies*, in Proc. of the Fifteenth Annual SPIE/ACM International Conference on Multimedia Computing and Networking (MMCN 2008), 2008.

[22] WU C., LI C., HO J., *Improving the Download Time of BitTorrent-like Systems*, In Proc. of IEEE International Conference on Communications, ICC 2007, pp. 1125-1129.

[23] WU G., TZI-CKER C., *Peer to Peer File Download and Streaming*, RPE report, TR-185, 2005.

[24] YAMAZAKI S., TODE H., MURAKAMI K., *CAT: A Cost-Aware BitTorrent*, In Proc. of 32nd IEEE Conference on Local Computer Networks, 2007, pp. 226-227.

[25] YANG X., DE VECIANA G., *Service Capacity of Peer to Peer Networks*, In Proc. of INFOCOM 2004, pp. 2242-2252.

[26] YING L., BASY A., *Traceroute-Based Fast Peer Selection without Offline Database*, in Proc. of Eighth IEEE International Symposium on Multimedia, ISM'06, 2006, pp. 609-614.

[27] ZHU Y, LI B., *Overlay Networks with Linear Capacity Constraints*, IEEE Transactions on Parallel and Distributed Systems, Vol. 19, No. 2, 2008, pp. 159-173.

AUTHOR INDEX